APPLIED OPTICS
and
OPTICAL ENGINEERING

VOLUME VII

CONSULTING EDITOR
RUDOLF KINGSLAKE

APPLIED OPTICS
and
OPTICAL ENGINEERING

EDITED BY

ROBERT R. SHANNON

and

JAMES C. WYANT

Optical Sciences Center
University of Arizona
Tucson, Arizona

VOLUME VII

ACADEMIC PRESS   New York   San Francisco   London   1979
A Subsidiary of Harcourt Brace Jovanovich, Publishers

COPYRIGHT © 1979, BY ACADEMIC PRESS, INC.
ALL RIGHTS RESERVED.
NO PART OF THIS PUBLICATION MAY BE REPRODUCED OR
TRANSMITTED IN ANY FORM OR BY ANY MEANS, ELECTRONIC
OR MECHANICAL, INCLUDING PHOTOCOPY, RECORDING, OR ANY
INFORMATION STORAGE AND RETRIEVAL SYSTEM, WITHOUT
PERMISSION IN WRITING FROM THE PUBLISHER.

ACADEMIC PRESS, INC.
111 Fifth Avenue, New York, New York 10003

United Kingdom Edition published by
ACADEMIC PRESS, INC. (LONDON) LTD.
24/28 Oval Road, London NW1 7DX

Library of Congress Cataloging in Publication Data

Kingslake, Rudolf, ed.
  Applied optics and optical engineering.

  Vol. 7–   edited by R. R. Shannon and J. C. Wyant.
  "Cumulative index": v. 5, p. 358–382.
  Includes bibliographical references.
  PARTIAL CONTENTS:––v. 1. Light, its generation and
modification.––v. 2. The detection of light and
infrared radiation––v. 3. Optical components.––
v. 4–5. Optical instruments.
  1. Optical instruments. 2. Optics. I. Shannon,
Robert Rennie, Date   II. Wyant, James C.
III. Title.
QC371.K5   621.36   65–17761
ISBN 0–12–408607–1

PRINTED IN THE UNITED STATES OF AMERICA

79 80 81 82   9 8 7 6 5 4 3 2 1

Contents

LIST OF CONTRIBUTORS	ix
FOREWORD	xi
PREFACE	xiii
CONTENTS OF OTHER VOLUMES	xv

CHAPTER 1

Incoherent Light Sources

J. E. Eby and R. E. Levin

I. INTRODUCTION	1
II. TUNGSTEN FILAMENT (INCANDESCENT) LAMPS	2
III. GASEOUS DISCHARGE LAMPS	10
IV. PHOTOFLASH LAMPS	33
V. LIGHT EMITTING DIODES	34
VI. NATURAL RADIATION	38
REFERENCES	45

CHAPTER 2

Optical Materials—Refractive

Charles J. Parker

I. INTRODUCTION	47
II. OPTICAL GLASS	48
III. VITREOUS SILICA GLASS	65
IV. OPTICAL CRYSTALS	68
V. PLASTIC OPTICAL MATERIALS	70
VI. OPTICAL MATERIALS FOR THE INFRARED	71
REFERENCES	76

CHAPTER 3
Plastic Optical Components
Brian Welham

I. Introduction	79
II. Materials	80
III. Design	85
IV. Manufacturing Methods	92
Reference	96

CHAPTER 4
Optical Materials—Reflective
William P. Barnes, Jr.

I. Introduction	97
II. Mirror Property Requirements	98
III. Property Comparisons	101
IV. Lightweight Mirrors	110
References	119

CHAPTER 5
Photographic Detectors
R. Shaw

I. Properties of Photographic Grains	121
II. Statistical Properties of Images	124
III. Spatial Frequency Analysis of Images	132
IV. Problems of Detection and Information Storage	141
V. Photographic Detection for Unconventional Exposures	148
References	152

CHAPTER 6
Propagation of Laser Beams
H. Kogelnik

I. Paraxial Ray Propagation	156
II. Propagation of Gaussian Laser Beams	175
III. Fresnel Diffraction	185
References	190

CHAPTER 7
Scattering from Optical Surfaces
J. M. Elson, H. E. Bennett, and J. M. Bennett

I. Introduction	191
II. Methods for Calculating Light Scattering	193
III. Total Integrated Scatter	199
IV. Angular Dependence of Scattering	216
V. Scattering from Dielectric Multilayers	236
VI. Summary	241
References	242

CHAPTER 8
Adaptive Optical Techniques for Wave-Front Correction
James E. Pearson, R. H. Freeman, and Harold C. Reynolds, Jr.

I. Introduction	246
II. Adaptive System Types	262
III. Phase Shifting Elements	281
IV. System Design Considerations	299
V. Compensation Performance	312
VI. Target Considerations	329
References	337

Index — 341

List of Contributors

Numbers in parentheses indicate the pages on which the authors' contributions begin.

WILLIAM P. BARNES, JR., *Optical Systems Division, Itek Corporation, Lexington, Massachusetts* 02173 (97)

H. E. BENNETT, *Michelson Laboratories, Naval Weapons Center, China Lake, California* (191)

J. M. BENNETT, *Michelson Laboratories, Naval Weapons Center, China Lake, California* (191)

J. E. EBY, *Tests and Measurements, Lighting Division, GTE Sylvania, Inc., Danvers, Massachusetts* 01923 (1)

J. M. ELSON, *Michelson Laboratories, Naval Weapons Center, China Lake, California* (191)

R. H. FREEMAN, *United Technologies Research Center, Optics and Applied Technology Laboratory, West Palm Beach, Florida* 33402 (245)

H. KOGELNIK, *Electronics Research Laboratory, Bell Laboratories, Holmdel, New Jersey* 07733 (155)

R. E. LEVIN, *Lighting Division, GTE Sylvania, Inc., Salem, Massachusetts* (1)

CHARLES J. PARKER, *Corning Glass Works, Corning, New York* 14830 (47)

JAMES E. PEARSON, *United Technologies Research Center, Optics and Applied Technology Laboratory, West Palm Beach, Florida* 33402 (245)

HAROLD C. REYNOLDS, JR., *United Technologies Research Center, Optics and Applied Technology Laboratory, West Palm Beach, Florida* 33402 (245)

R. SHAW, *Xerox Corporation, Webster, New York* 14580 (121)

BRIAN WELHAM, *U.S. Precision Lens, Cincinnati, Ohio* 45245 (79)

Foreword

It has been about ten years since Volume V completed the initial set of books on this topic which were edited by Rudolph Kingslake. During this period, these books have served as the major reference on the many topics associated with the design and use of optical systems. The treatise was intended for use as a reference by the practicing engineer and the use of mathematics and derivations was kept to a minimum for this purpose, with lucid explanation serving as the mechanism for communication rather than the exposition of theory.

Many changes have occurred in the field of optical engineering during the past decade. Lasers in various wavelength regions, widespread use of computers and digital processors, new optical materials, and widespread innovations in electro-optical sensors are but a few of the new topics that need to be reviewed. Dr. Kingslake has acknowledged the need for treatment of new topics and updating of some previous topics and he has collaborated with Brian Thompson on Volume VI, for which the Table of Contents can be found elsewhere in this volume.

This volume is the first of a planned set of successor volumes. We are initiating these with the intention of providing as useful a reference as that provided by the previous volumes. Two efforts are required here. First, much of the reference material on topics such as sources and detectors and materials must be updated because these items have been supplanted by newer devices or materials that were not available ten years ago. Second, many new concepts and topics are of interest, some of which were not even in the conceptual stage at that time. In a few cases, the understanding of some topics, such as the transmission of light through the atmosphere, has been improved to the extent that a new treatment of the subject is in order. We are planning to respond to all of these needs in the subsequent volumes.

Our philosophy is a bit different than that of the previous volumes. The field of applied optics moves so rapidly that it is not possible to establish a grand plan to cover several volumes, with each volume covering a specific topic. We have instead chosen to include a broad sample of topics from each of the above categories in each volume. This generates a continuing series of reviews on topics in applied optics that will be both as up to date and as comprehensive as possible. Some of the newer topics in optics are sufficiently complex that it is not possible to retain the mathematical level of complexity at as low a level as that which could be maintained in the previous volumes. In such cases, we have endeavored to have the authors provide key formulas and explanations insofar as possible.

We believe that this and succeeding volumes should serve as valuable general references for the practicing optical engineer. Further, we intend that the choice of topics will be such that an engineer can learn of some of the newer techniques and processes that will be of vital interest if he is to advance in his field. It is our hope to be able to produce about one volume a year which meets this criterion. We are attempting to obtain as authors leaders in the field who will write articles of lasting significance and interest. We believe that the contents of this seventh volume, the first under our editorship, meet this goal.

ROBERT R. SHANNON
JAMES C. WYANT

Preface

This is the first volume of "Applied Optics and Optical Engineering" under our editorship. The volume is made up of the following chapters.

The first chapter, by Eby and Levin, is an updating and replacement for the chapter on sources in Volume I of the original series. It does not repeat the information contained in the earlier chapter but, rather, discusses some of the innovations introduced into the subject of incoherent light sources, including semiconductor diode sources.

The chapters on materials by Parker, Welham, and Barnes discuss the present state of the supply of materials for optical systems. The present-day prevalence of plastics in precision optical systems was not forecast ten years ago. The existence of mirror materials which attained zero coefficient of expansion was mentioned only as an innovation at that time. The suppliers of optical glass have been reduced to a small number, although the need has increased for larger amounts of such materials. Computer aided optical design has, on the other hand, reduced the need for the stocking of hundreds of different optical glasses. In fact, there are only about thirty or so optical glasses in common use today.

Photographic detectors are treated as a member of the family of light detectors by Shaw. This chapter permits the optical engineer to treat electro-optical detectors and photographic systems, using equivalent concepts and terminology.

The chapter by Kogelnik on beam optics is significant in that the average optical engineer has probably already encountered situations in the use of lasers in which the apparently contradictory imaging properties of geometrical optics and diffraction of a Gaussian beam mode are present. The techniques of calculation described in this chapter should permit the engineer to understand this subtle and often confusing problem. The level of mathematics necessary, while not difficult, is somewhat higher than that of some of the topics discussed above. However, no background is required of the reader that would not normally be attained by a practicing engineer.

The chapter on scattering of light by optical surfaces, by Elson, Bennett, and Bennett, is a comprehensive survey of the state of the art in this currently important field. This chapter covers the significant parameters involved in describing the scattering that arises at optical surfaces and will serve to acquaint the reader with the concepts necessary to describe the important effects involved.

The chapter on adaptive optics, by Pearson, Freeman, and Reynolds, is a review of a totally new field, one with considerable promise. This application is closely tied to electronic and electro-optical device technology, but operates within concepts and limitations that are explained in the chapter. It is expected that this may become the basic review article on the subject for the practicing engineer.

In summary, the breadth of content and immediacy of the chapters in this volume are representative of those which we expect to be able to provide in future volumes. This volume stands on its own but will be enhanced by reference to previous articles on these topics in Volumes I–V. We wish to thank the authors whose work is included in this volume for their cooperation and enthusiasm. They have set a high standard which we shall try to maintain in successive volumes.

Contents of Other Volumes

Volume I: Light: Its Generation and Modification
Edited by *Rudolf Kingslake*

Photometry—*Ray P. Teele*
Light Sources for Optical Devices—*F. E. Carlson and C. N. Clark*
Filters—*Philip T. Scharf*
Atmospheric Effects—*Harold S. Stewart and Robert F. Hopfield*
Optical Materials—*Norbert J. Kreidl and Joseph L. Rood*
Basic Geometrical Optics—*Rudolf Kingslake*
Diffraction—*Adriaan Walther*
Interference, and Optical Interference Coatings—*P. Baumeister*
Polarization—*Robert J. Meltzer*
Projection Screens—*R. E. Jacobson*
Precision and Accuracy—*Ralph D. Geiser*
Author Index—Subject Index

Volume II: The Detection of Light and Infrared Radiation
Edited by *Rudolf Kingslake*

The Eye and Vision—*Glenn A. Fry*
Stereoscopy—*Leslie P. Dudley*
The Photographic Emulsion—*Fred H. Perrin*
Combination of Lens and Film—*E. W. H. Selwyn*
Illumination in Optical Images—*Rudolf Kingslake*
Electro-Optical Devices—*Benjamin H. Vine*
Television Optics—*Charles H. Evans*
Infrared Detectors—*Henry Levinstein*
Infrared Equipment—*Charles F. Gramm*
Author Index—Subject Index

Volume III: Optical Components
Edited by *Rudolf Kingslake*

Lens Design—*Rudolf Kingslake*
Optical Manufacturing—*R. M. Scott*
Photographic Objectives—*G. H. Cook*
Microscope Objectives—*James R. Benford*
The Testing of Complete Objectives—*R. R. Shannon*
Spectacle Lenses—*Irving B. Lueck*
Mirror and Prism Systems—*R. E. Hopkins*
Mirror Coatings—*Georg Hass*
Eyepieces and Magnifiers—*Seymour Rosin*
Author Index—Subject Index

Volume IV: Optical Instruments—Part I

Edited by *Rudolf Kingslake*

Fiber Optics—*Walter P. Siegmund*
Microscopes—*James R. Benford and Harold E. Rosenberger*
Camera Shutters—*Alfred Schwarz*
Still Cameras—*Arthur A. Magill*
Microfilm Equipment—*Odeen G. Olson*
High-Speed Photography—*T. E. Holland*
Optical Workshop Instruments—*A. W. Young*
Radiometry—*Fred E. Nicodemus*
Interferometers—*K. M. Baird and G. R. Hanes*
Refractometry—*G. E. Fishter*
Author Index—Subject Index

Volume V: Optical Instruments—Part II

Edited by *Rudolf Kingslake*

Dispersing Prisms—*Rudolf Kingslake*
Diffraction Gratings—*David Richardson*
Spectrographs and Monochromators—*Robert J. Meltzer*
Spectrophotometers—*Walter G. Driscoll*
Colorimeters—*Harry K. Hammond III*
Astronomical Telescopes—*A. B. Meinel*
Military Optical Instruments—*Francis B. Patrick*
Surveying and Tracking Instruments—*M. S. Dickson and D. Harkness*
Medical Optical Instruments—*John H. Hett*
Ophthalmic Instruments—*Henry A. Knoll*
Motion Picture Equipment—*Robert M. Corbin*
Author Index—Subject Index
Cumulative Index, Volumes I–V

Volume VI

Edited by *Rudolf Kingslake and Brian J. Thompson*

Solid State Lasers—*David C. Brown*
Gas Lasers—*J. M. Forsyth and J. Wilson*
Semiconductor Diode Lasers—*Jack F. Butler*
Acousto-Optics—*Adrianus Korpel*
Coherent Light Valves—*David Casasent*
Scanning Devices and Systems—*Gerald F. Marshall*
Coherent Optical Processing in Mapping—*N. Balasubramanian and Robert D. Leighty*
Infrared Detectors—*Donald E. Bode*
Principles and Applications of Optical Holography—*B. J. Thompson*
Image Intensifiers—*Herbert K. Pollehn*
Fiber Optics for Communication—*D. B. Keck and R. E. Love*
Index

APPLIED OPTICS
and
OPTICAL ENGINEERING

VOLUME VII

CHAPTER 1
Incoherent Light Sources

J. E. EBY
Tests and Measurements, Lighting Division
GTE Sylvania, Inc., Danvers, Massachusetts

R. E. LEVIN
Lighting Division
GTE Sylvania, Inc., Salem, Massachusetts

I. Introduction	1
A. Raison d'Etre	1
B. Scope	2
C. Overview	2
II. Tungsten Filament (Incandescent) Lamps	2
A. Polar Intensity Distribution	2
B. Tungsten–Halogen Lamps	3
III. Gaseous Discharge Lamps	10
A. Overview	10
B. Low Pressure Lamps	11
C. Medium Pressure Lamps	18
D. High Pressure Lamps	30
IV. Photoflash Lamps	33
V. Light Emitting Diodes	34
A. Overview	34
B. Operation	34
C. Optical Characteristics	35
VI. Natural Radiation	38
A. Overview	38
B. Spectral Distribution	38
C. Luminance	39
D. Illuminance	45
References	45

I. INTRODUCTION

A. Raison d'Etre

"Light Sources for Optical Devices" (Volume I, Chapter 2) introduces high luminance light sources especially suited for optical devices. The objectives of the present chapter are twofold. First, this previous material will be selectively updated. Second, coverage will be expanded to other sources that can be used for, or may indirectly affect, not only optical devices but also a range of technical applications such as film and television photography and reprographic and photochemical systems.

B. Scope

It is estimated that there are on the order of 10^4 different light sources available. Characteristics of each type may vary between manufacturers and with time as improvements are incorporated. Consequently, tabular listings of sources and their characteristics are not presented although selected exemplary data have been included. The reader is referred to manufacturer's literature and to industry summaries (Illuminating Engineering Society, 1972) for current light source data. In depth analysis of the operating principles for incandescent and arc discharge sources may be found in Waymouth (1971) and Elenbaas (1972).

C. Overview

The luminous performance of some optical systems is proportional to source efficacy (lumens per watt). For other systems where throughput (étendue) is satisfied,* performance is proportional to source luminance; in this case source efficacy is only related to the power required. Thus, two significant metrics of a luminous source are luminance and efficacy.

The spectral radiance at each wavelength of incandescent and arc discharge sources is bounded by the radiance of a complete radiator at the temperature of the effective source. This limit extends, albeit indirectly, even to fluorescent sources since it bounds the 253.7 nm fluorescent excitation line. When considering a light source of a specific type and mechanism, this limit can be extended to establish an order of magnitude estimate for the maximum luminance. Figure 1 summarizes the typical upper limiting range for luminance and luminous efficacy of a variety of light sources in the forms currently available.

II. TUNGSTEN FILAMENT (INCANDESCENT) LAMPS

A. Polar Intensity Distribution

A tungsten surface is nearly lambertian (Forsythe and Worthing, 1925). The far field effect of a filament expressed as an intensity distribution is

---

* This is sometimes said to be a completely filled or flashed system.

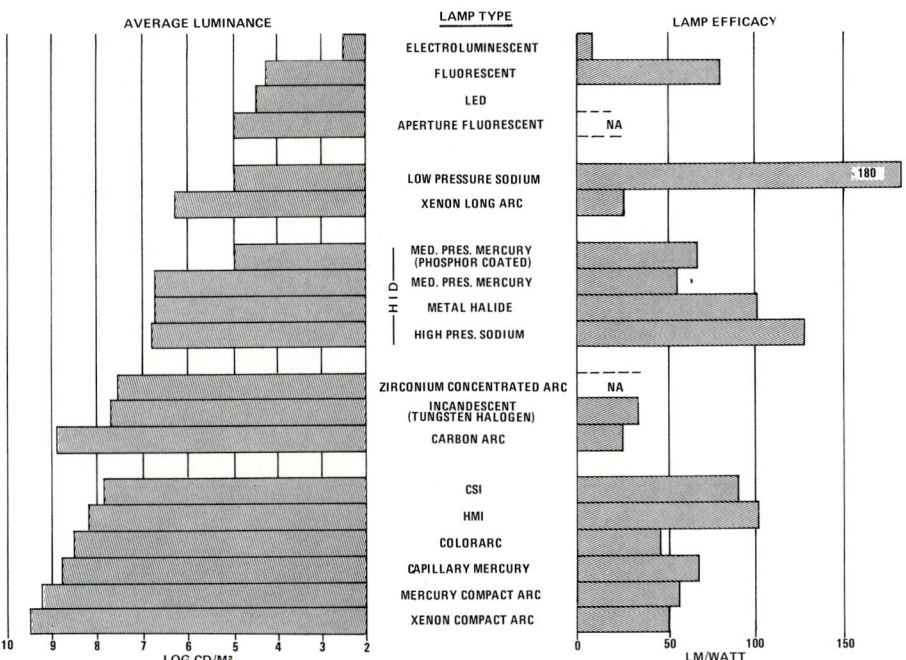

FIG. 1. Typical maximum luminance and efficacy available in current state-of-the-art light sources.

principally determined by the filament form with lamp envelope and base having a secondary effect. Figure 2 illustrates typical normalized intensity distributions for several filament forms.

B. Tungsten–Halogen Lamps

1. *Principles and Construction*

Tungsten–halogen incandescent lamps (formerly known as quartz–iodine lamps) are tungsten filament lamps in which a halogen gas has been added to the inert fill gas of the lamp (T'jampens and van de Weijer, 1966; Kopelman and Van Wormer, 1968; Burgin and Edwards, 1970; McHale, 1971). A halogen transport cycle returns evaporated tungsten to the filament thus preventing tungsten from depositing on the bulb wall. Consequently, these lamps maintain an essentially constant output during life. The bulb wall temperature required to maintain this halogen cycle, above 200–250 °C at all points, is achieved by using a sufficiently small bulb size for the lamp power, a bulb much smaller than for an ordinary incandescent lamp of the

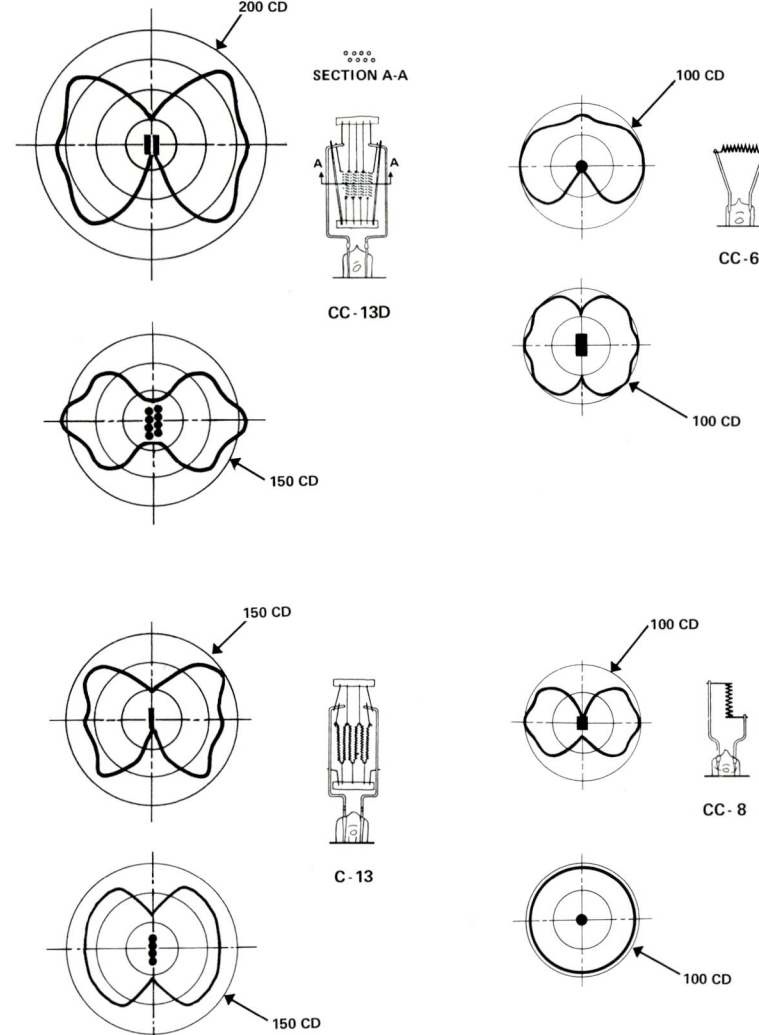

Fig. 2. Typical polar intensity distributions per 1000 lamp lumens for several filament forms.

same power. Due to the high temperatures involved, the bulb is of either quartz or fused silica.

If the lamp seal temperature exceeds 350 °C, oxidation of the molybdenum foil leads required for these low expansion bulb materials can cause early lamp failure. Since oxidation is a time–temperature function, seal temperatures as high as 450 °C may be permitted for some short design life lamps.

Many substances, including oils from the skin, will cause quartz to devitrify at elevated temperatures. For this reason these lamps should not be handled without gloves or the protective materials in the original packing. If the lamp is handled, it should be cleaned with a solvent before using.

2. *Operation*

In contrast to conventional incandescent lamps which operate with an internal gas pressure of about 1 atm, most tungsten–halogen lamps operate at several atmospheres internal pressure. This is acceptable due to the stronger bulb wall and the smaller gas volume. The pressure increase reduces the rate of tungsten evaporation. Thus, at a given filament temperature the tungsten–halogen lamp will have a significantly longer life than a comparable conventional incandescent lamp. Conversely, when designed for equal lives, the tungsten–halogen lamp will operate at a higher filament temperature with a consequential greater luminance and efficacy. With the exception of a few long, small diameter lamps which must be operated horizontally, tungsten–halogen lamps may be operated in any position.

3. *Exponential Relations*

Tungsten–halogen lamp operation can be approximated by the same exponential relations used for ordinary incandescent lamps (Volume I, Chapter 2, Section III.G) with the exception of the life relations. These relations are based on filament temperature change. The normal life of conventional lamps is principally dependent on tungsten evaporation, a function of filament temperature. Tungsten–halogen lamp life depends on other factors in addition to filament evaporation such as bulb temperature, quantity of halogen in relation to expected evaporation rate, filament etching (Lemons and Meyer, 1964), and operating gas pressure. The change in life due to voltage variation cannot be predicted accurately. The life of some tungsten–halogen lamps may be extended at reduced voltage. Generally, reduced voltage operation is not expected to decrease life below the rated value, and tungsten–halogen lamps can be operated on dimmers.

4. *Luminance and Filament Size*

The average filament luminance normal to the coil axis for single helix coils and coiled coils as determined from a variety of tungsten–halogen lamp types (Levin and Westlund, 1966) is given in Fig. 3. Interreflections in multiple coil filaments, e.g., C-13D filaments, can slightly increase these values. The necessary filament size for an optical device usually is established by characteristics of the device. Figure 4 permits an estimate of the required lamp power for single helix coils and coiled coils when the filament dimensions are known (Levin and Westlund, 1966).

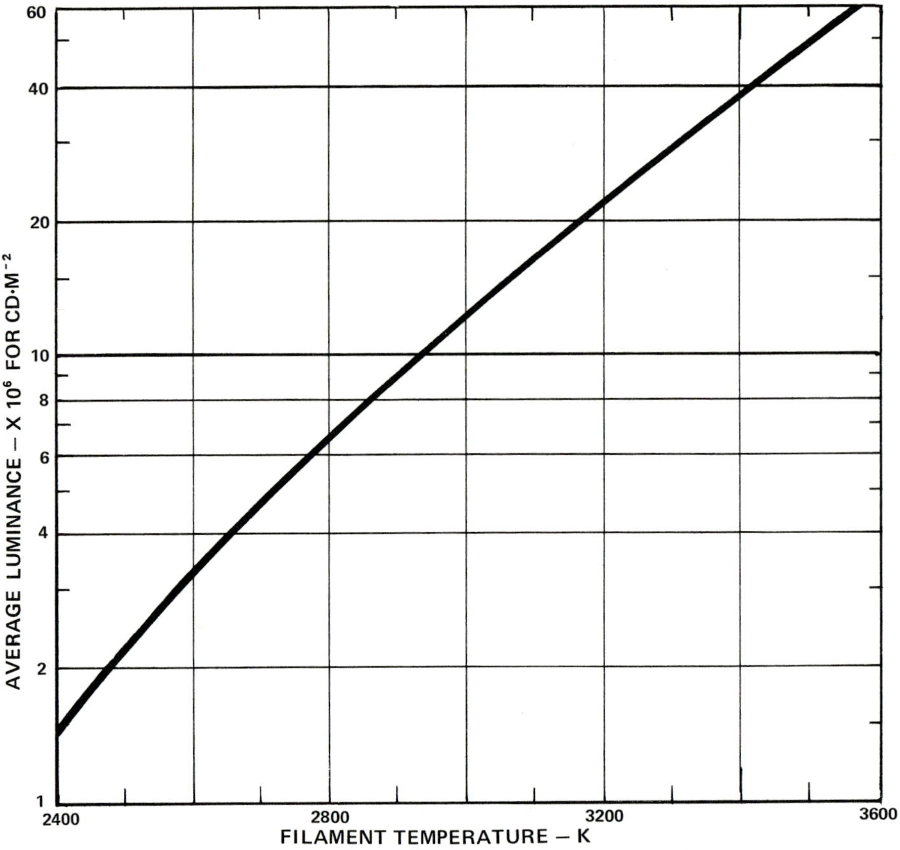

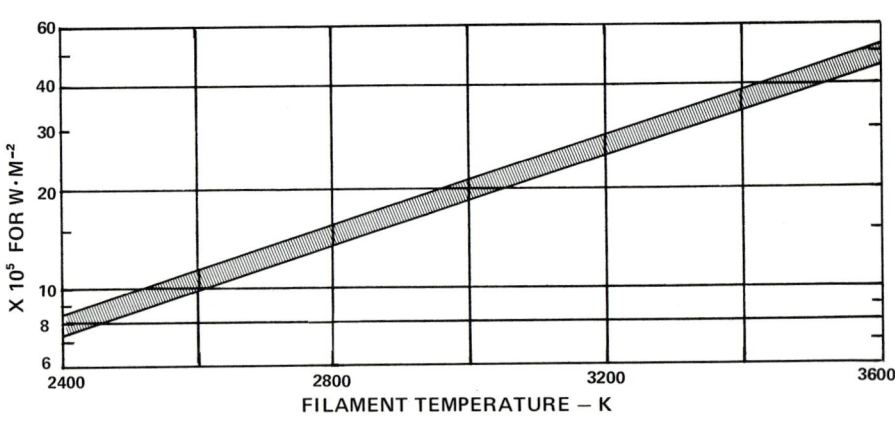

Fig. 3. Average luminance normal to axis of helical coil incandescent lamp filament.

Fig. 4. Lamp watts per unit filament area for helical coil tungsten–halogen lamp filaments. Area is defined as that of a closed end cylindrical boundary tangent to coil.

5. Spectral Power Distribution

Iodine and bromine are the halogens presently used. Iodine usually can be identified by the purplish tint observable immediately after the lamp is extinguished. Although bromine is colorless, iodine has a slight (a few percent) absorption, principally in the yellow-green part of the spectrum (Studer and Van Beers, 1964). The normalized spectral power of tungsten–halogen lamps is shown in Fig. 5. These are averaged data based on many lamp types. The integrals of these curves are given in Fig. 6. Certain high color temperature tungsten–halogen lamps, principally for photographic applications, use vanadium-doped fused silica bulbs to absorb radiation below about 350 nm, and the short wavelength end of these curves must be modified accordingly. These curves also can be applied to ordinary incandescent lamps providing the narrower pass band of the glass is taken into account.

6. Hard Glass Halogen Lamps

Both materials and fabrication technology contribute to the high cost of tungsten–halogen lamps. Tungsten–halogen lamp bulbs can be made of a hard glass, aluminosilicate. This allows the use of less expensive materials and manufacturing technology. At present, hard glass halogen lamps are limited to low voltage lamps below about 50 W.

7. Available Range

Tungsten–halogen lamps are made in both single-ended and double-ended form. Most bulbs are T-shaped (tubular) with the G-shaped (spherical) bulbs used occasionally. Several compact filament forms are common, e.g., the C-6, C-8, CC-6, CC-8, C-13, C-13D, C-Bar 6. Proximity reflectors are used with some C-13D filaments.

The majority of lamps range from 50 W to 2000 W although lamps from 4 W to 10,000 W are standard. Since low wattage, high voltage filaments are not practical in compact form, tungsten–halogen lamps less than about 250 W are commonly low voltage. The majority of lamps range from 3000 K color temperature (about 2000 hr life) to 3400 K color temperature (about 25 hr life), but lamps below 3000 K are available.

Some lamps for optical devices, e.g., photographic projectors, are comprised of a reflector with an integrally mounted tungsten–halogen capsule.* The reflector, which serves as a reflective condenser for the optical device, often is glass coated with a dichroic cold mirror to control system heat. The filament and the reflector of each particular lamp are designed to meet

---

* Generic types are "tru-beam" and "rim mount."

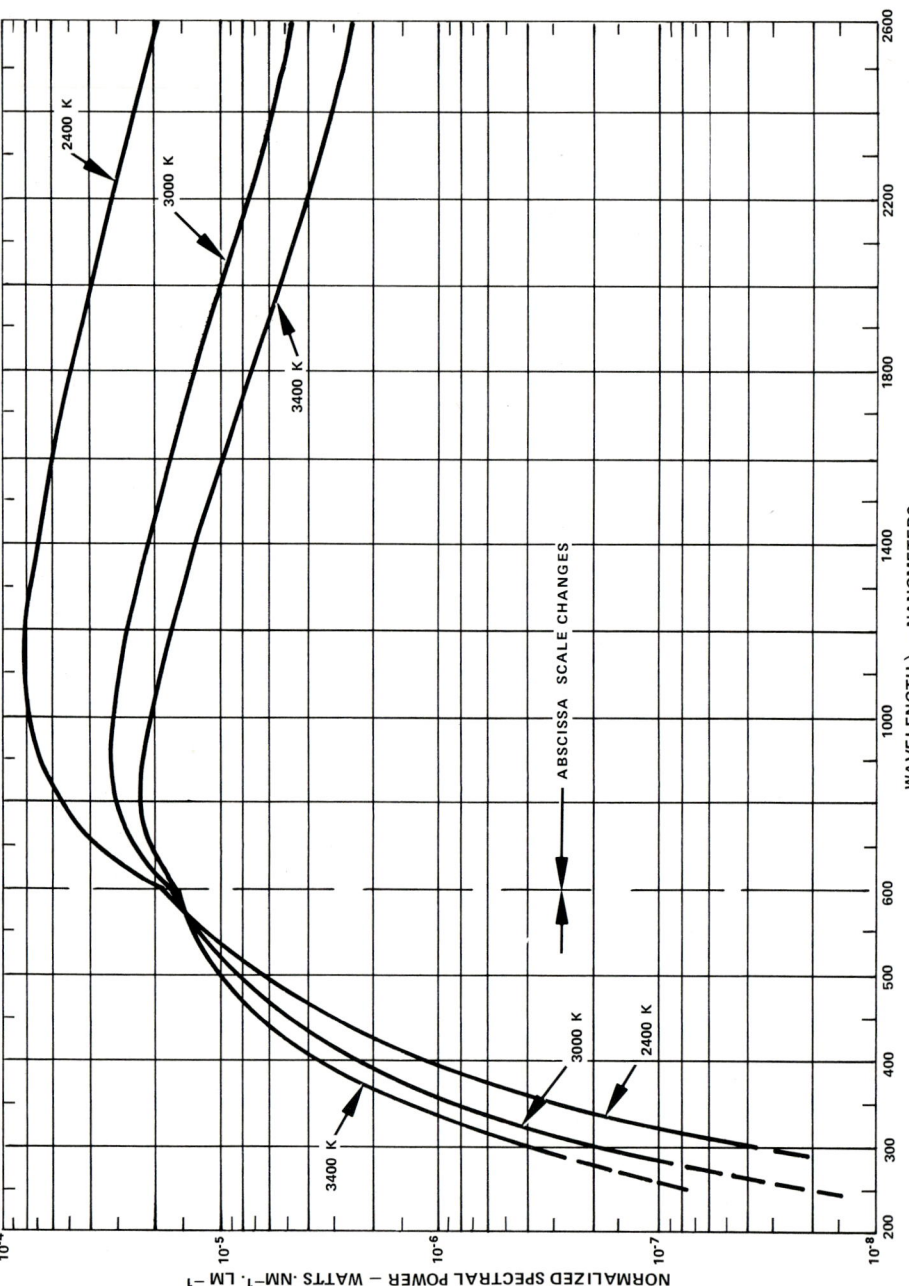

Fig. 5. Normalized spectral power distribution for tungsten–halogen lamps.

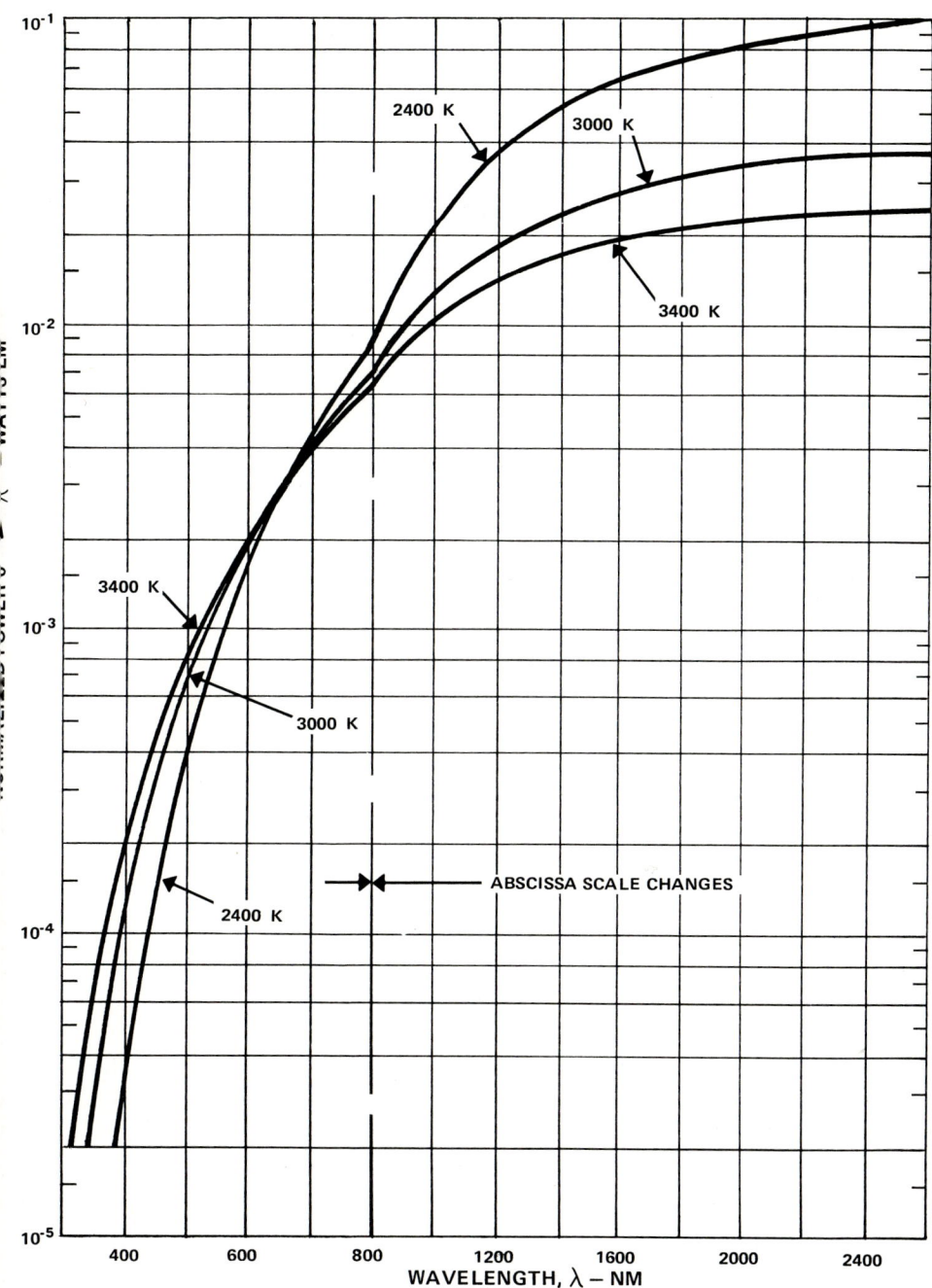

FIG. 6. Integral from 0 to λ for functions of Fig. 5.

the requirements of some specific optical system. Once available, however, a lamp can find use in other devices. The principal advantage of combining the filament and reflector into one unit is tolerance control, but other advantages exist.

III. GASEOUS DISCHARGE LAMPS

A. OVERVIEW

Over the years a wide variety of gaseous discharge lamps have evolved in parallel with the advancements in incandescent lamps. While developed primarily for use in the field of illumination, these discharge lamps also have their applications in optical instrumentation. In comparison with incandescent lamps, these sources tend to be long lived and highly efficient but require more complicated circuitry for operation and have higher initial costs.

All discharge lamps have a "negative resistance" characteristic in which the voltage across the arc drops as the current increases. This leads to runaway self-destruction unless the current is limited by an external ballast.

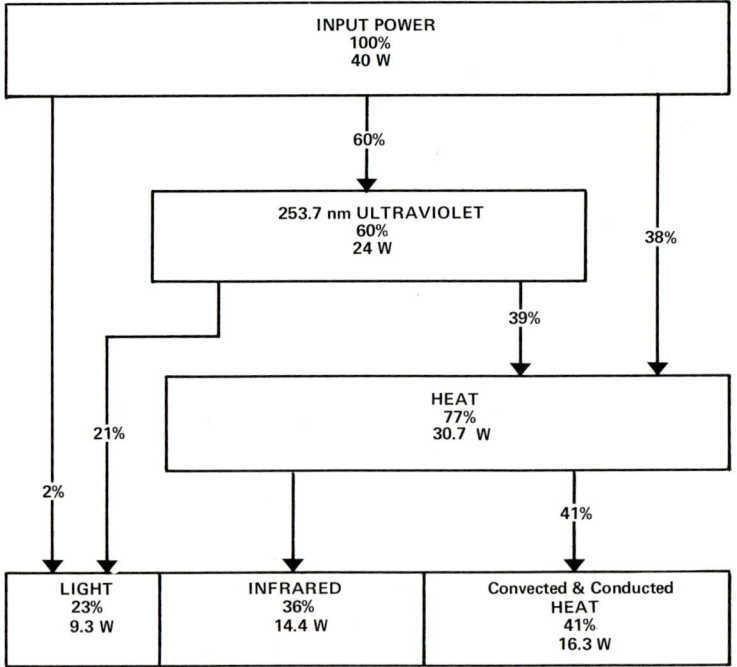

FIG. 7. Power distribution of typical 40 W cool white fluorescent lamp.

The presence of this external ballast, as well as possible starting aids, are the factors which account for the additional circuit complexity with these sources. The vast majority of commercial lamps are based on either mercury or sodium arcs. Generally speaking, the mercury lamps have been developed first due to the materials problems caused by the highly reactive nature of sodium.

As a matter of convenience, arc discharges can be divided into three categories based upon their operating pressures. Low pressure arcs operate in the 10^{-3} atm pressure range, have loadings on the order of $\frac{1}{2}$ W/cm of arc length, and have spectral outputs consisting of the narrow characteristic lines of the elements involved in the discharge. Medium pressure lamps, generically referred to as high intensity discharge or HID lamps, operate at pressures in the 1–10 atm range, have loadings in the area of 50 W/cm, and emit a small amount of continuum radiation in addition to spectral lines which can be pressure broadened or even self-reversed. High pressure lamps are those which operate above 10 atm, have loadings in the range of kilowatts per centimeter of arc length and emit large amounts of continuum radiation. This last class has been dealt with above (Volume I, Chapter 2, Section V) under the heading of "Compact-Source Arc Lamps."

B. Low Pressure Lamps

1. *Fluorescent*

a. Theory of Operation. With few exceptions the available low pressure mercury sources fall under the heading of fluorescent lamps (Waymouth, 1971; Elenbaas, 1971, 1972; Amick, 1960).

In these sources a low pressure mercury arc transforms electrical energy into 253.7 nm radiation with about 60% efficiency. This ultraviolet is then employed to excite phosphors which in turn emit the working radiation in the desired spectral bands. A diagram of the actual power usage is shown in Fig. 7.

b. Bulb Sizes and Shapes. Fluorescent lamps are predominantly tubular, ranging in length from 15 to 245 cm and in diameter from 16 to 54 mm. Wattages in the range from 4 to 215 are available in standard production.

c. Polar Distribution of Radiation. The normal tubular lamp has essentially uniform luminance around its circumference, but the center of the lamp has slightly higher luminance than the ends. This circumferential uniformity is modified in the reflector lamp (Balder and van der Weijer, 1956) by the presence of a white reflecting coat between the phosphor and the bulb glass over a portion of the circumference (see Fig. 8). The aperture lamp (Toomey, 1961; Spencer and Montgomery, 1961) carries this concept one step further

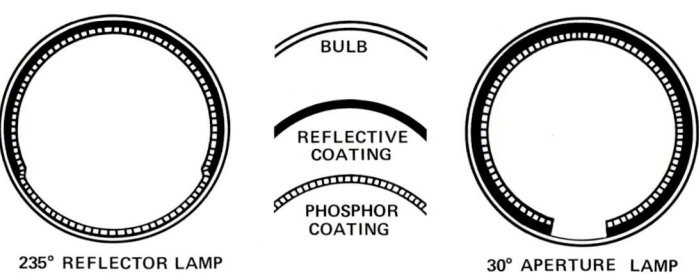

Fig. 8. Reflector and aperture fluorescent lamps.

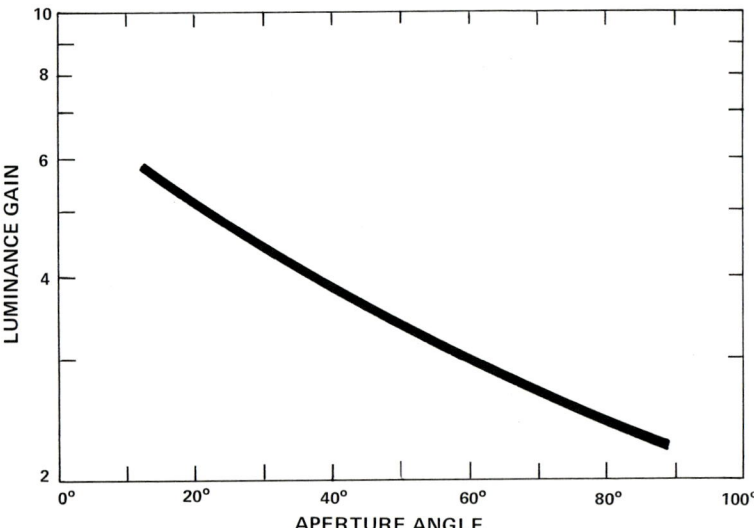

Fig. 9. Typical luminance gain of fluorescent aperture lamp.

by removing the phosphor from the bulb wall in the area of the window. While these alterations reduce the total radiation from the lamp, the luminance of the aperture area can be as much as six times that of the conventional lamp as shown in Fig. 9. For modelling purposes this aperture can be considered a lambertian source.

The intensity of linear lamps in a plane containing the tube centerline is nominally a cosine function but decreases somewhat faster at angles approaching 90° from the normal to the centerline. The total radiated flux can be approximated as 9.25 times the normal intensity (Baumgartner, 1941).

d. Spectral Distribution of Radiation. The spectral output of fluorescent lamps is characterized by one or more relatively broad phosphor emission bands in conjunction with the characteristic narrow spectral lines of the low pressure mercury arc. In conventional lamps the glass envelope absorbs strongly at wavelengths below 300 nm and very little radiation is emitted above 850 nm. Between these limits, a wide variety of distributions are possible. Figure 10 shows three of the many "white" lamps available. A number of distributions are available for nonvisual applications such as plant growth (Fig. 11) and photochemical activation (Fig. 12). Lamps designated as blacklight principally radiate within the 320–400 nm band, but the actual spectral distribution may differ between manufacturers since a variety of phosphors are available for this region.

e. Efficacy and Life. For many years fluorescent lamps were the most efficient visual sources in general use with efficacies as high as 85 lm/W excluding ballast losses. The life determining factor is usually a failure of the electron emitting coating on the electrodes. Since these electrodes are damaged during every startup, their useful life is a function of the average hours burned during an on–off cycle. An average life of 34,000 hr on continuous burning will drop to 18,000 hr when operated at 3 hr/start as shown in Fig. 13. Light output versus temperature and light output versus life, referred to as lumen maintenance, are also illustrated in Fig. 13. The principal values shown above are for standard 40 W rapid start cool white lamps which exhibit perhaps the best performance in all these areas. The choice of other phosphors or specialized lamp configurations such as aperture lamps may be expected to degrade these values.

f. Temporal Variation of Radiation. While maintenance may be regarded as a description of the long term behavior of the light output, there are short term variations as well. Radiation from the mercury arc in the interior of the lamp shows 100% modulation at twice the line frequency affecting both the visible mercury lines and the exciting radiation to the phosphors. Depending on their decay times, the phosphors will then exhibit this same modulation

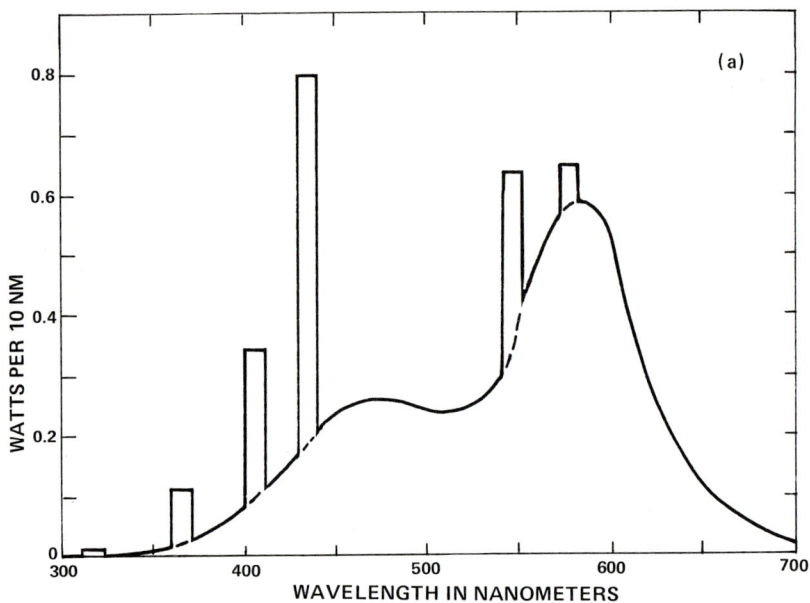
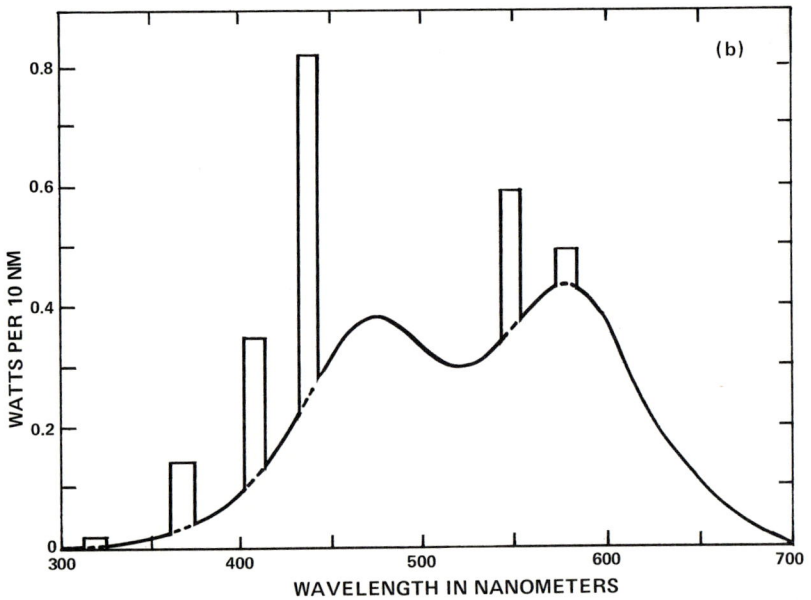

FIG. 10a,b. Spectral power distributions for 40 W rapid start "white" fluorescent lamps: (a) cool white, (b) daylight.

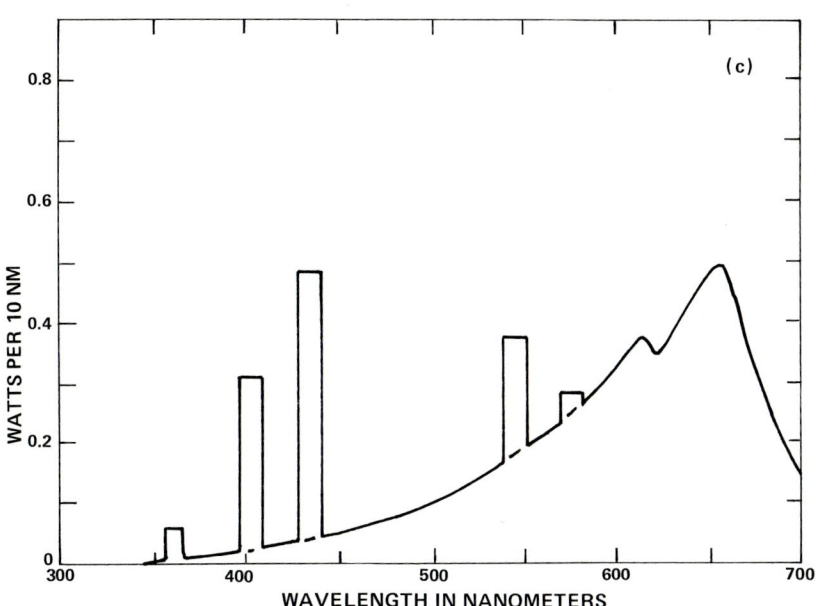

FIG. 10c. Incandescent fluorescent.

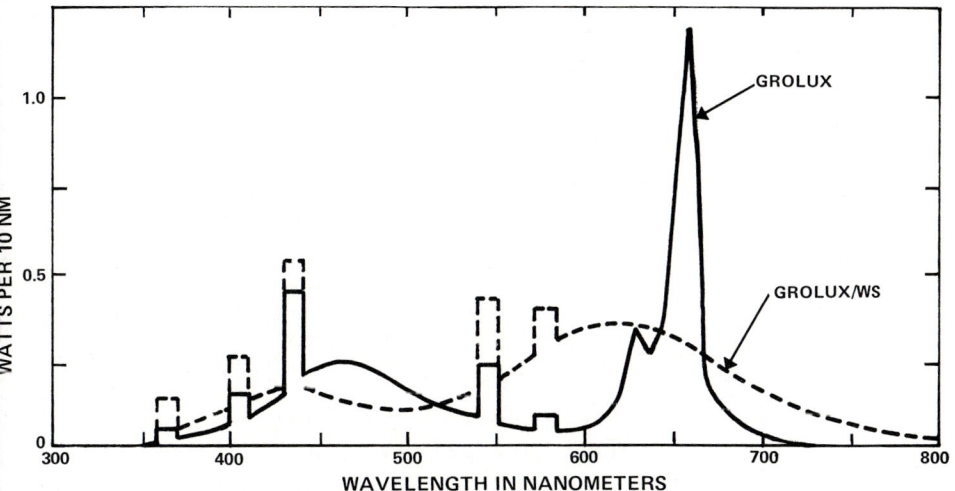

FIG. 11. Spectral power distribution for 40 W rapid start plant growth fluorescent lamps.

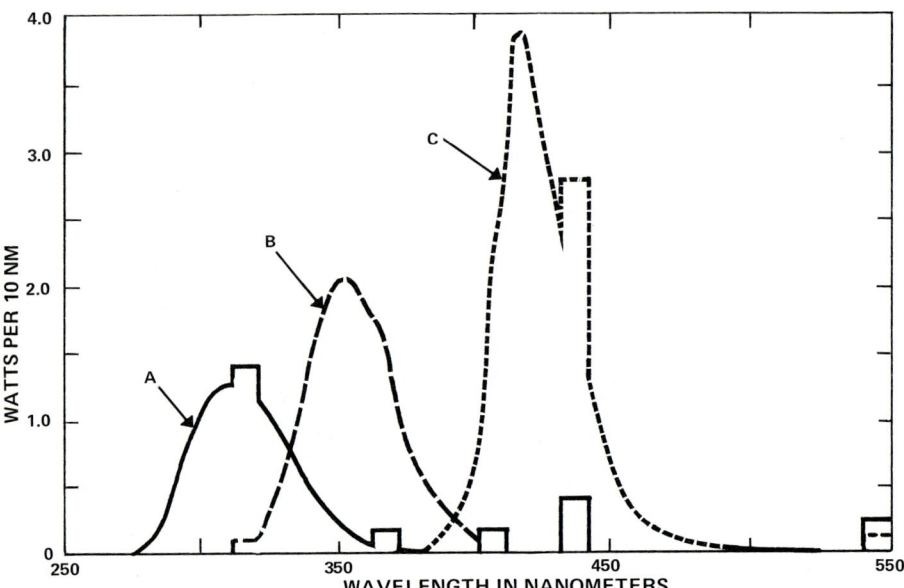

Fig. 12. Spectral power distribution for 40 W rapid start actinic fluorescent lamps: A, FS40 sunlamp; B, black light blue; C, super diazo blue (SDB).

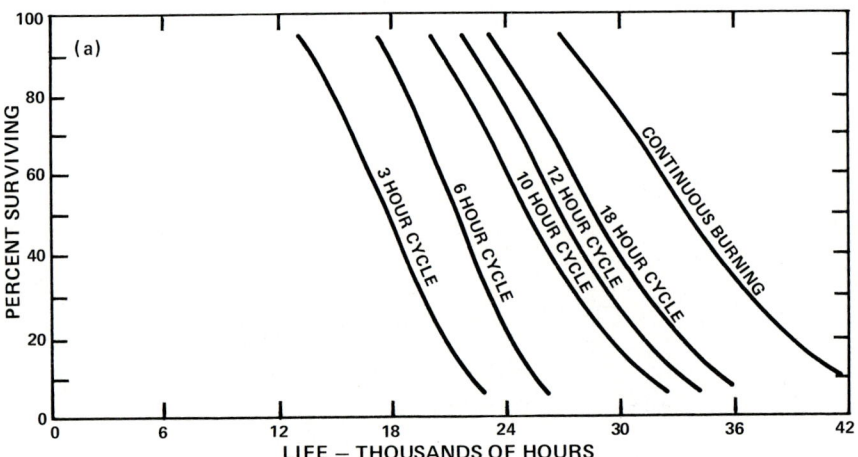

Fig. 13a. Fluorescent lamp operating characteristics: mortality curves for 40 W rapid start lamps with rated life of 18,000 hr.

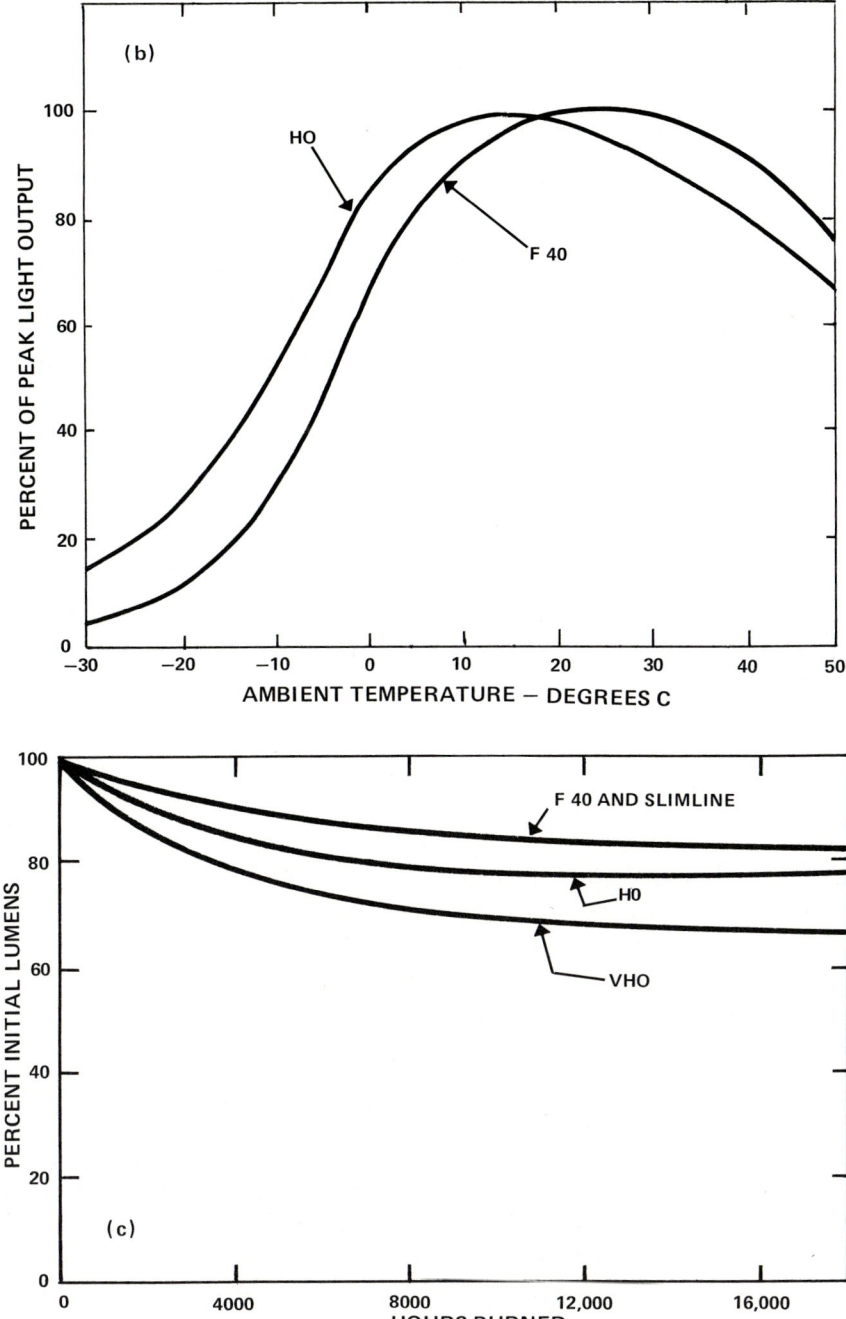

Fig. 13b,c. Fluorescent lamp operating characteristics: (b) change in output as function of ambient temperature in still air; typical lumen maintainance of cool white lamps [F40–40 W (430 mA) rapid start; slimline—instant start; HO—800 mA preheat; VHO—1500 mA preheat].

to a greater or lesser degree. An interesting consequence is that lamps containing one or more phosphors with appreciable decay times will show a detectable variation in color at twice the line frequency. This is due to the greater modulation of the visible mercury lines and any fast decay phosphors which may be present compared to the longer decay phosphors.

2. *Germicidal Lamps*

Germicidal lamps may be regarded as a specialized subset of the fluorescent family. In these lamps the envelope is formed of a UV transmitting glass, and no phosphor is employed. The result is simply a low pressure mercury arc lamp emitting primarily at 253.7 nm. These lamps are available in loadings up to 25 W/cm (Sadoski and Roche, 1976).

3. *Low Pressure Sodium*

Low pressure sodium lamps (Waymouth, 1971; Elenbass, 1972; Elenbass *et al.*, 1969; Beijer *et al.*, 1974) have not yet gained wide acceptance in the United States and are not readily available here. Their primary use has been in highway lighting where their efficacies of up to 185 lm/W are outstanding. Their spectral output is very nearly monochromatic, consisting primarily of the sodium doublet emission at 589 nm.

C. Medium Pressure Lamps

1. *Mercury**

a. Theory of Operation. A diagram of the electrical circuit of a typical medium pressure mercury lamp is shown in Fig. 14. When voltage is applied, a discharge is initiated between the starting electrode and the adjacent operating electrode with the discharge current being limited by the starting resistor. After sufficient amounts of ionized argon and mercury become available, the arc strikes between the two operating electrodes. Due to the increased current flow, the ballast output voltage drops causing the starting discharge to extinguish because of the starting resistor, and the lamp current now flows between the operating electrodes.

In normal lamp construction, the quartz arc tube and the starting resistor are enclosed within a nitrogen-filled outer glass envelope. This outer housing absorbs unwanted ultraviolet radiation, regulates the arc tube operating temperature, and protects it from atmospheric corrosion. When desired, however, this outer jacket can be removed for special applications. This will

---

* See Waymouth (1971), Elenbaas (1972), Noel and Martt (1956), Fraser *et al.* (1961), Levin and Lemons (1971).

1. INCOHERENT LIGHT SOURCES

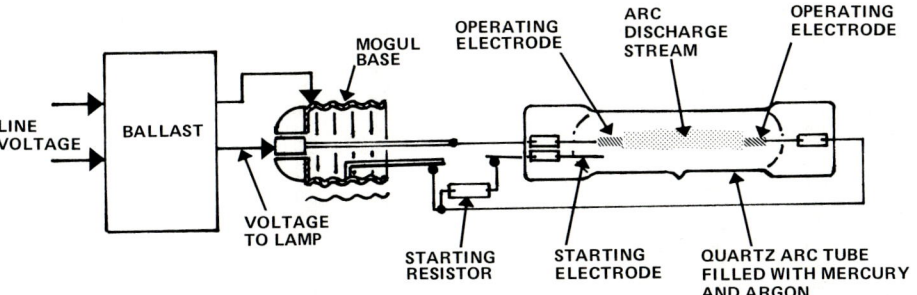

FIG. 14. Electrical circuit of typical medium pressure mercury vapor lamp.

result in the presence of hazardous ultraviolet radiation and lead to shortened lamp life due to oxidation of the quartz to metal seals of the arc tube.

Some self-ballasted lamps are available in which a tungsten filament located inside the outer jacket is employed as a resistive ballast for the discharge. Such lamps will have comparatively short lives and low efficiencies, but they have the advantage of operating directly from the available line voltage.

b. Spectral Distribution of Radiation. The clear mercury lamp has a spectrum consisting primarily of strong lines at 365.0, 404.7, 435.8, 546.1 and 578.0 nm, as illustrated in Fig. 15. This forms a bluish-white source in which there is virtually no red radiation and hence a low color rendering index. To improve the color rendering qualities, phosphors can be applied to the inner surface of the outer bulb wall. These transform some of the lamp's ultraviolet radiation into light in the red portion of the spectrum as shown in Fig. 15. This phosphor coating also has the effect of transforming a relatively small, intense source into a larger one with lower luminance. In the self-ballasted sunlamp, a hard glass outer jacket is employed to transmit the erythemal UV wavelengths as shown in Fig. 16. The radiation from a quartz tube medium pressure mercury arc in the absence of an outer jacket is also shown in Fig. 16.

c. Size and Range. Standard mercury lamps are available in wattages from 40 to 1000 with arc lengths from 16 mm to 150 mm, respectively. While the outer jacket is usually roughly ellipsoidal, there are a number of lamps where this jacket has been shaped and internally aluminized to form an integral reflector. These reflector lamps are available in a variety of beam patterns from narrow spots to wide beam floods and with the aperture area either clear or phosphor coated (see Fig. 17).

Unjacketed mercury lamps are also available, used primarily in the reprographic and photochemical fields. These are tubular lamps with diameters

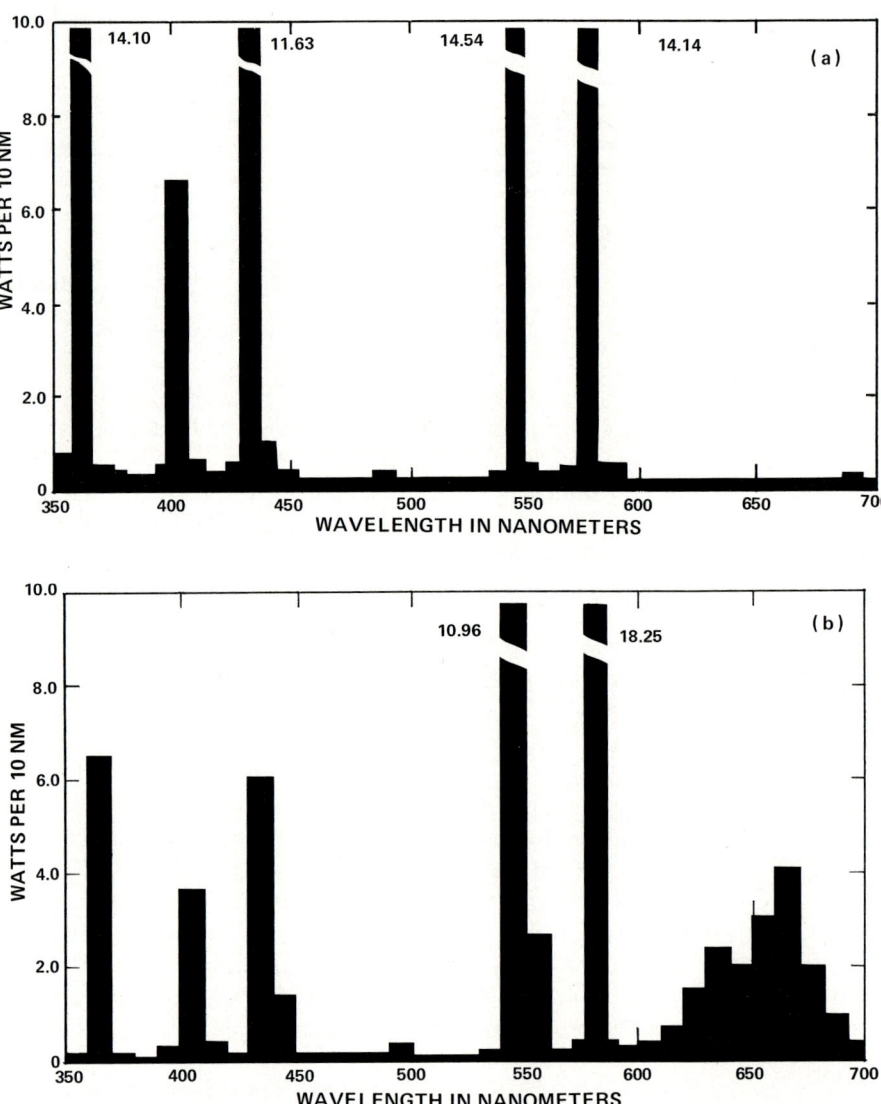

FIG. 15. Spectral power distributions of 400 W medium pressure mercury lamps: (a) clear; (b) phosphor coated (color improved).

from 9 to 30 mm, with arc lengths from 10 to 180 cm, and with wattages from 75 to 15,000.

d. Efficacy and Life. The mercury arc was for a long time the most efficient source available for outdoor lighting with outputs as high as 63 lm/W.

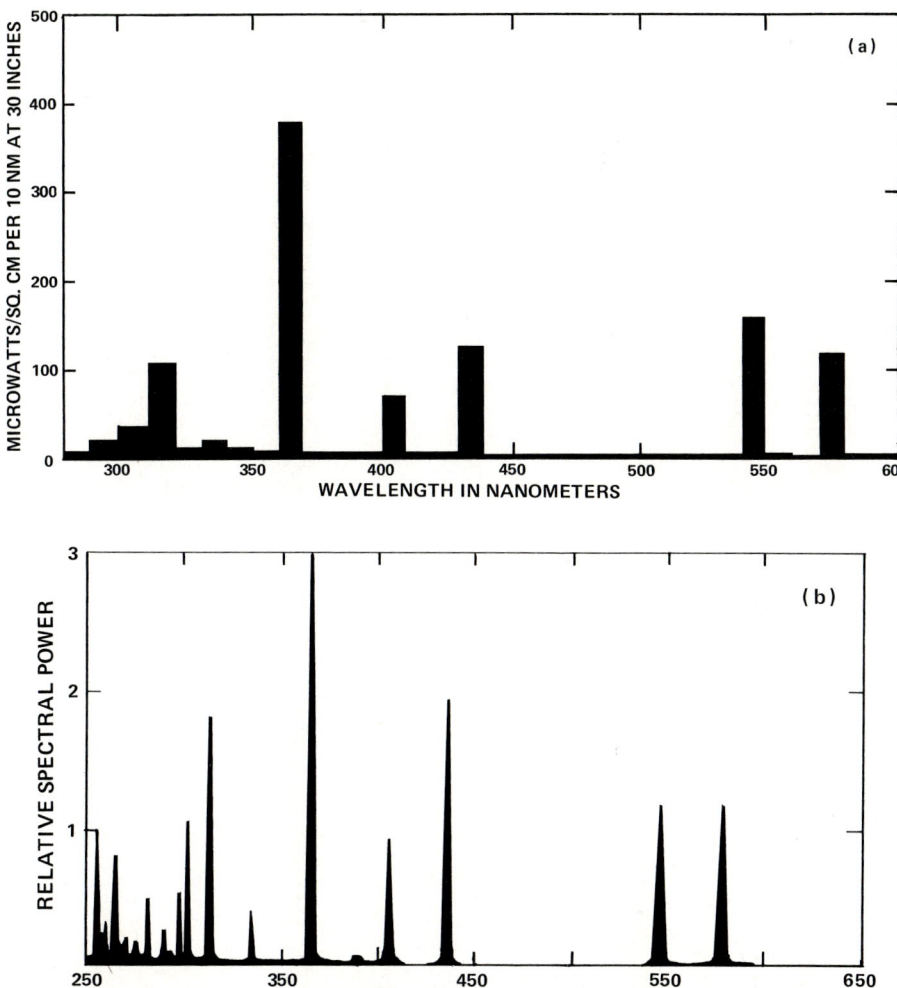

Fig. 16. Spectral power distributions of medium pressure mercury vapor arcs emitting ultraviolet radiation: (a) 275 W (arc plus ballast) RS sunlamp; (b) unjacketed quartz arc tube.

In sizes of 100 W and up, long life is one of their outstanding characteristics. These lamps have average life ratings of at least 24,000 hr; the luminous output decreases during life (see Fig. 18). This decrease in output is due principally to blackening of the arc tube walls with materials sputtered from the electrodes.

e. Temporal Variation of Output. In common with the other arc discharge sources, the medium pressure mercury arc exhibits modulation of its output

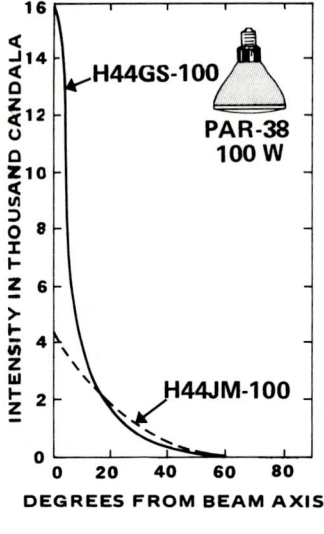
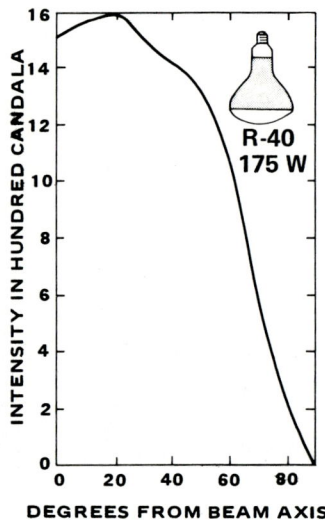
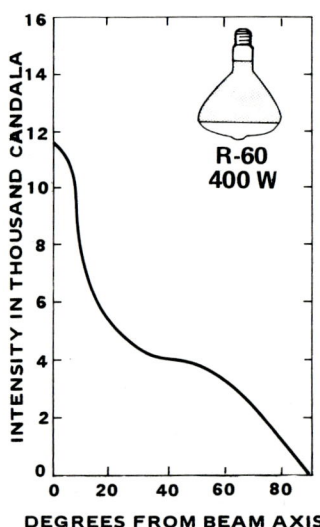

Fig. 17. Medium pressure mercury arcs in reflectorized jackets.

at twice the line frequency. For the clear mercury lamp this modulation is about 80%. In addition, there is a warm up period of about 5 min after the initial application of power (see Fig. 18). As the arc tube comes up to temperature and reaches equilibrium during this period, the light output will steadily increase. Once turned off, about 15 min cooling is required before the lamp will be ready to restart.

2. Metal Additive (Metal Halide) Lamps*

a. Theory of Operation. In order to overcome the spectral limitations of the mercury arc, it was long desired to incorporate other metals into the arc stream and benefit from their spectral characteristics. The major problem was the lower vapor pressures of these other metals, and this was finally overcome by incorporating them in the form of their iodides. While these iodides

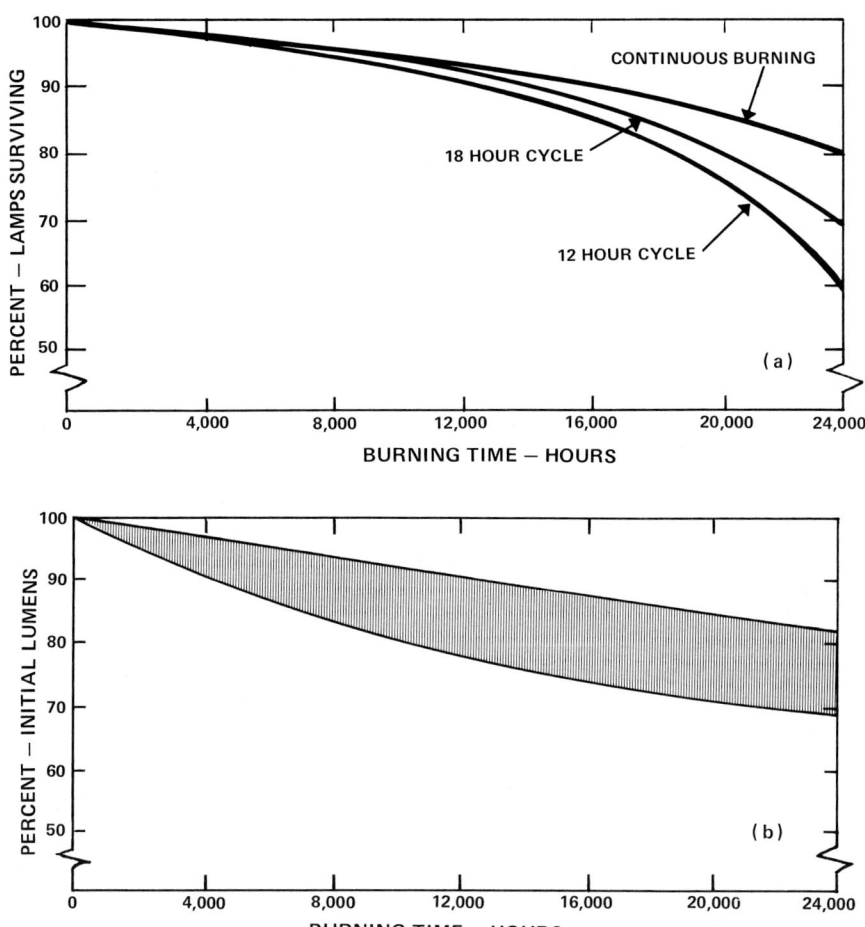

FIG. 18a,b. Operating characteristic of typical jacketed medium pressure mercury lamps (a) mortality curves; (b) maintenance (clear lamps).

* See Waymouth (1971), Elenbaas (1972), Levin and Lemons (1971), Larson et al. (1963), Waymouth et al. (1965), Tataronis (1972), Dobrusskin (1971).

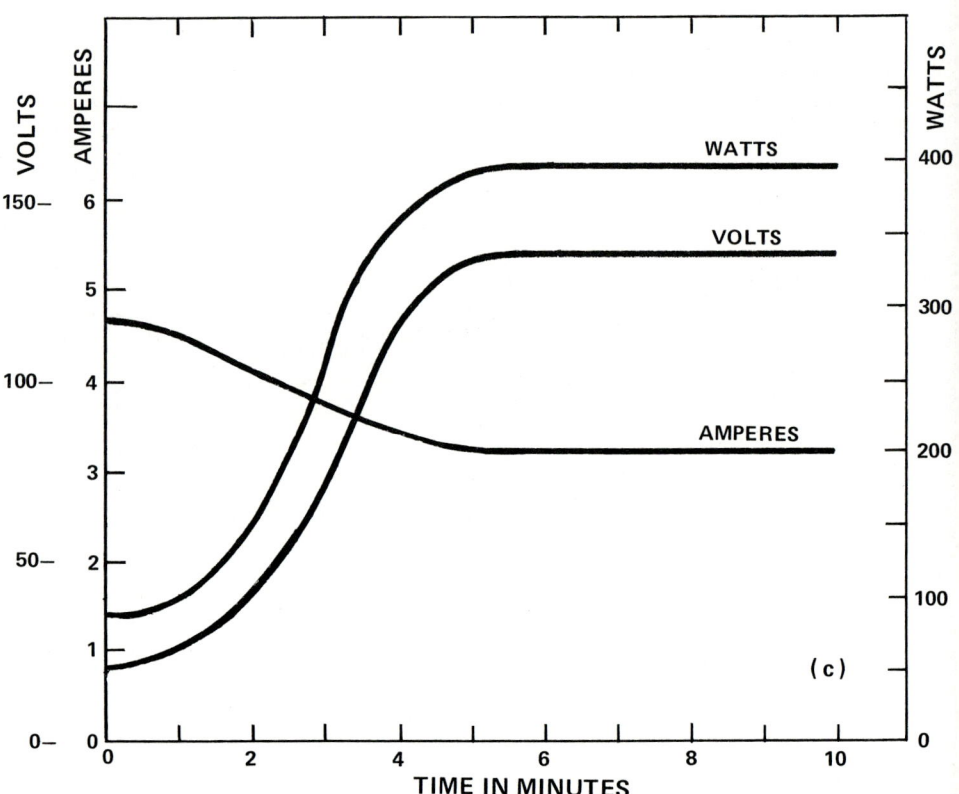

FIG. 18c. Electrical characteristics of 400 W lamp on reactor or autotransformer ballast.

condense on the arc tube walls when the lamp is cold, they evaporate once the arc is struck and the walls begin to warm up. After evaporating, they become disassociated in the region of the high arc stream, and the metals then produce their desired spectral contributions.

While these lamps closely resemble mercury lamps in their outer form, their inner construction has been somewhat modified. This is necessary to accomodate the complexities introduced by the presence of the metal iodides. The vapor pressure of the iodides in the arc tube is dertermined by the temperatures of the coolest points which are the tube ends. To maintain these ends at sufficiently elevated temperatures they are provided with a heat retaining outer coating. A bimetal shorting switch is also provided between the starting probe and its adjacent main electrode. This switch operates once the lamp has warmed up and precludes any flow of current through iodide deposits between these elements while the lamp is in operation. Such currents could lead to electrolytic failure of the vulnerable molybdenum foils in the end seals. Lamps manufactured in Europe frequently have no starting electrode and rely on an external starting aid to provide momentary high voltage pulses to initiate the arc.

The inclusion of the metal additives also change the electrical parameters of the lamp. When compared to the mercury family, these lamps require higher peak voltage for reliable starting and a higher lamp current crest factor (ratio of peak to rms) when operating. While some metal additive lamps will operate on certain mercury ballasts, this is not generally the case. Special ballasts are normally required for reliable operation.

b. Spectral Distribution of Radiation. By varying the choice and proportions of the additives which are incorporated, these lamps can be varied in their spectral output almost at will. For general lighting applications a combination of scandium and sodium iodides is widely used resulting in the spectral distribution shown in Fig. 19. While this spectrum affords vastly improved color rendition when compared to the mercury lamp, it can be improved still further by the addition of phosphors to the outer envelope. For special applications, lamps can be produced with the majority of their output restricted to fairly narrow regions of the visible or ultraviolet regions.

c. Size and Range. Standard metal additive lamps are currently available in wattages from 175 to 1500 with arc lengths from 28 to 95 mm, respectively.

As was the case for mercury, these lamps are also available in tubular forms, both jacketed and unjacketed (Ayotte and Tataronis, 1969), ranging

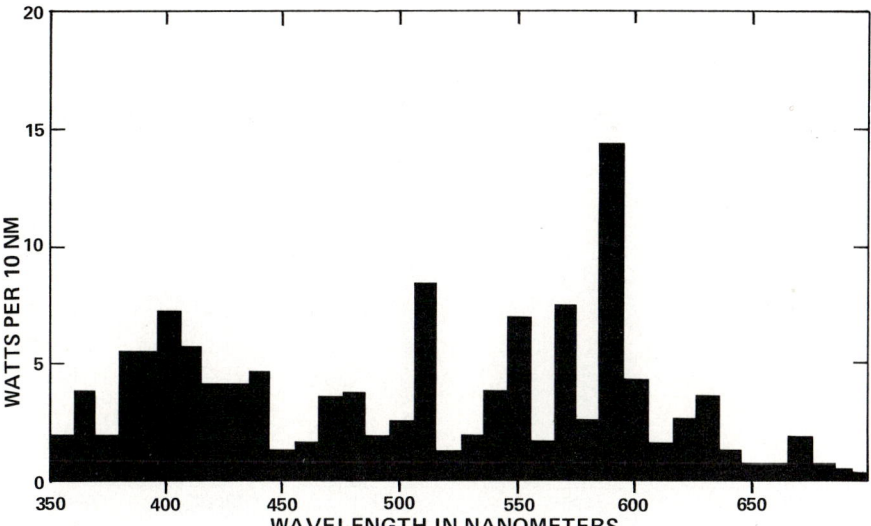

FIG. 19. Spectral power distribution of clear 400 W metal halide (Metalarc) lamp.

from about 10 to 100 cm in length and from 400 to 11,000 W. Used primarily in the reprographic photochemical fields, these lamps can be tailored to give energy in the desired spectral bands (Ayotte and Hale, 1967). A lamp with iron iodide additive (Gardner *et al.*, 1975), for example, produces from three to four times as much energy in the 300–400 nm region as a comparable mercury lamp.

d. Efficacy and Life. In addition to improved color rendering qualities, the standard metal additive lamps show significantly higher efficacies than mercury lamps with values on the order of 100 lm/W. The average life varies from 1,500 to 15,000 hr at present, depending on lamp wattage and burning cycle. These figures can be expected to increase as further development of these sources takes place. Identifying and optimizing the factors which lead to failure can be a tedious process for lamps which take a year or two to die.

e. Temporal Variation of Light Output. This family of lamps shows both long term maintenance and short term flicker values which are similar to those of mercury. The maintenance characteristics are illustrated in Fig. 20, and the flicker at twice line frequency is about 50%. As with mercury, these lamps take several minutes to warm up to full light output and require a cooling period before they can be restarted. When immediate restart capabilities are imperative, a high voltage pulse circuit can be employed to restrike the arc. Pulses up to tens of kilovolts are required, so this technique can only be

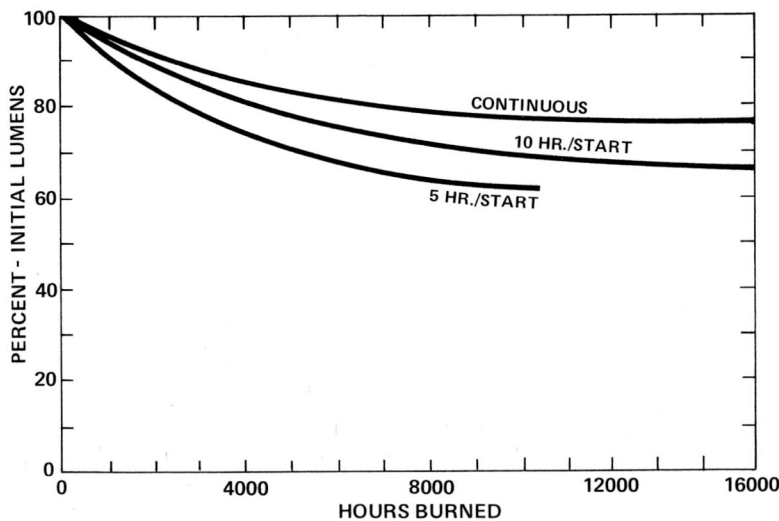

FIG. 20. Maintenance characteristics of 400 W metal halide lamps.

used with double ended lamps where the external leads are sufficiently separated to prevent arcing.

3. *Sodium**

a. Theory of Operation. Development of a practical medium pressure sodium lamp[†] had to wait until arc tube materials could be found which would withstand the attack of high temperature sodium. This problem was finally overcome with a combination of translucent sintered polycrystalline alumina for the wall material and niobium metal for the electrical feedthroughs. A schematic of a typical lamp is shown in Fig. 21. Since it is not feasible to incorporate a third electrode for starting purposes, an external circuit to provide high voltage starting pulses is normally required. This circuit is usually incorporated into the ballast and provides a single microsecond wide pulse on the order of 2–4 kV once every cycle or half cycle. By replacing the normal xenon gas fill with a Penning mixture (Penning, 1957) of rare gases, it is possible to reduce the starting voltage requirements considerably. This development has lead to a family of medium pressure sodium lamps[†] which can be substituted directly for mercury lamps on certain ballasts (Cohen and Richardson, 1975). While having a lower efficacy than regular sodium lamps, they offer a significant improvement over the mercury lamps they replace.

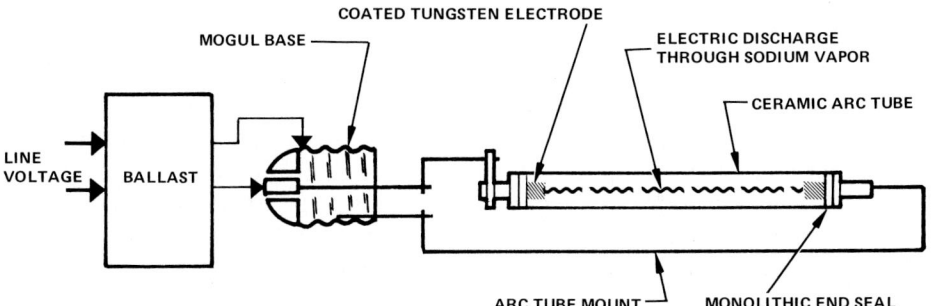

FIG. 21. Electrical circuit of typical medium pressure sodium arc lamp.

b. Spectral Distribution. In going from the low pressure to the medium pressure sodium arc, the characteristic sodium emission doublet becomes enormously broadened and self-reversed as shown in Fig. 22. While deficient

---

* See Waymouth (1971, 1977), Elenbass (1972), Lin (1970), Lin and Knochel (1974).
† In the lighting industry, these are usually referred to as high pressure sodium lamps in contrast to the older low pressure sodium lamps.

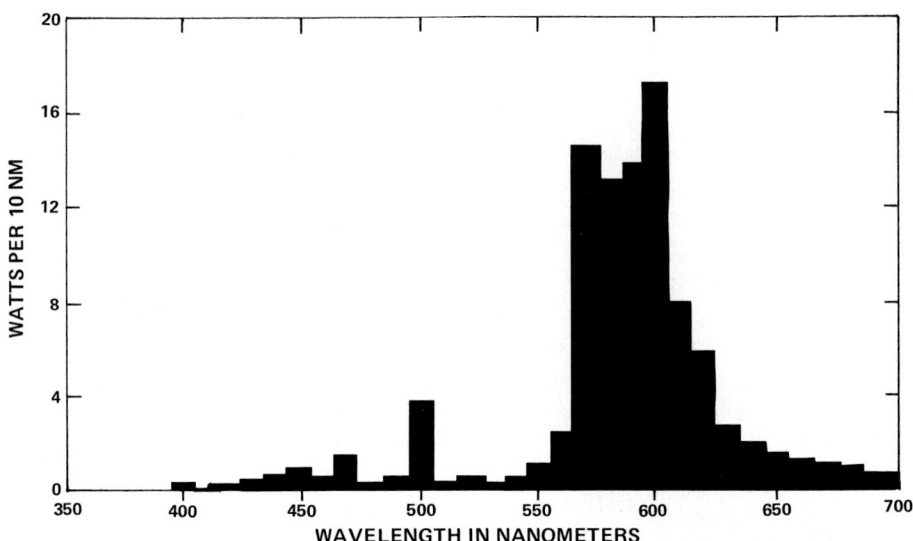

FIG. 22. Spectral power distribution of 400 W medium pressure sodium lamp.

in the blue portion of the spectrum, this source does a moderately good job of color rendering and it is much more satisfactory than the medium pressure mercury lamp.

   c. Size and Shape. At this time these lamps are available in wattage from 150 to 1000 with arc lengths from 4 to 21 cm, respectively. Arc tube diameters range from about 6 to 9 mm. It is expected that lower wattage lamps will be available in the future. These sources are all enclosed in outer jackets to provide protection from the atmosphere and are available with diffusing coatings on the jackets if larger effective source sizes of lower luminance are desired. Phosphors are not used in this case due to the low levels of ultraviolet radiation which are produced by the arc.

   d. Efficacy and Life. Sodium lamps have the highest efficacies of the medium pressure family with initial values as high as 140 lm/W. The maintenance of this light output with life is excellent; life is also quite long with values in the vicinity of 20,000 hr (see Fig. 23). Under normal conditions the lamp requires about four minutes to warm up and will restrike after about 1 min of cooling following a power interruption. The normal failure mode is rather unusual in that the lamp will start but then extinguishes during the warmup process. After cooling down, the lamp restrikes and the process repeats itself. For obvious reasons this failure mode is referred to as "cycling."

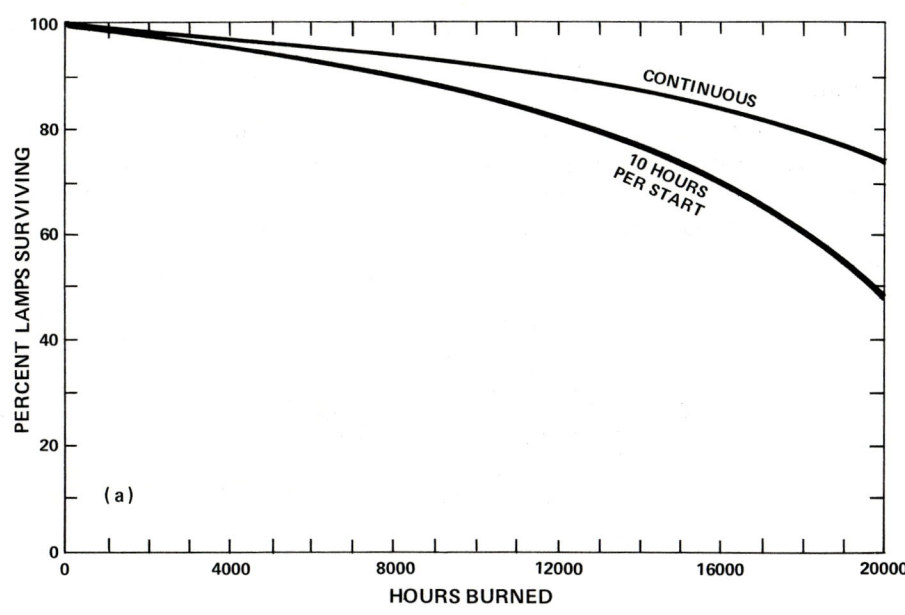
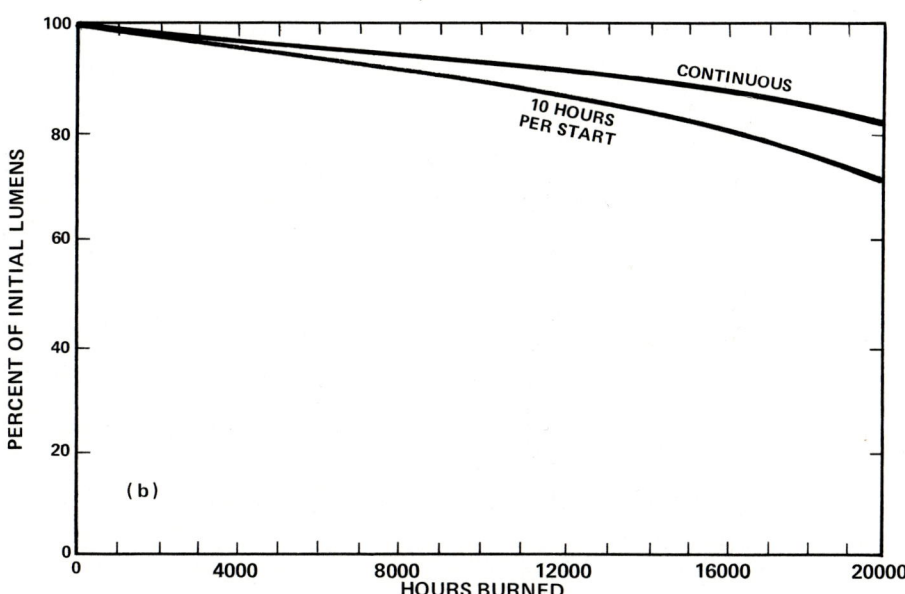

FIG. 23. Operating characteristics of typical 400 W medium pressure sodium lamp: (a) mortality curves; (b) lumen maintenance curves.

D. High Pressure Lamps

1. *HMI Metal Additive Lamp*

Just as metal iodides are added to the medium pressure arc lamps, they have been added to the high pressure compact arc as well. Developed in Germany primarily for film and television lighting, Osram's HMI lamps (Block *et al.*, 1974) are available in wattages from 575 to 4000 with arc lengths from

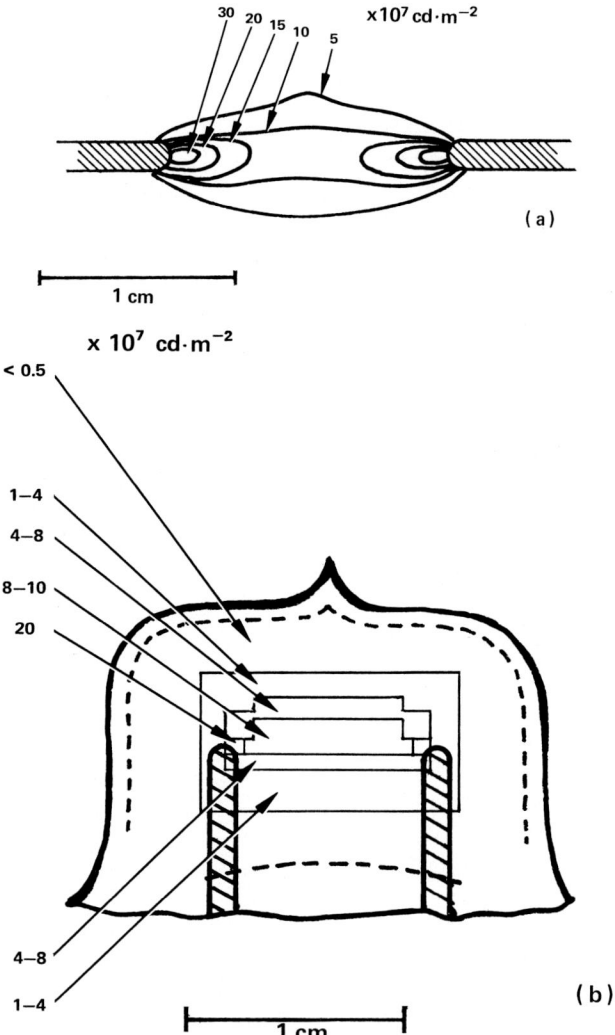

FIG. 24. Typical arc luminance of high pressure lamps: (a) 1200 W HMI lamp; (b) 400 W CSI lamp.

12 to 35 mm, respectively. They are characterized by excellent color rendering qualities and efficacies on the order of 80–100 lm/W. For color critical applications their useful life is limited to about 200 hr since the color temperature decreases about 1 K/hr. However, the lamps themselves will last on the order of 1000 hr with 90% lumen maintenance at 600 hr. High voltage pulses in the 20–40 kV range are required to start these lamps and are of sufficient magnitude to offer instant restrike capability after the lamp has been momentarily extinguished. On initial lightup, two to three minutes are required for the lamp to stabilize. Figures 24 and 25 illustrate typical luminance and spectral characteristics.

2. *CSI Metal Additive Lamp*

The CSI lamp (Ainsworth and Beeson, 1967), developed by Thorn in Great Britain, is somewhat similar to the HMI lamp in concept although not in form. Available in 400 and 1000 W sizes with 9 and 15 mm arc lengths, respectively, these lamps are of compact single-ended construction with the electrical leads brought out side by side. When enclosed within a protective outer envelope these lamps will have lives on the order of 1000 hr. When exposed to the atmosphere, this figure drops to about 200 hr. Their efficacy and color rendering are similar to the HMI lamp, and their light output wave form shows about 40% modulation when operated on 50 Hz power. This figure should be slightly smaller for 60 Hz operation. These lamps are commonly started with high voltage pulses in the 10 kV region which are not sufficient to give instant restrike capabilities. If the normal cooling period of from 2 to 10 min is not acceptable before restarting, pulses in the 30–40 kV region can be used. Due to the reduced spacing between the external electrodes, however, considerable care has to be used in the layout of the electrical connections for this situation. Figures 24 and 25 illustrate typical luminance and spectral characteristics.

3. *Xenon*

a. Overview. Compact-source xenon arc lamps (Volume I, Chapter 2, Section V) traditionally are made with roughly ellipsoidal quartz bulbs. A new version has been developed (Varian/Eimac Division) using an integral reflector in a ceramic body with a sapphire window. The reflector surface is silver or, in some cases (UV), aluminum to provide increased ultraviolet output. Reflectors are either paraboloidal to produce a nearly collimated beam or an ellipsoidal/spherical combination to concentrate the flux on a small spot. The spectral distribution is similar to other compact-source xenon lamps and has a color temperature of 6000 K. Horizontal operation is recommended, but the lamps may be operated in any position except where the lamp axis is within 45° from vertical with the window up.

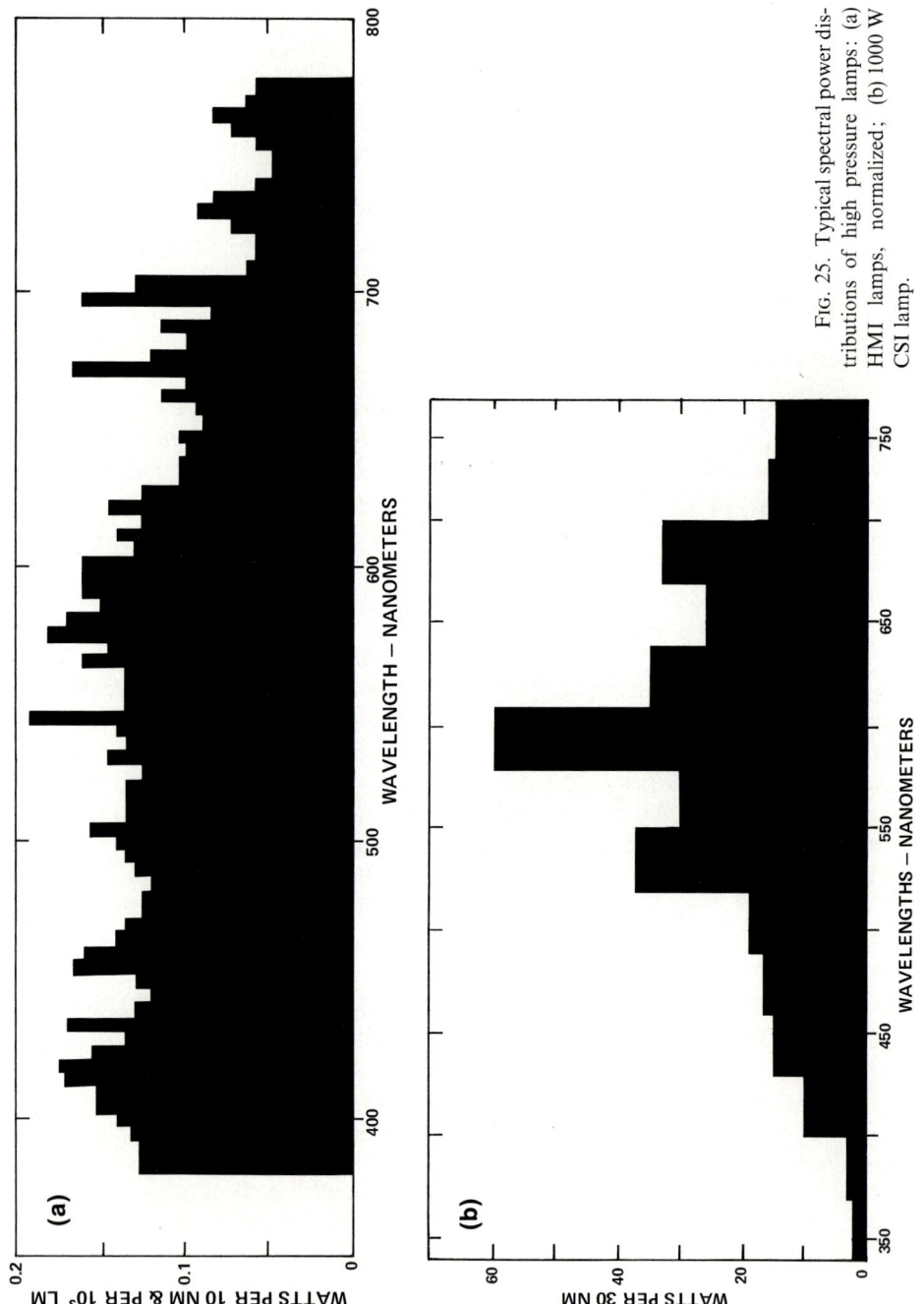

Fig. 25. Typical spectral power distributions of high pressure lamps: (a) HMI lamps, normalized; (b) 1000 W CSI lamp.

TABLE I

INTEGRAL REFLECTOR COMPACT-SOURCE XENON LAMPS (10 HR VALUES)

Type	Input power (watts)	Total luminous flux (lumens)	Reflector	Window diameter (mm)	Focal[a] ratio	Focal[b] distance (mm)	Peak intensity (candela)	Beam[c] angle (deg)	Radiant[d] power (watts) Total	UV	IR
VIX-150	150	1750	Parabola	25.4			200,000	8°	22	1	14
VIX-150UV	150	1500	Parabola	25.4			170,000	8	20	2	12
VIX-300	300	5000	Parabola	25.4			300,000	10	60	2	40
VIX-300UV	300	4000	Parabola	25.4			250,000	10	50	4	30
VIX-500	500	8000	Parabola	48.0			1,500,000	6	100	5	65
VIX-500UV	500	6000	Parabola	48.0			1,100,000	6	75	6	45
VIX-501	500	8000	Ellipse/sphere	48.0	1.33	48.3			100	5	65
VIX-800	800	15,000	Parabola	48.0			5,000,000	6	215	9	145
VIX-801	800	15,000	Ellipse/sphere	48.0	1.33	48.3			215	9	145

[a] Focal length/reflector aperture.
[b] Focal point from window mounting flange.
[c] Total spread to 10% of peak intensity.
[d] UV: 200–400 nm; IR: 800–2100 nm.

b. Operating Characteristics. At recommended power and approximately 1 hr/start, satisfactory operation is possible in excess of 1000 hr. Table I summarizes initial operating characteristics. Stability and change of performance characteristics determine end of useful life. At 1000 hr the color temperature can drop 200 K, the luminous flux can drop 25–50%, and the peak intensity can drop 60–80%. Also, the beam spread will increase and the focal distance will change due to an increase in arc gap dimensions.

IV. PHOTOFLASH LAMPS

Subsequent to the earlier discussion (Volume I, Chapter 2, Section VI), this family of lamps has shown evolution in two directions. First, the development of smaller bulb sizes has lead to the concept of self-contained packages of multiple bulb–reflector combinations. Secondly, the traditional problems associated with battery ignition of the lamps has encouraged development of alternative triggering methods.

To date, the multiple flash packages which have appeared are the flashcube, hi-power flashcube, magicube, flash bar, and flip flash. About 3 cm on an edge, the cubes all contain four lamps and reflectors, one behind each side face of the cube. The hi-power cube provides roughly twice the peak intensity and total light output of the other two cubes. The flash bar contains two side by side arrays of five lamps each, the two arrays being placed back to back in a package about 11 cm wide by 4 cm high.

The flip flash has a total of eight lamps in an array four lamps high by two lamps wide with package dimensions of about 14 cm high by 5 cm wide. The light output of the flash bar is about 80% that of the hi-power cube, while the flip flash is comparable to the flashcube. All of these packages incorporate blue filters which result in a color temperature of about 5500 K for use with color film.

The innovations in lamp ignition have consisted of a mechanically activated lamp, used in the magicube, and a lamp designed for piezoelectric firing, used in the flip flash.

The magicube lamp has a metal tube extending from its base containing a percussive primer which is fired by releasing a cocked spring in the cube base. The flip flash lamp contains a primer designed for the extremely low energy but very high voltage pulse which is obtained by mechanically shocking a piezoelectric element located in the camera body. Both of these approaches increase system reliability by avoiding the problems of battery life and contact resistance in low voltage circuits.

V. LIGHT EMITTING DIODES

A. Overview

The light emitting diode (LED) or solid state lamp is a p–n junction semiconductor which operates on the principle of junction luminescence (Bergh and Dean, 1972). Under forward bias, it conducts current by the same mechanism as other p–n junction devices and radiates in the visible or infrared spectral region. Light emitting diodes are formed both as integral parts of solid state displays and as discrete solid state lamps.

B. Operation

Radiative recombination occurs when photons are generated by the energy of recombing holes and electrons. In direct gap semiconductor material, minority carrier recombination may occur in a single step directly between the conduction band and the valence band across the forbidden gap. Temporary trapping may occur in the forbidden gap of indirect gap semiconductor material and result in a two- or three-step recombination process. The direct gap materials intrinsically are more efficient as light producers.

When electron energy is released by a direct transition from the conduction band to the valence band, that energy is radiated as a photon. Two-step electrical transitions due to crystal flaws and impurities cause nonradiative recombinations. This contributes to the total diode current thus affecting the radiation efficiency.

One measure of LED efficiency is the internal quantum efficiency, i.e., the ratio of the number of photons generated at the junction to the number of electrons through the junction. It is possible for this value to approach unity. However, the user generally is more interested in the external efficiency. Contact resistance and bulk resistance constitute the electrical losses which reduce external efficiency.

External efficiency is also reduced by optical losses. There is significant internal absorption as photons travel from the junction to the semiconductor material surface. The high refractive index of this material causes a Fresnel loss and a loss due to total internal reflection at the semiconductor surface since reflected photons are rapidly absorbed. Further losses may occur in the lenses and/or diffusing materials above the diode chip.

The common commercially available LED's are formed of a ternary compound of gallium arsenide and gallium phosphide ($GaAs_{1-x}P_x$). Pure gallium phosphide radiates at about 565 nm in the yellow-green while pure gallium arsenide radiates at about 900 nm in the near infrared. By changing the relative percentages of the two in the ternary compound, intermediate wavelengths are obtained.

LED's using other materials are in various states of development and availability (RCA Electro-Optical Handbook, 1974). In some instances, phosphors may be used to convert infrared radiation to visible emission. Fig. 26 shows typical spectral power distributions of LED's.

C. Optical Characteristics

The intrinsic Lambertian intensity distribution of a diode chip is almost always modified to a more desirable form. A diode may be hermetically sealed directly below a lens using the lens spacing and power to modify the intensity

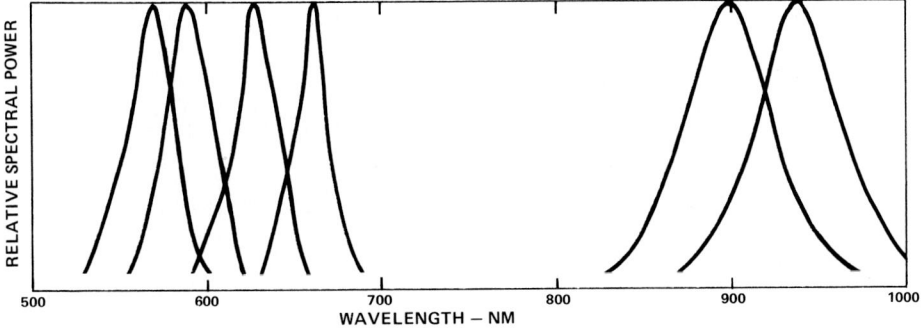

Fig. 26. Typical relative spectral power distributions of light emitting diodes.

distribution. However, the diode and its mount are most frequently sealed directly in plastic or an epoxy whose surface forms the lens; this now is an immersion lens. By thus increasing the refractive index of the material in contact with the high index semiconductor material, both the Fresnel loss and the loss due to total internal reflection will be reduced allowing more light to escape from the diode. With proper lens surface shapes, much of this additional light can be emitted from the solid state lamp increasing its external efficiency.

The luminance of the lamp is established by junction luminescence as modified by losses. It is nearly constant at all angles. Projected area of the emitter as a function of direction is modified by the lens to control the intensity distribution. Although luminance relates to contrast, photographic exposure, etc., intensity is generally considered to be a better overall metric for viewing information. Intensity in various directions relates quite closely to the LED effectiveness for viewing.

In some LED's, diffusing material is incorporated in the encapsulation to increase the apparent size of the source and/or to give a very wide intensity distribution. Also, the encapsulation may have a wavelength selective filtering action either to modify the emission spectrum or to enhance contrast by decreasing reflection of ambient incident light.

A large variety of solid state lamps exist in terms of electrical parameters, optical characteristics, and packaging. As a somewhat tenuous generalization, the typical solid state lamp can be characterized as having a forward voltage of about 2 V, a forward current on the order of tens of milliamperes, a minimum reverse breakdown voltage of 3–5 V, a response time in tens of nanoseconds, an axial intensity of a few millicandela, and a half power bandwidth on the order of 30 nm.

The spatial intensity distribution virtually always has a nominal axial symmetry; it usually decreases monotonically with increasing angle. Many

TABLE II
Typical Parameters for Some Commercially Available Solid State Lamps

Crystal	GaAsP	GaAsP	GaP
Color	Red	Yellow	Green (diffuse)
Peak wavelength	660 nm	585 nm	565 nm
Bandwidth ($\frac{1}{2}$ power)	20 nm	35 nm	35 nm
Maximum ratings (25°C)			
Forward Current (continuous)	70 mA	35 mA	35 mA
Power	175 mW	105 mW	105 mW
Reverse voltage	5 V	5 V	5 V
Typical Operation (25°C)			
Forward current	50 mA	20 mA	20 mA
Forward voltage	1.65 V	2.1 V	2.2 V
Axial intensity	4 mcd	45 mcd	1.0 mcd
50% intensity angle	±35°	±10°	±55°

designs tend to be somewhat narrow for better utilization of the limited available power. Howerever, broad intensity distributions for wide angle viewings are available.

The maximum permissible power must be derated several fold as ambient temperature increases from 25 °C to the upper operating limit, usually about 80–100 °C. Most characteristics are highly dependent on operating temperature. The spectral distribution is sensitive to temperature and driving conditions; shifts as great as 10 nm may be encountered.

Table II presents typical operating parameters for three commercially available solid state lamps. As an example of functional relations, characteristic variations for the red lamp (column 1, Table II) are presented in Fig. 27. The inherent variation in performance parameters between individual lamps

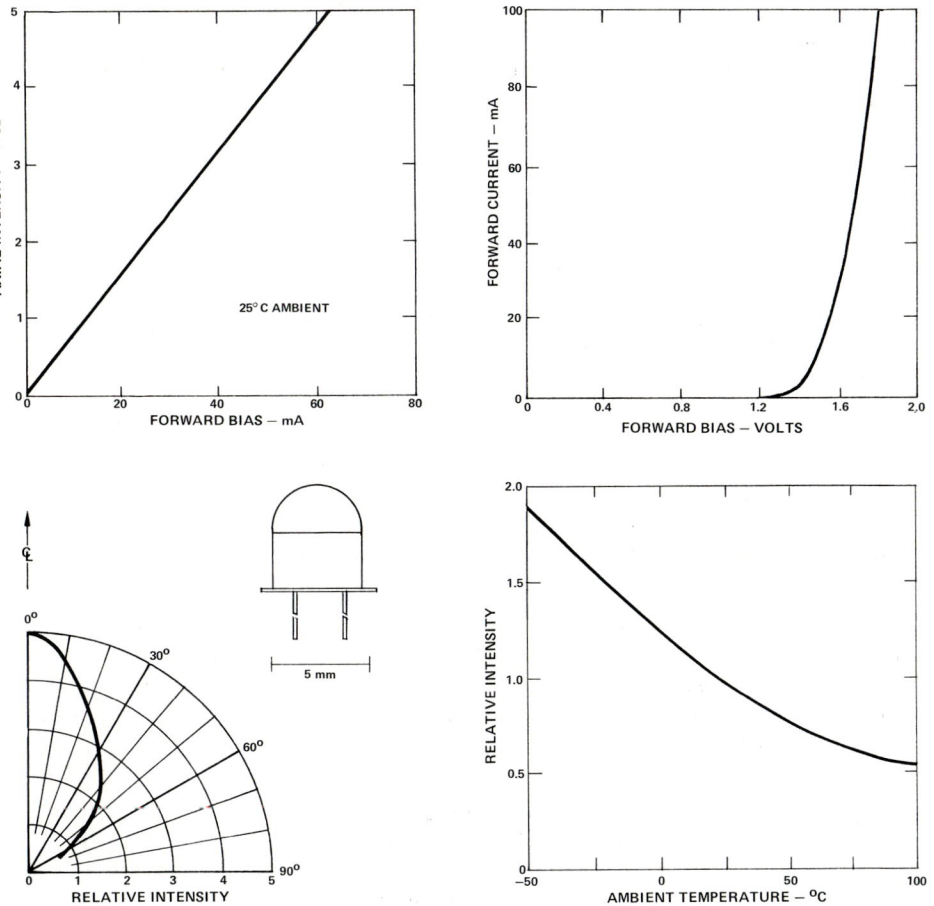

FIG. 27. Typical solid state lamp characteristics (red lamp, column 1, Table II).

of a given type may be quite large. In some cases, the maximum intensity may vary five to one; the peak wavelength may vary ±20 nm. The optical and mechanical axis may have as much as a ±10° tolerance. Since tolerances vary considerably for the various types of LED's, it is generally important to refer to such information in the manufacturer's specifications.

VI. NATURAL RADIATION

A. Overview

Natural solar radiation at the earth's surface is a function of meteorological and atmospheric conditions, seasonal variations, terrestrial location, solar elevation and azimuth, and orientation of the receiving plane. This radiation may be resolved into three components: direct solar radiation, sky radiation (the component scattered by the earth's atmosphere), and radiation reflected from the earth's surface. Irradiance on a horizontal plane is independent of the reflected component and of the solar azimuth; it is termed "total radiation" in geophysics.*

Atmospheric scattering is a function of wavelength and principally affects the ratio of solar to sky irradiance. Atmospheric attenuation also is a function of wavelength, but it changes the spectral composition of natural radiation. Absorption is a function of time of day, season of year, latitude, and elevation above sea level[†] as well as turbidity and aerosols (clouds, fog, haze, dust, smoke, etc.) and ozone layer thickness. This dependence increases as wavelength decreases.

B. Spectral Distribution

Judd *et al.* (1964) subjected over 600 daylight spectral distribution functions taken under a variety of conditions to eigenvector analysis. It was determined that correlated color temperature and spectral distribution in the visible spectrum are well correlated. Representative relative spectral power distribution functions as illustrated in Fig. 28 were reconstituted from the mean and the first and second eigenvectors. The extremely fine structure of the daylight was eliminated, with the values being taken as the true average of spectral power over wavelength intervals of 10 nm (Judd *et al.*, 1964; Wysecki and Stiles, 1967).

---

* In some disciplines, this is also known as "global radiation."

[†] These first four effects are reasonable correlates of air mass. Air mass is a measure of the relative amount of atmosphere traveled by direct solar radiation with unity at the zenith; it varies approximately as the secant of the solar angle from the zenith.

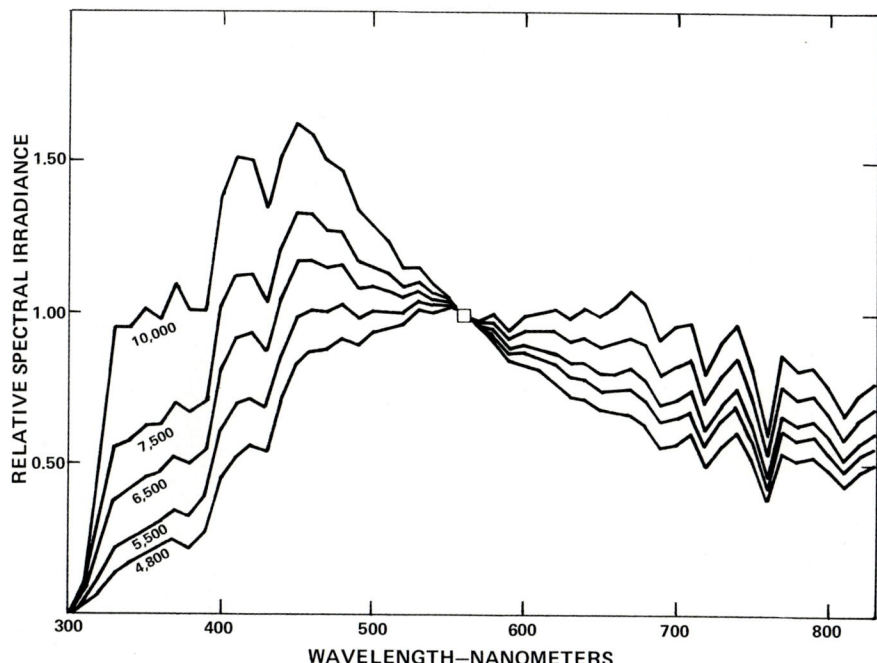

FIG. 28. Relative spectral distributions for five phases of daylight specified in terms of correlated color temperature (K).

As a basis for standardized simulation and uniform calculation, the Commission Internationale de l' Éclairage (C.I.E.) (1972) has recommended values of the total radiation in spectral bands for zenith sun, air mass = 1, and cloudless days. This (geometric) total spectral irradiance on a horizontal plane at the earth's surface is shown in Fig. 29 and Table III. Figure 30 shows the fraction of total radiation due to the sky radiation as a function of solar altitude and wavelength as well as total irradiance as a function of solar altitude. Table IV shows the effect of cloud cover on total radiation, while Table V shows the luminance of natural sources.

C. Luminance

1. Overcast Sky

The luminance $L(\beta)$ at point P with zenith angle β for an overcast sky has been standardized by the C.I.E. (1955) as a function of zenith luminance L_0 (see Fig. 31 for geometry).

$$L(\beta) = (L_0/3)(1 + 2\cos\beta) \qquad (1)$$

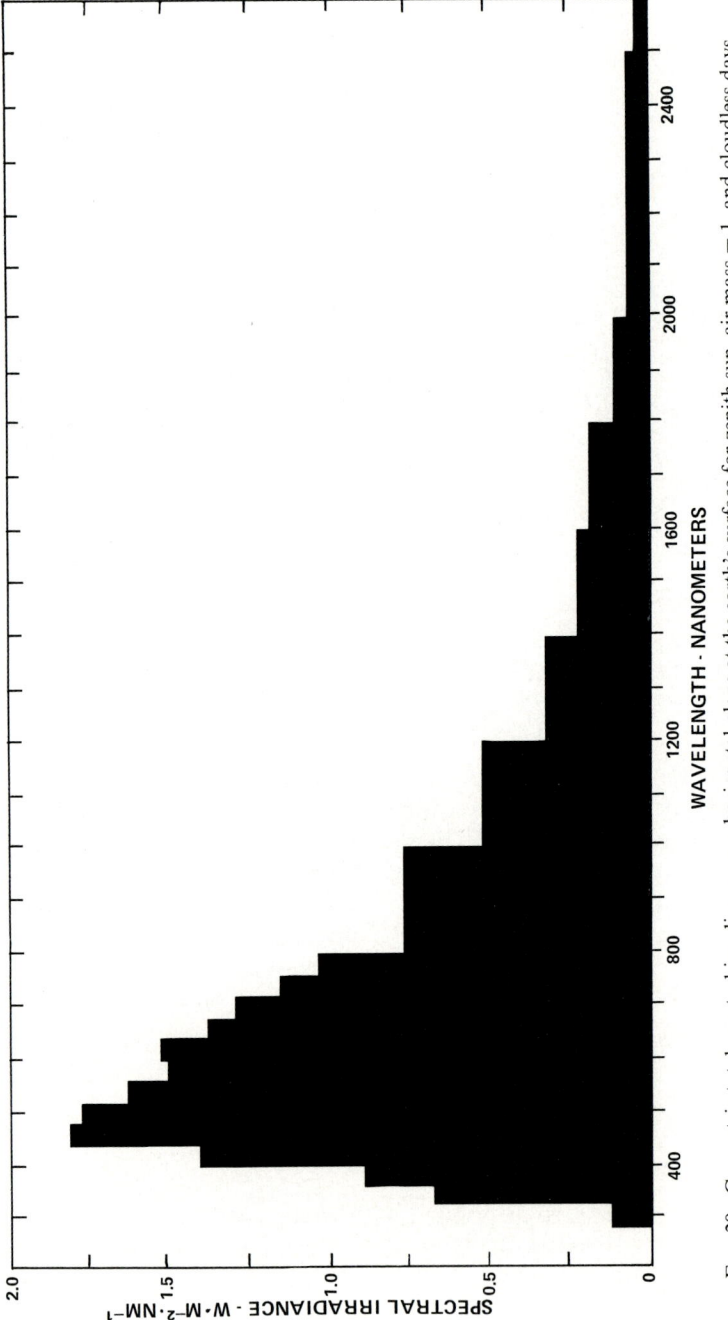

FIG. 29. Geometric total spectral irradiance on a horizontal plane at the earth's surface for zenith sun, air mass = 1, and cloudless days.

TABLE III
IRRADIANCE OF THE TOTAL RADIATION IN SPECTRAL BANDS[a]

Range	Wavelength (nanometers)	Irradiance (watts/meter^2)	Range	Wavelength (nanometers)	Irradiance (watts/meter^2)
0	<280	0	3	800–1000	156
				1000–1200	108
	280–320[b]	5		1200–1400	65
1	320–360	27			
	360–400	36		1400–1600	44
				1600–1800	29
	400–440	56	4	1800–2000	20
	440–480	73		2000–2500	35
	480–520	71		2500–3000	15
	520–560	65			
2	560–600	60	5	3000[b]	—
	600–640	61			
	640–680	55	0–5	Σ	1120
	680–720	52			
	720–760	46			
	760–800	41			

[a] Spectral irradiance on horizontal plane due to direct solar radiation plus sky radiation for zenith sun (altitude = 90°), air mass = 1, and cloudless day. Commission Internationale de l'Éclairage (1972).

[b] Radiation below 300 nm does not reach the surface of the earth; radiation above 3000 nm is negligible.

Krochmann (1963; Krochmann and Seidl, 1974) has suggested a function for the zenith luminance of an overcast sky L_0 in terms of the solar zenith angle δ.

$$L_0 = 123 + 8600 \cos \delta \quad \text{cd/m}^2 \tag{2}$$

2. Clear Sky

The luminance $L(\alpha, \beta)$ at a point P for an absolutely cloudless sky has been standardized by the C.I.E. (1973) as a function of the zenith luminance L_z (see Fig. 31 for geometry).

$$L(\alpha, \beta) = [L_z f(\varepsilon) \varphi(\beta)][f(\delta) \varphi(0°)]^{-1} \tag{3}$$

where

$$f(x) = 0.91 + 10 \exp(-3x) + 0.45 \cos^2 x \tag{3a}$$
$$\varphi(x) = 1 - \exp(-0.32 \sec x) \tag{3b}$$

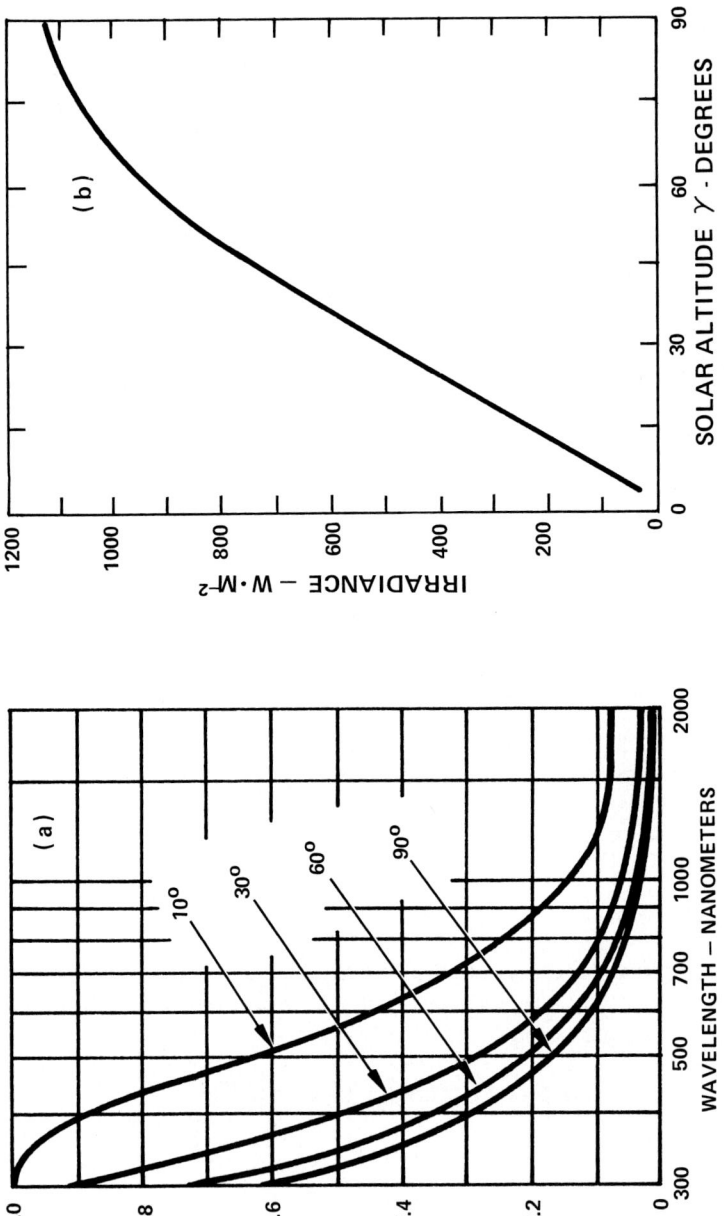

FIG. 30. (a) Fraction of total radiation due to sky radiation as a function of solar altitude and wavelength; (b) total irradiance as a function of solar altitude.

TABLE IV
EFFECT OF CLOUDS ON
TOTAL RADIATION[a]

Clouds	Total radiation (%)
Cloudless	100
$\frac{1}{5}$ cloudiness	89
$\frac{2}{5}$ cloudiness	77
$\frac{3}{5}$ cloudiness	64
$\frac{4}{5}$ cloudiness	46
Total cloudiness	20

[a] Commission International de l'Éclairage (1972).

TABLE V
LUMINANCE, CHROMATICITY COORDINATES, AND CORRELATED COLOR
TEMPERATURE OF NATURAL SOURCES[a]

Source	Luminance (cd/cm^2)	Chromaticity coordinates		Correlated color temperature (K)
		x	y	
Sun's disk				
Measured above atmosphere (m = 0)	200,000	0.318	0.330	6200
Measured near sea level for air masses				
m = 1	150,000	0.331	0.344	5600
m = 2	125,000	0.343	0.357	5100
m = 3	100,000	0.356	0.369	4700
m = 4	80,000	0.368	0.379	4400
m = 5	65,000	0.380	0.388	4100
Sky				
Clear blue sky	0.06–0.4	0.262	0.270	15,000
		0.247	0.251	30,000
Partly cloudy sky	0.1–0.4	0.294	0.309	8000
		0.279	0.291	10,000
Overcast sky	0.2–0.5	0.313	0.329	6500

[a] Wyszecki and Stiles (1967).

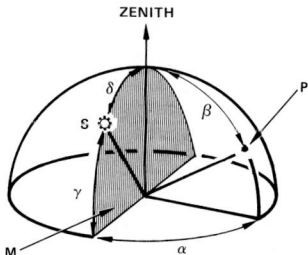

FIG. 31. Geometrical relations: S, sun; M, meridian plane of sun; δ, solar zenith angle; γ, solar elevation.

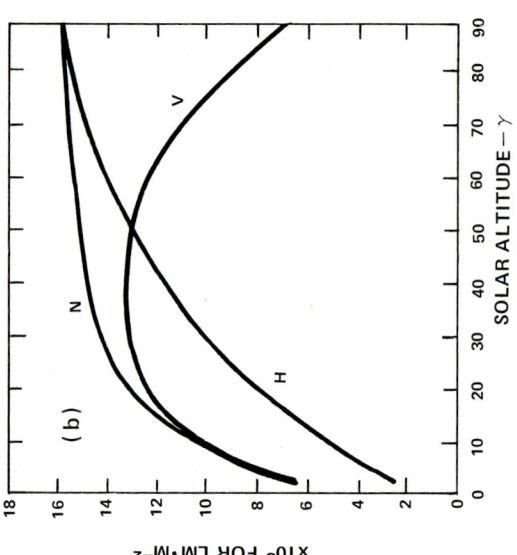
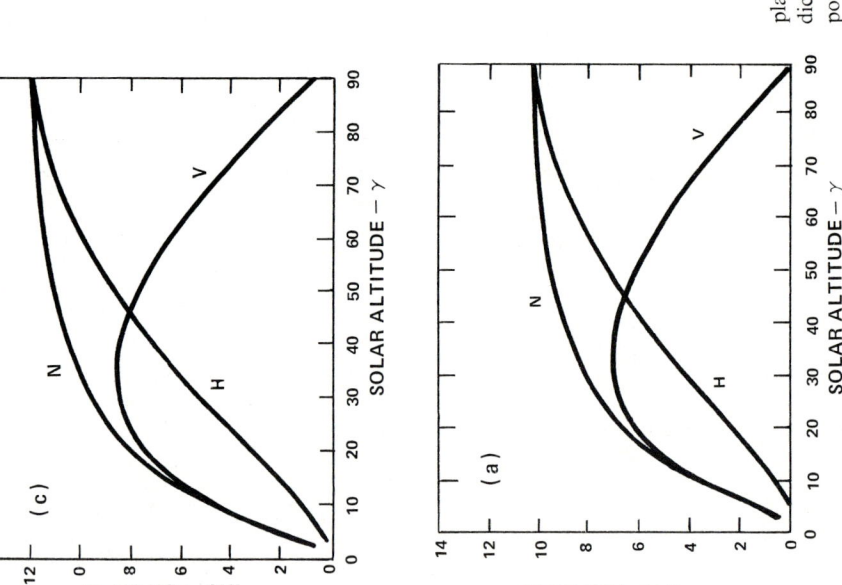

Fig. 32. Illuminance under clear sky on a horizontal plane (H), a plane normal to the direction of the sun (N), and a vertical plane perpendicular to the sun's meridian plane and facing the sun (V). (a) Solar component; (b) sky component; (c) daylight (sun plus sky).

Under the polluted atmosphere of large cities of industrial areas, replace Eq. (3a) by

$$f'(x) = 0.856 + 16\exp(-3x) + 0.3\cos^2 x \tag{3c}$$

Angle ε may be determined from

$$\cos\varepsilon = \cos\delta\cos\beta + \sin\delta\sin\beta\cos\alpha \tag{4}$$

Krochmann (Krochmann and Seidl, 1974; Krochmann et al., 1970) has suggested a function for the zenith luminance of a clear sky L_z in terms of the solar elevation angle γ expressed in degrees,

$$L_z = 100 + 63\gamma + \gamma(\gamma - 30)\exp[0.0346(\gamma - 68)] \quad \text{cd/m}^2 \tag{5}$$

D. Illuminance

1. Overcast Sky

The illuminance due to an overcast sky expressed in terms of zenith luminance (cd/m^2) can be determined (Krochmann and Seidl, 1974) by integrating Eq. (11). The illuminance on a horizontal plane is given by

$$E_h = 2.44 L_0 \quad \text{lm/m}^2 \tag{6}$$

and that on a vertical plane (neglecting direct ground reflection) is

$$E_v = 0.969 L_0 \quad \text{lm/m}^2 \tag{7}$$

2. Clear Sky

Illuminance under clear sky conditions can be divided into the direct solar component and the sky component. These values (Jones and Condit, 1948) plus their sum, daylight, are shown in Fig. 32 as a function of solar altitude γ. Values are given on three planes: (a) a horizontal plane, (b) a plane normal to the direction of the sun,* and (c) a vertical plane perpendicular to the sun's meridian plane and facing the sun.

REFERENCES

Ainsworth, T. S., and Beeson, E. J. G. (1967). *Photogr. J.* p. 324 (October).
Amick, C. L. (1960). "Fluorescent Lighting Manual." McGraw-Hill, New York.
Ayotte, R. D., and Hale, R. R. (1967). *Illum. Eng.* (*N.Y.*) **62**, 221.
Ayotte, R. D., and Tataronis, R. T. (1969). *Illum. Eng.* (*N.Y.*) **64**, 103.
Balder, J. J., and van de Weijer, M. H. A. (1956). *Philips Tech. Rev.* **17**, 198.
Baumgartner, R. G. (1941). *Illum. Eng.* (*N.Y.*) **36**, 1340.

* This plane is tipped at angle δ from the vertical.

Beijer, L. B., van Boort, H. J. J., and Koedam, M. (1974). *Light. Des. Appl.* **4**, 15 (July).
Bergh, A. A., and Dean, P. J. (1972). *Proc. IEEE* **60**, 156.
Block, W., McGovern, M. J., and Lemons, T. M. (1974). *J. Soc. Motion Pict. Telev. Eng.* **83**, 725.
Burgin, R., and Edwards, E. F. (1970). *Light. Res. Technol.* **2**, 95.
Cohen, S., and Richardson, D. A. (1975). *Light. Des. Appl.* **5**, 12 (September).
Commission Internationale de l'Éclairage (1955). *Proc. CIE, 13th Sess.*, Zurich **2**, Part 3-2.
Commission Internationale de l'Éclairage (1972). "Recommendations for the Integrated Irradiance and the Spectral Distribution of Simulated Solar Radiation for Testing Purposes." Publ. CIE No. 20 (TC-2.2). Comm. Int. Éclair.
Commission Internationale de l'Éclairage (1973). "Standardization of Luminance Distribution on Clear Skies." Publ. CIE No. 22 (TC-4.2) 1973. Comm. Int. Éclair.
Dobrusskin, A. (1971). *Light. Res. Technol.* **3**, 125.
Elenbaas, W. (1971). "Fluorescent Lamps." Macmillan, New York.
Elenbaas, W. (1972). "Light Sources." Crane, Russak, New York.
Elenbaas, W., van Boort, H. J. J., and Spiessens, R. (1969). *Illum. Eng. (N.Y.)* **64**, 94.
Forsythe, W. E., and Worthing, A. G. (1925). *Astrophys. J.* **61**, 146.
Fraser, H. D., Waldbauer, W. M., Unglert, M. C., and Walick, J. A. (1961). *Illum. Eng. (N.Y.)* **56**, 215.
Gardner, P. J., Morris, J. C., Watson, W. R., Silver, H. G., and Scholz, J. A. (1975). *J. Illum. Eng. Soc.* **5**, 45.
Illuminating Engineering Society (1972). "I.E.S. Lighting Handbook," 5th Ed., Sec. 8. Illum. Eng. Soc., New York.
Jones, L. A., and Condit, H. R. (1948). *J. Opt. Soc. Am.* **38**, 123.
Judd, D. B., MacAdam, D. L., and Wyszecki, G. W. (1964). *J. Opt. Soc. Am.* **54**, 1031.
Kopelman, B., and Van Wormer, K. A., Jr. (1968). *Illum. Eng. (N.Y.)* **63**, 176.
Krochmann, J. (1963). *Lichttechnik* **15**, 559.
Krochmann, J., and Seidl, M. (1974). *Light. Res. Technol.* **6**, 165.
Krochmann, J., Müller, J., and Retzow, V. (1970). *Lichttechnik* **22**, 551.
Larson, D. A., Fraser, H. D., Cushing, W. V., and Unglert, M. C. (1963). *Illum. Eng. (N.Y.)* **58**, 434.
Lemons, T. M., and Meyer, E. R. (1964). *Illum. Eng. (N.Y.)* **59**, 723.
Levin, R. E., and Lemons, T. M. (1971). *IEEE Trans. Ind. Gen. Appl.* **IGA-7**, 218.
Levin, R. E., and Westlund, A. E. (1966). *J. Soc. Motion Pict. Telev. Eng.* **75**, 589.
Lin, F. C. (1970). *Illum. Eng. (N.Y.)* **65**, 250.
Lin, F. C., and Knochel, W. J. (1974). *J. Illum. Eng. Soc.* **3**, 303.
McHale, J. J. (1971). *Illum. Eng. (N.Y.)* **66**, 280.
Noel, E. B., and Martt, E. C. (1956). *Illum. Eng. (N.Y.)* **51**, 513.
Penning, F. M. (1957). "Electrical Discharges In Gases," Ch. 6. Cleaver-Hume Press, London.
"RCA Electro-Optical Handbook" (1974). EOH-11, Sect. 9. RCA/Commercial Eng., Harrison, New Jersey.
Sadoski, T. T., and Roche, W. J. (1976). *J. Illum. Eng. Soc.* **5**, 143.
Spencer, D. E., and Montgomery, L. L. (1961). *J. Opt. Soc. Am.* **51**, 727.
Studer, F. J., and Van Beers, R. F. (1964). *J. Opt. Soc. Am.* **54**, 945.
Tataronis, R. T. (1972). *J. Illum. Eng. Soc.* **1**, 191.
T'jampens, G. R., and van de Weijer, M. H. A. (1966). *Philips Tech. Rev.* **27**, 173.
Toomey, C. L. (1961). *Illum. Eng. (N.Y.)* **56**, 227.
Waymouth, J. F. (1971). "Electric Discharge Lamps." MIT Press, Cambridge, Massachusetts.
Waymouth, J. F. (1977). *J. Illum. Eng. Soc.* **6**, 131.
Waymouth, J. F., Gungle, W. C., Harris, J. M., and Koury, F. (1965). *Illum. Eng. (N.Y.)* **60**, 85.
Wysecki, G. W., and Stiles, W. S. (1967). "Color Science." Wiley, New York.

CHAPTER 2

Optical Materials—Refractive

CHARLES J. PARKER

Corning Glass Works, Corning, New York

I. Introduction	47
II. Optical Glass	48
A. Description and Terminology	48
B. Optical Properties	49
C. Chemical Properties	59
D. Physical Properties	62
III. Vitreous Silica Glass	65
A. Terminology	65
B. Properties	66
IV. Optical Crystals	68
V. Plastic Optical Materials	70
VI. Optical Materials for the Infrared	71
A. Glasses	72
B. Crystalline Materials	74
C. Metallic Materials	75
References	76

I. INTRODUCTION

Refractive optical materials include broad classes of solids with closely specified optical properties that are useful for control of light in the ultraviolet, visible, and infrared spectral regions. The prime optical properties are refractive index and transmittance (i.e., absorption) and the wavelength dependence of these properties through the region of interest. In the visible region, where most optical materials are transparent, selection is made primarily according to index characteristics. Outside the visible, transparency is usually the critical property and the user fits his application to the index available.

There is a variety of further properties, optical and otherwise, whose order of importance and specification depend strongly on the end use. These include, for example, inclusion quality, index homogeneity, scattering, chemical reactivity, thermo-optic and stress-optic coefficients, hardness, density, and strength. This chapter does not discuss polarizing or optically active refractive materials, or materials designed for optical thin films.

II. OPTICAL GLASS

A. Description and Terminology

Although it is one of the oldest optical materials, glass still offers the widest range of optical properties, and is the primary type of material used in the visible and near-visible region. Optical glass can be considered as ordinary glass that is made by processes that make it usable for precise control of light by refraction. Nearly every element in the periodic table is used in the compositions of the many optical glasses to manipulate the optical properties to meet the needs of the optical designers. There are hundreds of optical glasses available, and the catalogs of the manufacturers should be consulted. They contain extensive tabulations and graphs of optical and other properties, along with useful descriptive matter to assist in selection.

The critical properties of optical glass are the index and its variation with wavelength, which is called dispersion. Thus, optical glasses are commonly identified by the index at the wavelength of the helium d line (or the sodium D line), along with the Abbé number v_d (nu value), which is one measure of the dispersion. The Abbé number is a recurring fraction in optical design that gives in a convenient number the reciprocal of a dispersion in relation to the refraction:

$$v_d = (n_d - 1)/(n_F - n_C) \tag{1}$$

Most catalogs also list

$$v_e = (n_e - 1)/(n_{F'} - n_{C'}) \tag{2}$$

(See Table II for wavelengths of the lines described above by subscript letters.)

Figure 1 illustrates one version of the index versus Abbé number (n_d versus v_d) chart commonly used to tabulate available glasses. Traditionally, most glasses with v_d greater than 50 are called crowns and those with v_d less than 50, flints, although all modern glasses cannot be classified by this rule. Glasses are available with n_d from 1.45 to 1.96 and v_d from 20 to 82.

The large number of optical glasses has led to a somewhat confusing terminology situation among different manufacturers, although all agree that the six-digit designation is the most meaningful. This is a numeral consisting of the first three digits after the decimal of the refractive index, followed by the first three digits of the Abbé number. Thus the six-digit code immediately gives the index level and an indication of the dispersion. For example, glass type 613586 has $n_d = 1.613$, $v_d = 58.6$.

The main difficulty in nomenclature arises when descriptive names and abbreviations are used to designate the many glass types. Table I illustrates several of the names and letter designations used for the same glass by

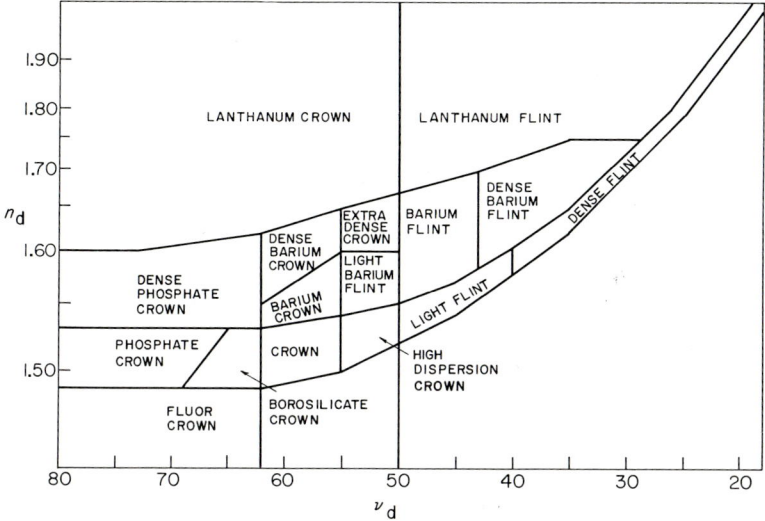

FIG. 1. Types of optical glasses.

TABLE I
Some Terminology Differences

	529517	658509	744448
Corning–Sovirel[a]	CHD B29-52 High dispersion crown	BCDD C58-51 Very dense barium crown	FBS D44-45 Special baryte flint
Schott	KzF2 529517 Short flint	SSKN5 658509 Extra dense crown	LaFN2 744448 Lanthanum flint
Chance–Pilkington	TF 530512 Telescope flint	DBC 658509 Dense barium crown	SBF 744447 Special barium flint
Hoya	SbF2 529516 Antimony flint	BaCED5 658509 Extra dense barium crown	LaF2 744449 Lanthanum flint

[a] This manufacturer also lists by six-digit number.

different manufacturers. Only the six-digit code is universally useful in designating optical glasses.

B. Optical Properties

1. *Index and Dispersion*

Index and dispersion of optical materials, including glasses, will increase as the wavelength decreases. Glasses of higher refractive index generally have greater dispersion. Index usually, but not always, increases as the density of the glass increases. The index and dispersion of optical glass derive primarily from the oscillator strengths and frequencies of the fundamental absorption bands in the far-ultraviolet region, with added contributions from the bands in the infrared. Figure 2 pictures the complete dispersion curve for a material such as glass that will have fundamental absorption bands at several wavelength positions across the electromagnetic spectrum, as indicated by the λ's.

From basic considerations of the interaction of light with glass, equations have been developed over the years that more or less closely represent the course of the index versus wavelength; i.e., the dispersion, in the optical region. The Sellmeier dispersion equation (1871) is one of the best known of these, and is of the form

$$n^2 = 1 + \Sigma_s \frac{a_s \lambda^2}{\lambda^2 - \lambda_s'^2} \tag{3}$$

where n is the index, λ the wavelength, and a a constant. This was modified by Ketteler and Helmholtz (1875–1885) to improve its representation of the dispersion curve close to absorption frequencies. A widely used formula that is similar to the Ketteler–Helmholtz equation is that proposed by Herzberger in 1942:

$$n = A + B\lambda^2 + \frac{C}{(\lambda^2 - 0.035)} + \frac{D}{(\lambda^2 - 0.035)^2} \tag{4}$$

where A, B, C, and D are constants. Other formulas by Cauchy and by Hartmann have some usefulness over relatively restricted spectral ranges.

Mathematical manipulation and series expansion of theoretical equations selected from the foregoing lead to the series form relating index n to the wavelength λ,

$$n^2 = A_0 + A_1\lambda^2 + (A_2/\lambda^2) + (A_3/\lambda^4) + \cdots \tag{5}$$

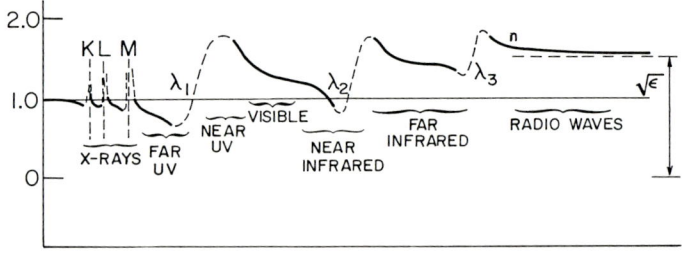

FIG. 2. The complete dispersion curve for a substance, such as glass, transparent in the visible region. (From Jenkins and White, 1976; by permission of McGraw-Hill Book Company.)

that represents the true dispersion curve of a glass as closely as desired, depending on the number and accuracy of the index measurements made on the glass and on the number of terms retained in the least-squares fit to the data. Equation (5) is the dispersion equation found in manufacturers' catalogs. The power terms in λ represent the contributions of the infrared absorptions, and the squared term only is retained because the fundamental absorption frequencies in the infrared are quite distant from visible region frequencies. The terms containing reciprocal powers of λ represent the all-important ultraviolet absorptions.

Table II lists 19 spectral lines from ultraviolet to infrared that are used for index specification and measurement. Those marked with an asterisk (∗) appear in most catalogs. The F and C lines are rarely measured nowadays due to the cumbersome nature of hydrogen discharge sources and the ready availability of the convenient cadmium lamp. Measurements at the D line are also less common due to the uncertainty arising from the doublet nature of this line. It should be noted that index is normally measured and reported relative to air at 20 °C and 760 Torr. Absolute values can be obtained using published data on the refractive properties of air (Owens, 1967; Werner, 1968).

Index measurements must be made at a minimum of six and preferably more, well-spaced wavelengths to provide reasonable accuracy for the six

TABLE II
SPECTRUM LINES FOR REFRACTIVE INDEX

Letter designation	Source	Wavelength in nanometers
i	Hg	365.015*
—	He	388.865
h	Hg	404.656*
g	Hg	435.835*
f	Cd	467.816
F	H	486.133*
F'	Cd	479.992*
E	Cd	508.582
e	Hg	546.074*
d	He	587.561*
D	Na	589.294*
C'	Cd	643.847*
p	He	667.815
C	H	656.272*
r	He	706.519*
R	Rb	780.023
s	Cs	852.110*
t	Hg	1013.98*
—	He	1083.03

constants for Eq. (5) that are listed by most manufacturers for each glass. The cataloged index data or the constants of the dispersion curve can then be used to obtain the various index values, partial dispersions $(n_x - n_y)$, and relative partial dispersions $(n_x - n_y)/(n_a - n_b)$ needed for optical glass selection and optical system design. Actually, in computer-assisted design only the constants need be stored and all index and dispersion data are then called out as needed.

The dispersion formula, Eq. (5), should be used only for interpolation. Extrapolation beyond the spectral region covered by the index measurements is not recommended and can lead to gross errors. If extrapolated data are needed they should be obtained only after measurements at suitable additional wavelengths.

a. Secondary Spectrum Correction. For most glasses the rate of change of dispersion is similar for the same v_d value; i.e., the relative partial dispersions, such as $(n_x - n_y)/(n_F - n_C)$, are linearly related to the v_d values. In lens design good correction of the second-order chromatic aberration, the so-called secondary spectrum, is not possible from this selection of glasses. Thus new glasses were sought and were developed with different dispersion ratios. One such glass, a lead borate glass, is called short flint because the important blue partial $(n_g - n_{F'})$ is "shorter" than normal. Other new glasses such as the lanthanum-containing types provide further "non-normal" glasses. Considerable space in most catalogs is devoted to presenting the glasses with appreciable deviation from the "normal" relative partial dispersions at all levels of v_d.

b. Effect of Temperature on Refractive Index. The refractive index of glass changes as its temperature is changed. Two effects contribute to the change in index. For glass with a positive coefficient of thermal expansion, a rise in temperature causes a decrease in density which would correspond to a decrease in index. But also as temperature increases, the fundamental ultraviolet absorption edge shifts to longer wavelengths, tending to increase the index. The observed change results from the combination of these two effects and the magnitude and direction of the index change depend upon the type of glass and its properties.

One must distinguish between the relative and absolute temperature coefficients of index. The index relative to air changes more with temperature than does the absolute index because of the significant temperature sensitivity of the index of air. Catalogs and publications do not always make clear which refractive index is involved in the reported data on temperature effects. Table III shows the temperature dependence of both absolute and relative index at 546.1 nm wavelength for a variety of optical glasses (Parker and Popov, 1971). It is apparent from Table III that users should be aware

2. OPTICAL MATERIALS—REFRACTIVE

TABLE III
Some Temperature Coefficients of Index[a]

Type	Absolute dn/dT	Relative dn/dT	Type	Absolute dn/dT	Relative dn/dT
Crowns			Flints (continued)		
511604	−0.03[b]	1.39[b]	581409	0.99	2.48
517642	1.54	2.96	620364	2.96	4.48
522595	0.34	1.77	648339	3.18	4.73
523602	2.59	4.02	699301	4.46	6.06
Barium crowns			755276	7.24	8.89
573575	1.08	2.55	805254	9.21	10.9
613586	1.14	2.65	Barium flints		
613574	1.57	3.08	650392	1.80	3.35
618551	3.09	4.61	670471	2.42	3.99
620603	0.75	2.27	702410	6.67	8.27
Lanthanum crowns			Borate flints		
652585	−0.74	0.81	613443	4.14	5.66
691547	2.15	3.73	653397	1.72	3.27
720504	3.56	5.18	Lanthanum flints		
Flints			717480	0.60	2.21
526510	1.95	3.38	744448	0.43	2.07
548458	1.12	2.57			

[a] $\lambda = 546.1$ nm (e line) $T = 22$ °C.
[b] Units in the sixth decimal, per degree Celsius.

of possible index changes of the order of several units in the sixth decimal per degree celsius for conventional optical glasses (Baak, 1969).

The effect of temperature on dispersion is found by measuring the temperature coefficients for a range of wavelengths. The coefficient generally increases at shorter wavelengths so that the dispersion generally increases with temperature rise for glasses with positive temperature coefficients of index. This is illustrated by Fig. 3 that shows the wavelength dependence of the slopes of the absolute dispersion curves for one glass at three different temperatures (Parker and Popov, 1971).

c. Effect of Stress on Refractive Index. The refractive properties of optical glass—and of any optical material, for that matter—are sensitive to stress within the material. Consideration must be given to two types of stress: (a) applied stress from an external load; and (b) residual stress resulting from thermal treatment during the fabrication process (Spinner and Waxler, 1966), which leads further to consideration of effects of variations in annealing heat treatments.

While glass is normally considered to be an isotropic material, under stress it exhibits double refraction, called birefringence, in an amount directly

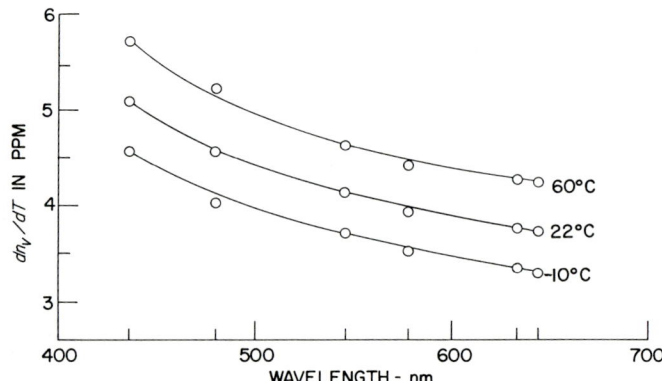

FIG. 3. Wavelength and temperature dependence of the temperature coefficient of absolute index, type 613443 borate flint glass. (From Parker and Popov, 1971.)

proportional to the stress. It behaves as a uniaxial optically negative crystal with the optical axis parallel to the stress. This affords a means for measurement and control of the residual annealing stresses in the glass, which must be kept very low to minimize the birefringence that might adversely affect image quality. This measurement is made with a compensating polarimeter such as the de Sénarmont arrangement (ASTM Standard F-218).

The birefringence resulting from the stress gives no information on the absolute change in the refractive index. Although the index change induced by residual annealing stress is very small, it is measurable, and in high-performance systems requiring maximum optical homogeneity in the glass the effect of stress on homogeneity should be recognized. The relative change in index is directly proportional to the relative change in density caused by the stress, and since it is therefore related to the elastic properties, the magnitude of the change will vary with glass type.

As an example, in a borosilicate crown glass a unidirectional stress giving a birefringence of about 10 nm/cm would be about 5 kg/cm^2; this would change the index by about 4 in the sixth decimal place. The change of index is generally greater in high index glasses even though these do exhibit lower birefringence under stress, as pointed out in the following paragraphs. Measurement of the refractive index change resulting from the direct application of hydrostatic pressure affords one method of evaluating this effect (Waxler and Cleek, 1973).

d. Stress–Optical Coefficient. The proportionality constant between stress in the glass, whether applied or residual and the resulting birefringence or double refraction, is variously called the stress–optical coefficient, Brewster's constant, or birefringence constant.

2. OPTICAL MATERIALS—REFRACTIVE

$$\Delta\lambda = K\sigma P \tag{6}$$

where $\Delta\lambda$ is the measured optical retardation, K the stress optical coefficient, σ the stress, and P the path length. A common unit for K is the Brewster—nm/cm/kg/mm².

The coefficient is a strong function of the chemical composition of the glass and must be known for each glass if stress is to be evaluated from observation of birefringence (ASTM Standard C-770). Figure 4 shows the wide variation in the stress–optical coefficient among the 27 widely used glasses listed in Table III. Their refractive indices n_d are also plotted in Fig. 4 from which it is evident that the effect of stress on index does generally, but not always, increase with the index.

It is customary in the optical industry to specify certain low, viz., 5 or 10 nm/cm, values of stress-induced optical retardation in the optical materials used, generally with the assumption that this is specifying residual stress. (The presence of residual stress may cause difficulties in figuring, and may cause refractive inhomogeneity.) However, as suggested in Fig. 4 it is possible for some, viz., high index flint, glasses to have low, or zero, or even negative indication of stress in a photoelastic measurement (Waxler and Napolitano, 1957). The user can be led into misuse of the glass if he assumes that low

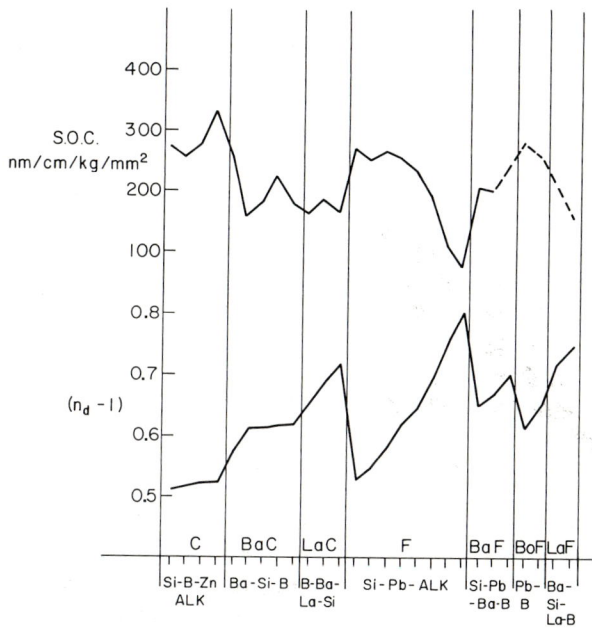

FIG. 4. Stress–optical coefficient and refractive index for the 27 glasses in Table III.

birefringence always indicates low stress and care must be taken in selection and use of the low-coefficient glasses.

e. Annealing. Refractive index is a strong function of thermal history, not only because of the induced changes due to unannealed residual stresses, but also because glass at room temperature retains structural characteristics of a higher temperature state, to a degree that relates to the rate at which the glass was cooled from the molten state. Thus, a quenched or less "fine-annealed" glass has a lower index than one of the same composition that has undergone a slow-cooling fine-annealing treatment. Index differences resulting from different treatments of the same glass can be of the order of 0.001—a large difference. Thus, proper annealing (and definition of the annealing) for optical glasses is a critical factor in their manufacture and use and considerable understanding of the effects of various annealing processes has been obtained in the industry (Spinner and Waxler, 1966). Optical glass annealing is as critical in achieving quality as the composition or the melting of the glass. Proper annealing is needed to: (a) arrive at the required index; (b) give low enough double refraction to prevent image degradation; (c) give low enough residual stress to prevent distortion during figuring; and (d) provide homogeneity; i.e., index uniformity.

Manufacturers' catalogs give useful information on product annealing and on various annealing quality grades available.

f. Optical Homogeneity; Inclusions. The term "optical homogeneity," as used in reference to optical glass, has come to mean the degree of overall refractive index variation within a blank or a melt. Strictly speaking, however, such defects as striae, bubbles, or inclusions are also inhomogeneities, but these are generally discussed and specified separately from bulk index variations.

In Section II.B.1.*e* it was pointed out that proper annealing is required for optimum homogeneity, referring to the need to have all elements of a piece undergo the same heat treatment schedule to arrive at the same refractive index at room temperature, as well as to minimize stress–optical effects. An even more critical problem in glass making is chemical inhomogeneity, caused by variation in chemical composition, that cannot be erased by careful heat treatment.

Striae are one type of undesirable chemical inhomogeneity. These are internal threads or veins of glass of different composition (and thus optical properties) from that of the bulk glass. They are very localized and feature sharp gradients of index that may be visible to the eye, or may require a sensitive shadowgraph or Foucault knife-edge test for detection. These are rarely seen nowadays in most optical glasses and those that do appear in some glasses are carefully specified and controlled. Striae have little or no effect on the bulk index of a piece away from a focal plane, since they constitute only a small fraction of the volume or of the optical path. Their chief optical effect is generally a scattering that can degrade contrast or reduce transmitted power.

Excessive variation in the bulk refractive index of a piece from chemical inhomogeneity or from improper annealing can have a more severe effect on optical performance than striae. Such variations are not seen by the eye or by shadowgraph methods, but require knife-edge methods to detect and interferometry to measure. Normal optical glass processes commonly yield glass with maximum index variation within one melt near $\pm 1 \times 10^{-4}$, or $\pm 2 \times 10^{-5}$ for $(n_F - n_C)$. Homogeneity within one piece will be considerably better than that; for example, $\pm 5 \times 10^{-5}$. On special request glass can be obtained with higher homogeneity within a melt or a piece, up to $\pm 1 \times 10^{-6}$ in one blank. Cost of the glass increases rapidly with increase in homogeneity, however.

An important third class of inhomogeneities is small bubbles—sometimes called seeds—and inclusions. These are also chemical inhomogeneities and total absence of these is really not attainable in glassmaking processes. Bubbles are gaseous inclusions and the term "inclusion" is usually reserved for a solid particle. Like striae, these defects away from a focal plane generally have a minor optical effect affecting only the contrast, and even then to a minor degree. However the ready visibility of these defects does lead to a "cosmetic" problem such that users of optical glass tend to specify much better bubble and inclusion quality than needed, for appearance's sake.

Optical glasses are provided in several classes of "bubble quality." This is specified by the total cross-section area of bubbles per 100 cm^3 of glass. Solid inclusions are here treated as bubbles. Bubbles or inclusions with a diameter equal to or less than 0.05 mm are generally disregarded. Table IV lists six classes of bubble quality used by some suppliers. Others may list

TABLE IV
BUBBLE CLASSIFICATION BY AREAS

Class	Total bubble cross section per 100 cm^3 of glass (mm^2)
0	0–0.029
1	0.03–0.10
2	0.11–0.25
3	0.26–0.50
4	0.51–1.00
5	1.01–2.00

number per unit volume in specified diameter ranges, a specification that is easier to apply in practice.

2. Transmittance

Optical glass must permit maximum possible transmittance of light in the spectral regions in which it is used. Since certain trace impurities in concentrations as low as ten parts per billion will noticeably affect the transmittance, absorption of light in optical glass is minimized by strict control of raw materials used in its manufacture (Campbell and Adams, 1969) and of the materials used in contact with the glass, including the atmosphere, during melting and forming. Many of the high-index glasses, however, may have a yellow tint even at highest chemical purity levels due to intrinsic absorption in the ultraviolet region by the heavy elements required to achieve high index.

Transmittance is defined as the ratio of radiant flux transmitted by a body to that incident upon it. For a plane parallel plate of glass with polished surfaces, the transmittance is given by

$$T_\lambda = K_\lambda 10^{-\beta \lambda t} \tag{7}$$

where λ indicates the wavelength, β is the absorption coefficient in cm^{-1}, a characteristic constant of the glass, and t is the thickness in centimeters.

The factor $10^{-\beta \lambda t}$ is the *internal* transmittance, the ratio of radiant flux incident on the second surface to that leaving the first surface. For zero absorption, $\beta_\lambda = 0$, the internal transmittance is 100%. Manufacturers' catalogs give extensive tabulations of internal transmittance for their optical glasses, often for several thicknesses. It is evident that optical glasses do have very low absorptions over most of their usable spectral ranges, but care should be used in selecting glasses at either the short- or long-wave extremes—absorption could become a problem. For very long path lengths, as in fiber optics, the transmittance can be low even in the lowest absorbing regions and care must also be used in selecting glass for long-path uses. Calorimetric methods must be resorted to for absorption measurements in low-absorbing glasses for long-path uses (Mitra and Bendow, 1975).

The factor K_λ in Eq. 7 represents surface reflection losses

$$K_\lambda = (1 - \rho_\lambda)^2 \tag{8}$$

where ρ_λ is the reflectance of a single surface of the glass. Note that K_λ is actually the maximum attainable transmittance; i.e., the transmittance for nonabsorbing glass.

For normal incidence the surface reflectance depends on the refractive index, and for a glass–air interface (index of air = 1.0) it is given by

$$\rho_\lambda = [(n_\lambda - 1)/(n_\lambda + 1)]^2 \tag{9}$$

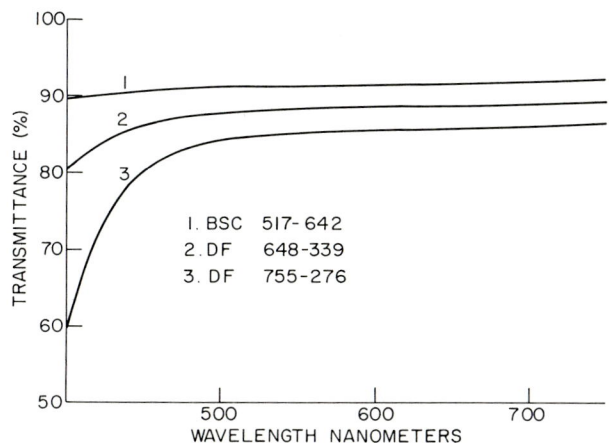

FIG. 5. Visible transmittance of thick (5.5 cm, $2\frac{3}{16}$ in.) specimens of three types of optical glass.

Thus, ρ and therefore K are functions of the index of the glass. Surface reflectance thus decreases with increasing wavelength, which means that even a glass with zero absorption would have a spectral variation in transmittance that would affect the color of the light passing through it.

Figure 5 illustrate the visible transmittance of 5.5 cm plates of three types of glass. This shows the yellowing effect for the flint glasses and also shows how the maximum transmittance is lower for the higher-index glasses, because of increased reflectance. The latter observation demonstrates why reflection-reducing films are so important in increasing throughput of optical systems and in reducing flare and ghosts resulting from multiple reflections.

C. Chemical Properties

The chemical compositions of optical glasses are designed primarily to give optimum refractive properties needed by designers and users. This has led, for many glasses, to the use of compositions with decreased resistance to chemical corrosion. This is manifested by increased staining or clouding of the glass surface and care must be taken with the less resistant glass types in handling and exposure during fabrication and use. In hot, humid climates fungus may grow on the surface leaving permanent marks. Attempts are continually being made by the glass chemists to improve chemical properties and from time to time new glasses appear with improved corrosion resistance that also retain the refractive properties of the older glasses that they replace. However, the required optical properties place severe limitations on what can be accomplished in these programs.

Manufacturers use a variety of standard test methods to evaluate chemical resistance of their glasses and include such information in their catalogs. Such tests, of course, cannot duplicate service conditions, but are intended to assess the relative resistance of the various optical glasses under the standard test conditions. Some of the test procedures and results are discussed in the following paragraphs.

1. *Resistance to Weathering*

Weathering is defined as corrosion by the atmosphere. Under conditions of moderate or high humidity, a reaction can occur between the glass and water vapor. This usually appears as a cloudy film (the phenomenon called "dimming" in the optical trade), which can sometimes be wiped off if discovered in the early stages. However, permanent damage can occur in which case an iridescent, darkened, or cloudy film will remain after cleaning. The effect can usually be avoided if the humidity at which the glass is stored does not exceed 20 or 30%.

To compare glasses for their relative stability, an accelerated weathering test is performed in which polished, carefully cleaned glass plates are exposed to a 98% relative humidity atmosphere at 50 °C. Specimens are removed periodically and visually inspected. Some of the specimens are washed at each inspection to remove accumulated weathering products and some remain unwashed during the entire test period of 28 days. Samples are given ratings from A—no change, to E—very poor. The bottom curve in Fig. 6 shows the course of such ratings for the 27 glasses in Table III.

A somewhat similar accelerated weathering test is also used in which the exposure time is up to 7 days in a saturated atmosphere that cycles hourly between 40 and 50 °C to provide cyclic condensation on the surface. Diffuse light photometry of surface scattering is used to quantify the results. Glasses are thereby graded from group 1—no attack after 7 days, to group 4—marked attack in a few hours. This test has been found to rank the same 27 glasses in essentially the same order.

2. *Resistance to Staining*

Several methods are also used by various manufacturers to measure and report resistance to staining, specifically by acids. Staining can occur because of the leaching action of the acids in polishing or cleaning compounds or in atmospheric pollutants. The "stain" may be a slight cloudiness, darkening, or iridescence. Under extreme reactions the surface may become frosted. It is also possible for the glass surface to be uniformly dissolved, leaving the visual appearance unchanged!

2. OPTICAL MATERIALS—REFRACTIVE

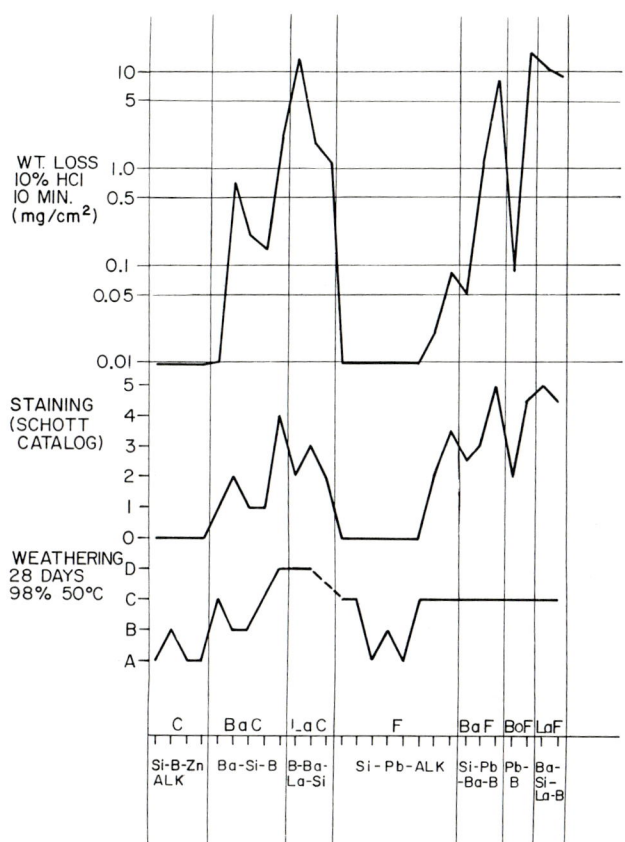

Fig. 6. Results from three methods for measuring chemical stability of the 27 glasses in Table III.

One company evaluates staining on a flat, polished sample. This is pressed onto a cuvette with a spherical hollow 0.25 mm deep that contains a few drops of the test solution. Two different solutions are used in this test:

Solution I: Standard acetate, pH 4.6
Solution II: Sodium acetate buffer, pH 5.6

Stains of interference colors are formed more or less rapidly from the reaction at 25 °C. Glasses are classified according to the time required for a stain to appear: group 0—no stain after 100 hr, solution I; groups 1–5— stain appears before 100 hr, solution I, with No. 1 being the slowest; group 5—less than 12 min by exposure to both solutions. The middle curve of Fig. 6 shows the stain ratings by this method for the Table III glasses.

Another method for rating relative acid stability of glasses uses a flat, polished, carefully washed specimen that is dipped halfway into a 10% hydrochloric acid solution at 25 °C. Weight loss per unit surface area and visual appearance are recorded after 10-min and 2-hr exposures. The top curve in Fig. 6 shows the 10-min results for the glasses of Table III. Glasses showing a visual change after a 10-min exposure should be processed with care. When weight loss at 10 min is high—over 0.1 mg/cm^2—there is no need to proceed to the 2-hr test. For comparison of very resistant glasses with negligible weight loss in 2 hr, chemical analysis of the test solution for alkali, alkaline earth, or other glass constituents can be used.

Inspection of Fig. 6 suggests that the several methods for measuring chemical stability generally agree as to the stable and unstable classes of optical glass. The two staining tests correlate very closely, even though different solutions and different end-point evaluation systems are used. The major cation constituents of the seven classes in Fig. 6 are listed at the bottom of the chart. It is evident that glasses containing large percentages of barium and/or boron oxides are chemically unstable. Perhaps surprisingly, the crowns and flints are the most stable of the glasses represented here even though they contain major amounts of the alkalies, sodium and potassium. Table III and Fig. 6, of course, contain only a fraction of available optical glasses. Manufacturers' catalogs should be consulted for additional chemical corrosion information and ratings.

D. Physical Properties

1. *Thermal Expansion*

The wide variety of glass compositions developed to meet the refractive properties needs of optical designers has resulted in glasses with a wide range of mechanical and thermal properties, not all of which are optimal or desirable but, as with chemical properties, all of which must be accepted and suitably dealt with in design and fabrication. Again referring to the 27 widely used glasses of Table III, these types have a wide variety of physical as well as chemical properties and these can be used to illustrate some characteristic properties, examine the differences between glass types, and observe possible correlation among the properties.

Figure 7 shows plots of the indicated properties for the 27 glasses arranged in the seven groupings of Table III. The expansion coefficient α that is given in all suppliers' catalogs and that is shown plotted for the room-temperature region is the most critical nonoptical physical property of optical glass. This is a prime parameter in consideration of thermal stressing and awareness of this factor is needed in handling or using these glasses where temperature changes or high temperatures are involved, as in

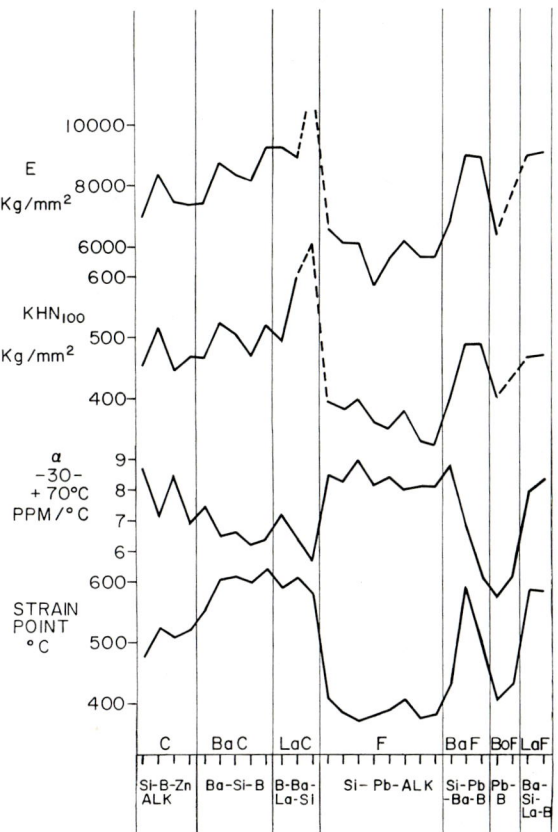

FIG. 7. Some physical properties of the 27 glasses in Table III. E is Young's modulus; α is the thermal expansion coefficient; KHN_{100} is the Knoop hardness, 100 gm load.

waxing to a lap or mounting in a metal cell to be used in extreme environments. Thermal expansion is also critical to optical performance with changing temperature because of the effect of density change on the index and because of the contribution of temperature-caused physical path change to the total optical path change in an optical part (Reitmayer and Schroeder, 1975). Most optical glasses, as shown in Fig. 7, have relatively high thermal expansions and must be carefully handled. It is interesting to see, however, that three of the glasses have usefully low room-temperature expansions—at or below $50 \times 10^{-7}/°C$. These are among the glasses that were developed specifically to meet the needs of designers for secondary spectrum correction. (See Section II.B.1.a. The borate flint glasses are the so-called short flints.) A high percentage of boric oxide was required in these compositions to give the desired optical properties and this resulted in their traditionally inferior chemical properties but it also imparted low thermal expansion. This is a good illustrative example of how the physical (and chemical) properties must necessarily depend on the optical properties designed into the glass.

2. Strain Point

The strain point is a temperature point on the viscosity curve of glass. At this temperature internal stress is substantially relieved by true viscous flow in a few hours. As a glass reaches the strain point on cooling from forming temperatures it becomes virtually an elastic solid and can develop stresses from thermal or mechanical loading. Thus, the strain point represents an absolute maximum temperature to be applied in use and practical temperature excursions should actually be held to 100–200 °C below this, particularly if the optical properties are to remain unaffected. The plot on Fig. 7, then, is a measure of the relative melting range of the glasses and an indicator of usable service temperatures. The sharp dip in the curve for the lead-based flints gives physical evidence of the well-recognized limited temperature capabilities of these glasses.

3. Mechanical and Elastic Properties; Strength

The Knoop hardness number (KHN) is one of various hardness measures (ASTM Standard C-730) and is the load per unit unrecovered projected area of an identation made in the surface by a shaped diamond indenter under a small load—100 gm for Fig. 7. Studies at Hoya Glass Works have shown that the KHN is a reasonable estimator for lapping hardness, a more directly useful characteristic for opticians (Izumitani and Suzuki, 1973). It was demonstrated there that both diamond penetration and lapping speed depend eventually on the same factor—the yield stress of the glass. Again we see a wide range of hardness in optical glasses, including the indication of the well-known relative softness of the flint glass family.

Young's modulus E is another critical parameter in thermal and mechanical stress considerations (Shand, 1958, p. 111). Comparing the E graph with the one below it, it is evident that Knoop hardness correlates closely with Young's modulus, an observation that should be useful in estimating relative workability of glasses. Most manufacturers list Young's modulus, while indentation hardness information is infrequently given.

The practical tensile strength of glass is several orders of magnitude below theoretically derived instrinsic strength. This arises from the completely brittle behavior of glass and the resultant controlling effects of surface flaws or defects in the surfaces created in the manufacturing process or in use (Tooley, 1974). Thus, all glasses in the annealed state have essen-

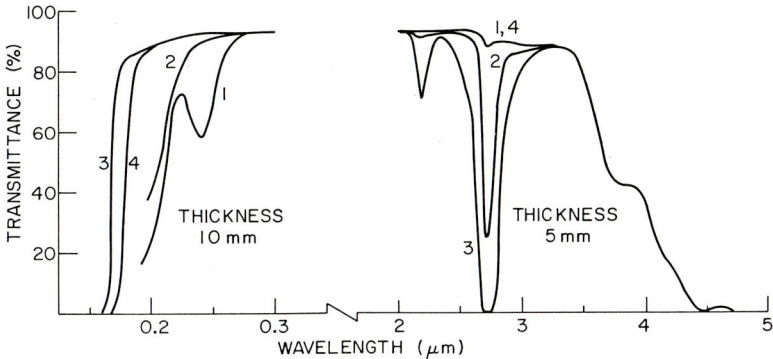

FIG. 8. Approximate ultraviolet and infrared transmittance of four types of vitreous silica. 1 = fused quartz; 2 = Heraeus process; 3 = synthetic; 4 = water-free synthetic.

2. Refractive Properties

The Bureau of Standards has published results of careful measurements of refractive index relative to air for vitreous silica from several manufacturers at wavelengths from 0.214 to 3.707 nm (Malitson, 1965). The index values were fitted to a three-term Sellmeier dispersion equation (see Section II.B.1) to give the following:

$$(n^2 - 1) = \frac{0.696\,1663\,\lambda^2}{\lambda^2 - (0.068\,4043)^2} + \frac{0.407\,9426\,\lambda^2}{\lambda^2 - (0.116\,2414)^2}$$
$$+ \frac{0.8974794\,\lambda^2}{\lambda^2 - (9.896\,161)^2} \qquad (10)$$

The agreement between observed and computed values is shown by an average absolute residual Δn of 10.5×10^{-6}. Considering the whole range of wavelengths this indicates that index values may be interpolated to five significant decimal places. The nominal index for vitreous silica at the d line is 1.45846. Later, Brixner presented a four-term Sellmeier equation fit to the same Bureau data that decreased the average residual to 4×10^{-6}. He pointed out, however, that three terms are still sufficient if certain questionable data points are rejected (Brixner, 1967).

The v value for vitreous silica is 67.8 (67.77–67.88 for the various Bureau program specimens). This shows that the dispersion is quite low in the visible range. It passes through a minimum near 1.3 μm and is relatively high in the 2–4 μm range. Dispersion and index also increase rapidly below 300 nm in the ultraviolet.

In view of the well-recognized thermal and chemical stability of vitreous silica, it comes as a surprise to note that it has one of the largest temperature coefficients of index among glasses. At ambient temperature the change in relative index is about $10 \times 10^{-6}/°C$ for the visible region, which is comparable to values assigned to high-index glasses (see Table III). It was pointed out in Section II.B.1.*b* that the thermal coefficient of index results from the combination of two factors. In vitreous silica the short-wave absorption does shift towards the visible region with increasing temperature, but one of this glass' most unique properties—its very low thermal expansion coefficient—permits only negligible density decrease with temperature so the net result is a large increase in index. At the Bureau of Standards, one of the glasses in the refractive index program was later measured at temperatures down to -200 °C, showing a monotonically decreasing index at all wavelengths, with the temperature coefficient at the d line going from 10×10^{-6} to 2×10^{-6} over that range (Waxler and Cleek, 1971).

3. *Other Properties*

The various types of vitreous silica share its most generally useful property of low thermal expansion, $5.5 \times 10^{-7}/°C$ for the range 0 to 300 °C. The slope of the expansion curve shows an anomalous (for glass) *decrease* with temperature up to the strain point region, such that the 0 to 1000 °C coefficient is about $5.0 \times 10^{-7}/°C$. The expansion curve also passes through a minimum near $-120°$ C. The low expansion gives high thermal shock resistance, leading to many product uses. It also imparts temperature insensitivity to optical mirrors, as discussed in Chapter 4. Adding about 7.5 wt % of titania to the silica will, in fact, yield the so-called ULE™ (ultralow expansion) mirror substrate glass that has essentially zero expansion at ambient temperatures.

The annealing range viscosities, which control the upper use temperatures, do vary among the glass types, being dependent largely on the hydroxyl and metallic impurity levels (Hetherington *et al.*, 1964). The synthetic materials, with the highest purity and water content, have the lowest annealing point, approximately 1020 °C. Fused quartz types have lowest purity and water content but the highest annealing point, near 1180 °C.

The literature and catalogs should be consulted for further information on properties. Some of the additional dimensions of uniqueness for this material are extreme resistance to radiation darkening (for the synthetics), unusually low loss for high-frequency ultrasonic waves, high electrical resistivity and low dielectric loss, highest permeability of any glass to helium and other gases, and generally high chemical inertness.

IV. OPTICAL CRYSTALS

Most refractive uses for crystalline materials—both as single crystals or in polycrystalline form—are found in infrared technology (see Section VI and Table VII). However, there are a number of broad-spectrum crystals that should be noted as a special class of refractive materials because of their versatility in uses from ultraviolet to infrared. Some of these are recorded in Table V. Further refractive and other property data are available from the various suppliers (Kebler, 1954; Optical Crystals, 1967).

Before synthetic crystals were available the use of crystal optics was limited to the minerals rock salt (NaCl), fluorite (CaF_2), sylvite (KCl), and quartz (SiO_2). Synthetic growth of the halides, as well as of hard oxides and other crystals, has accelerated in the past 50 yr, so a wide variety of high-quality optical crystals are now available. Section VI includes further information on crystalline materials for the infrared.

The fluoride crystals, as grown in single crystal form, transmit well from the vacuum ultraviolet region to long-wave infrared. In particular, CaF_2 has long been used in achromatized ultraviolet lenses. It is the most durable and moisture resistant of the three fluorides listed in Table V. LiF can be used in a wide range achromatized lens useful from far ultraviolet into the infrared (Cartwright, 1939).

Sapphire is a transparent single crystal form of aluminum oxide and is synthetically grown from the melt in the Verneuil and other fusion processes. It is a very hard material with good strength and thermal shock resistance, usable to quite high temperatures (its melting point is 2040 °C). The unique combination of wide range and extreme durability makes it a widely used material for use in severe environments and in specialized optical applications where previously conventional materials fail or are short lived. Its relatively high index, however, does lead to appreciable loss by reflection (see Section II.B.2) and this must be recognized and, if necessary, corrected by antireflection films. Sapphire is also available in a polycrystalline form made by ceramic processes. Although the polycrystalline alumina (PCA) does scatter and is thus only translucent, its absorption is very low and it is widely used for discharge lamp tubing.

TABLE V
Some Wide Range Optical Crystals

Crystal	Useful range (μm)	n_d	dn/dT ($10^{-6}/$°C)	n at (λ)	α^a ($10^{-6}/$°C)	E^b (10^6 psi)
BaF_2	0.13–15	1.474	−15	1.451(5.0)	18	7.7
CaF_2	0.13–10	1.434	−10	1.399(5.0)	24	14
LiF	0.11–7	1.392	−12	1.34(4.3)	36	9
Sapphire	0.15–6.5	1.769	+13	1.738(2.0)	5.8	50
Quartz	0.16–4	1.5534(e)	−6(abs)	—	14.4($\perp$)	13.5($\perp$)
		1.5443(o)	−6(abs)	—	8($\parallel$)	11.4($\parallel$)

[a] α is the thermal expansion coefficient.
[b] E is Young's modulus.

Until the development of synthetic fused silica (Section III.B), quartz was the material of choice for durable ultraviolet prisms and windows, and for wide range spectrometric systems usable to beyond 3 μm in the infrared. In addition to its availability as a natural mineral, quartz is also artificially grown, providing good control over its impurities, and thus its transparency. Birefringent and optically active properties of quartz require that optical elements be carefully prepared with respect to the optical axes. Refractive indices are given in Table V for both the ordinary and extraordinary rays. The thermal expansion coefficient of quartz varies considerably with direction, so that quartz is sensitive to thermal shock and must be properly handled to minimize this problem.

V. PLASTIC OPTICAL MATERIALS

Although plastics have long been used for refractive optical elements, recent advances in the technology of these materials have prompted their increased acceptance. Progress in materials science has broadened the choice of properties available to the designer and this has been accompanied by improvements in fabrication techniques, particularly in injection molding technology (Handbook of Plastic Optics, 1973). A simple litany of advantages and disadvantages of these materials compared with glass, reads as follows (Weeks, 1975).

Advantages: Lightweight ($<\frac{1}{2}$ that of glass); not fragile; easy to make aspherics; simplified mounting and handling; configuration flexibility; low cost (in volume).

Disadvantages: Low heat tolerance; high temperature coefficients of expansion and index; low abrasion resistance; too small selection of optical plastics; lack of reliable properties data; expensive tooling; haze and solarization problems.

Table VI lists properties for the principal optical plastics. The first three thermoplastics are by far the most-used types. The acrylic material is the most popular in the industry. Styrene is important for lowest cost and for its high index in lens designs. Polycarbonate has high heat and impact resistance, although poor scratch resistance. The highest scratch resistance is shown by the thermosetting plastic CR39, that is used for a large percentage of the prescription ophthalmic lenses currently produced. It should be noted from Table VI that the n_d versus v curve for these plastics would fall below the optical glass region, across the lower section of Fig. 1. Figure 9 illustrates the transmittance characteristics of the first three types in Table VI (Handbook of Plastics Optics, 1973).

TABLE VI
Optical Plastics

Material[a]	Common name	n_d	v_d	Approximate α $10^{-6}/°C$	Maximum use T (°C)	Maximum haze (%)
AC	Acrylic	1.491	57.2	65	90	2
PS	Styrene	1.590	30.9	65–75	80	3
PC	Lexan®	1.586	29.8	70	120	4
SAN	SAN	1.571	35.3	70	95	4
NAS	NAS	1.562	34.7	65	95	4
CR39	CR39	1.499	57.8	120	100	1
TPX	TPX	1.466	56.4	117	180	—

[a] AC—acrylic polymers, methylmethacrylate; PS—polystyrene; PC—polycarbonate; SAN—styrene acrylonitrile; NAS—methylmethacrylate-styrene copolymer; CR39—allyl diglycol carbonate; TPX—polymethylpentane.

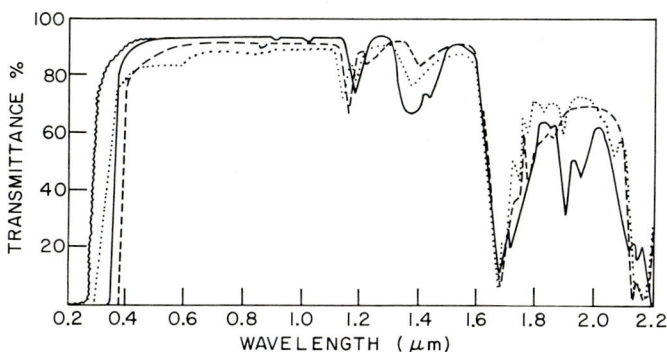

FIG. 9. Transmittance of four plastic optical materials. Dashed curve: polystyrene; solid: acrylic; dotted: polycarbonate; wiggly: ultraviolet acrylic. Thickness = 3.22 mm ($\frac{1}{8}$ in.).

VI. OPTICAL MATERIALS FOR THE INFRARED

Infrared materials include glasses, crystals, semiconductors, metals, and hot-pressed materials. As implied at the beginning of this chapter transmittance is the critical property by which infrared optical materials are identified and selected, although refractive, chemical, and mechanical properties must also be known and properly applied in designing for infrared optical uses.

Much of infrared technology, which has developed quite gradually over the years, is aimed at military applications. With the advent of new and increasingly complex devices involving the use of infrared materials, the materials problems themselves became quite complex, particularly as to the necessary tradeoffs among wavelength range, properties, fabrication, size, cost, etc. Because the variety of materials available is relatively limited, the infrared community eagerly procures specimens of every new material offered, studies it, measures it, and tries it. Selection is a difficult task although one that is eased if definitive knowledge of properties requirements is at hand, based on complete knowledge of service profiles and environments.

Table VII lists a variety of materials selected from the many more that are available. In virtually every case detailed properties data are found in suppliers' or published, literature (Optical Crystals, 1967; Kruse et al., 1962; Irtran Infrared Optical Materials, 1971). Two indices are given. The d line index is helpful in establishing the approximate level of the dispersion curve and an infrared index is given for somewhat better guidance. As pointed out in Section II.D., the thermal expansion coefficient α, and Young's modulus E, are important physical properties in designing for thermal and mechanical stress environments.

Many military infrared systems are airborne and choice of window materials must be based on such considerations as rapid temperature and humidity changes, abrasion resistance, absorption, and emissivity when heated. While refractive properties are of secondary importance in windows, infrared lenses and prisms are widely used in severe environments as well as in laboratories, so that the infrared-dispersion characteristic is an important material property. Although available materials do provide a fairly wide range of optical constants, it usually turns out that for reasons not primarily optical only a few media are available for a given application.

A. Glasses

Glass is widely used in the near infrared (Cleek et al., 1959; Malitson et al., 1963), but ordinary oxide glasses begin to absorb beyond 2.0 μm and absorption becomes very high in the OH^- ("water") band region, 2.7–3.0 μm. However, many military uses are found for high silica glasses such as fused quartz or VYCOR® brand glasses that are made with very low water content and that thus have useful transmittance beyond 3.0 μm (Fig. 10).

The infrared-transmittance cutoff for silicate glasses is limited to a maximum of about 5 μm and the absolute limit for oxide glass is estimated to be about 7.5 μm. The limits are established mainly by the fundamental vibration frequencies of the constituent cation–oxygen bonds. If one considers the oversimplified picture that frequency depends only on the force constant between ions and on the ionic masses, it is seen that for absorption to occur at lower frequency the ionic masses should be high and the attractive forces between ions should be low (other factors such as nearest neighbors and coordination numbers also influence frequency). However, lowering of force

TABLE VII
Selected Materials for the Infrared

Material	Useful trans. limit (μm)	Visible n_d	Visible dn/dT ($10^{-6}/°C$)	Infrared (λ) n	Infrared dn/dT ($10^{-6}/°C$)	α^a ($10^{-6}/°C$)	E^b (10^6 psi)
A. Glasses[c]							
Vycor® brand	3.5	1.457	+10	1.437(2.0)	—	0.8	9.6
(Ca,Al)SiO$_4$	4.7	1.605	—	1.580(2.0)	—	6.0	14.3
Germanate	5.7	1.660	—	1.637(2.0)	—	6.2	12.2
CaAl$_2$O$_4$	5.8	1.669	—	1.642(2.0)	—	8.4	15.2
As$_2$S$_3$	12.5	2.65	+95	2.411(3.3)	+3	26.0	2.3
B. Crystals[d]							
MgF$_2$	7.5	1.378(0)	+2(0)	1.376(3.3)	—	8.8–13.1	—
NaCl	22	1.544	−36	1.500(3.3)	−25	40	5.8
KCl	26	1.490	−33	1.470(3.4)	−34	36	4.5
KBr	35	1.560	−39	1.535(4.3)	−40	43	3.9
AgBr	35	2.253	—	—	—	35	4.6
KRS-5	40	2.625	−253	2.380(5.0)	−237	58	2.3
CsBr	45	1.698	−79	1.667(5.0)	−79	48	2.3
CsI	60	1.788	−99.2	1.739(10.0)	−91.7	50	0.8
C. Hot-pressed polycrystalline compacts							
Irtran 1 MgF$_2$	7.5	1.389	+1.5	1.349(4.3)	+1.5	10.7	16.6
Irtran 5 MgO	8	1.737	+15.4	1.660(4.3)	—	12	48.2
Irtran 3 CaF$_2$	10	1.434	−11	1.407(4.3)	−7.5	18.9	14.3
Irtran 2 ZnS	14	2.37	+64	2.249(4.3)	+75	6.6	14.0
D. Semiconductors; intermetallics							
Silicon	15	3.5	—	3.424(4.3)	+162	2.7	19
Germanium	23	—	—	4.022(4.3)	+280	5.5	14.9
GaAs	11	—	—	3.31(4.0)	+190	5.2	12
InSb	16	—	—	3.99(8.0)	+156	4.9	6.2
ZnSe (CVD)	22	2.59	+91	2.44(3.4)	+70	8.0	10.0

[a] α is the thermal expansion coefficient.
[b] E is Young's modulus.
[c] See Fig. 10 for transmittance curves.
[d] See also Table V for four additional crystals of great infrared interest.

constants leads to less stable glasses; i.e., inferior thermal and mechanical properties. It is generally not possible to make a hard glass with a wide transmission range. A hard glass requires strong bonds; an infrared transmitter requires weak bonds.

Since the use of glass for infrared optics is highly desirable because of homogeneity, ease of fabrication, increased element size, and economy, efforts have continued to develop oxide glasses with the longest possible

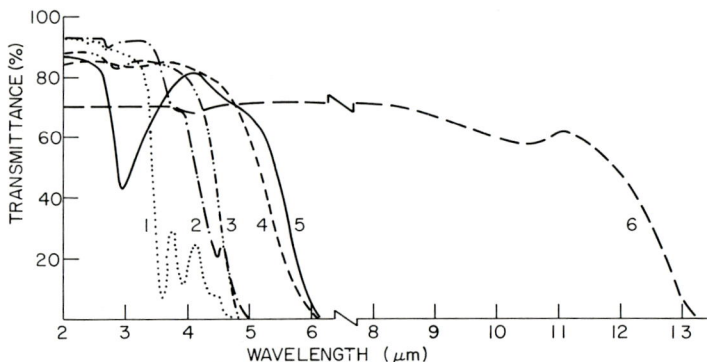

FIG. 10. Infrared transmittance of several types of glass. Thickness is 2 mm. 1 = VYCOR® brand Code 7905; 2 = fused quartz; 3 = calcium aluminosilicate; 4 = silica-free germanate; 5 = calcium aluminate; 6 = arsenic trisulfide.

infrared cutoff wavelengths, and some are now available commercially with useful transmittance beyond 5.0 μm. Figure 10 shows transmittance for five oxide glasses along with the representative nonoxide-type, arsenic trisulfide glass. These include the first five entries in Table VII. The limit for oxide glasses, as mentioned above, is about 7.5–8 μm. The calcium aluminate and germanate glasses are silica free, which is a necessary factor in maximizing long-wave transmittance in oxide glass. The strong absorption band at 3 μm, due to water content, in the curve for calcium aluminate glass in Fig. 10 is typical for early versions of this glass (Hafner *et al.*, 1958). At least one company now makes this material by a new process that eliminates water and its absorption so that the curve is virtually flat out to the cutoff region (Levengood, 1966). Elimination of the water band was also a key factor in successful development of the other glasses represented in Fig. 10.

A number of nonoxide chalcogenide systems have been developed in addition to the arsenic sulfide glass (Hilton, 1966). Germanium, selenium, and tellurium are also used to form various two to four-component glasses with a variety of properties. The glass with probably the best overall physical and optical properties for infrared use is in the system Ge–As–Se. Although its index is high, at 3.55, a Ge–As–Te glass can be formulated with useful transmittance to 20 μm.

B. CRYSTALLINE MATERIALS

The halides such as those listed in Tables V and VII, synthetically grown in single crystal form, have long been the primary materials for infrared instrumentation and experimentation. They traditionally exhibit two general shortcomings for uses such as in military systems—poor mechanical and chemical durability and lack of easy formability. Magnesium and calcium fluorides are the most rugged of these and they have had some limited systems use. With recently developed new fabrication techniques single crystal halides can now be shaped by extrusion or forging into new geometries such as tubing, domes, and lenses, adding a new flexibility in manufacturing ionic crystals (Optical Crystals, 1967). The molding process also imparts increased yield strength to these materials.

The development of the hot-pressed polycrystalline compacts was largely motivated by needs, such as military, to form various shapes from materials with good infrared properties. The rugged crystal, magnesium fluoride, was used for the first successful Irtran material (Irtran Infrared Optical Materials, 1971) and Irtran 1 is extremely important in infrared technology. Irtran 2, zinc sulfide, is probably second in popularity, although its high index does often dictate use of an antireflection coating. Irtran 5, magnesium oxide (periclase), is the hardest of the polycrystalline group and has the highest melting point. It is preferred where abrasion resistance or good thermal performance are primary requirements.

Because these compacts are made from powders that are densified by plastic flow to yield a mass of tiny crystalline aggregates, there is some light scattering from those grain faces that do not achieve optical contact with adjacent grains. The scattering increases at shorter wavelengths and is visually observed as haze. However, scatter is not a problem for most infrared applications of pressed compacts because it is insignificant beyond $2-3$ μm (Irtran Infrared Optical Materials, 1971).

C. Metallic Materials

The semiconducting metals silicon and germanium, in high-purity single crystal form, are widely used in a variety of infrared applications for lenses, domes, and windows (Kruse et al., 1962). They are both insoluble in water. Silicon has good resistance to mechanical and thermal shock and is excellent for guided missile domes. Silicon is also available in pressed polycrystalline form, with optical properties close to those for the single crystal. Germanium is fairly inert chemically and while hard, it is brittle and subject to mechanical fractures.

The high indices of refraction of these metals make them attractive for incorporation in single- or multi-element infrared lenses. Because of their semiconductor nature, both are quite temperature sensitive in absorption coefficient and in index, germanium much more so than silicon. Efficient antireflection coatings are known and are used on these high-index materials to increase their infrared throughput.

The first two intermetallics listed in Table VII, both III–V types, have seen some use in conventional infrared optics, but are probably most useful in optical components of integrated electro-optic systems. Zinc selenide, a II–VI compound, is available in polycrystal form as Irtran 4. However, the new form made by chemical vapor deposition (CVD) has superior optical properties (Lipson, 1976) and is now the preferred form, particularly for use with high-powered infrared sources such as the carbon dioxide laser.

REFERENCES

ASTM Standard C-730. Knoop indentation hardness of glass. *ASTM 1977 Annu. Book Stand.* Part 17, p. 754.
ASTM Standard C-770. Measurement of glass stress-optical coefficient. *ASTM 1977 Annu. Book Stand.* Part 17, p. 813.
ASTM Standard F-218. Analyzing stress in glass. *ASTM 1977 Annu. Book Stand.* Part 17, p. 987.
Baak, T. (1969). Thermal coefficient of refractive index of optical glasses. *J. Opt. Soc. Am.* **59**, 851.
Brixner, B. (1967). Refractive index interpolation for fused silica. *J. Opt. Soc. Am.* **57**, 674.
Campbell, D. E., and Adams, P. B. (1969). An evaluation of the problem of the chemical analysis of trace coloring oxides in optical glasses. *Glass Technol.* **10**, 29.
Cartwright, C. H. (1939). Lithium fluoride—quartz apochromat. *J. Opt. Soc. Am.* **29**, 350.
Cleek, G. W., Villa, J. J., and Hahner, C. H. (1959). Refractive indices and transmittances of several optical glasses in the infrared. *J. Opt. Soc. Am.* **49**, 1090.
Dumbaugh, W. H., and Schultz, P. C. (1969). Vitreous silica. *In* "Kirk–Othmer Encyclopedia of Chemical Technology," 2nd Ed., Vol. 18, pp. 73–105. Wiley, New York.
Hafner, H. C., Kreidl, N. J., and Weidel, R. A. (1958). Optical and physical properties of some calcium aluminate glasses. *J. Am. Ceram. Soc.* **41**, 315.
"Handbook of Plastic Optics" (1973). U.S. Precision Lens, Cincinnati, Ohio.
Hetherington, G., Jack, K. H., and Kennedy, J. C. (1964). The viscosity of vitreous silica. *Phys. Chem. Glasses* **5**, 130.
Hilton, A. R. (1966). Nonoxide chalcogenide glasses as infrared optical materials. *Appl. Opt.* **5**, 1877.
"Irtran Infrared Optical Materials" (1971). Eastman Kodak Co., Rochester, New York.
Izumitani, T., and Suzuki, I. (1973). Indentation hardness and lapping hardness of optical glass. *Glass Technol.* **14**, 35.
Jenkins, F. A., and White, H. E. (1976). "Fundamentals of Optics," 4th Ed., p. 489. McGraw-Hill, New York.
Kebler, R. W. (1954). Optical properties of synthetic sapphire. *J. Am. Instrum. Soc.* **1**, 69.
Kruse, P. W., McGlauchlin, L. B., and McQuistan, R. B. (1962). "Elements of Infrared Technology," Ch. 4. Wiley, New York.
Levengood, W. C. (1966). Stress-induced defects in vitreous calcium aluminates. *Appl. Opt.* **5**, 1906.
Lipson, H. G. (1976). Impurity absorption in CVD zinc selenide. *Tech. Dig. Pap. Meet., Opt. Phenom. Infrared Mater., Opt. Soc. Am.*
Malitson, I. H. (1965). Interspecimen comparison of the refractive index of fused silica. *J. Opt. Soc. Am.* **55**, 1205.
Malitson, I. H., Cleek, G. W., Stavroudis, O. N., and Sutton, L. E. (1963). Infrared dispersion of some oxide glasses. *Appl. Opt.* **2**, 741.

Maurer, R. D. (1973). Glass fibers for optical communication. *Proc. IEEE* **61**, 452.
Mitra, S. S., and Bendow, B., eds. (1975). "Optical Properties of Highly Transparent Solids," Plenum, New York.
"Optical Crystals" (1967). Harshaw Chem. Co., Cleveland, Ohio.
Owens, J. C. (1967). Optical refractive index of air: Dependence on pressure, temperature and composition. *Appl. Opt.* **6**, 51.
Parker, C. J., and Popov, W. A. (1971). Experimental determination of the effect of temperature on refractive index and optical path length of glass. *Appl. Opt.* **10**, 2137.
Reitmayer, F., and Schroeder, H. (1975). Effect of temperature gradients on the wave aberration in athermal optical glasses. *Appl. Opt.* **14**, 716.
Shand, E. B. (1958). "Glass Engineering Handbook," 2nd. Ed. McGraw-Hill, New York.
Spinner, S., and Waxler, R. M. (1966). Relation between refractive index and density of glasses resulting from annealing compared with corresponding relation resulting from compression. *Appl. Opt.* **5**, 1887.
Stoll, R., Forman, P. F., and Edelman, J. (1962). The effect of different grinding procedures on the strength of scratched and unscratched fused silica. *Proc. Symp. Mech. Prop. Glass Means Improv. It, Florence, 1961* p. 377. Cont. Sci. Union Glass, Charleroi, Belgium.
Tooley, F. V., ed. (1974). Physical properties of glass. *In* "The Handbook of Glass Manufacture," Vol. II, Sect. 17. Books for Industry, New York.
Waxler, R. M., and Cleek, G. W. (1971). Refractive indices of fused silica at low temperatures. *J. Res. Natl. Bur. Stand., Sect. A* **75**, 279.
Waxler, R. M., and Cleek, G. W. (1973). The effect of temperature and pressure on the refractive index of some oxide glasses. *J. Res. Natl. Bur. Stand., Sect. A* **77**, 755.
Waxler, R. M., and Napolitano, A. (1957). Relative stress—Optical coefficients of some National Bureau of Standards optical glasses. *J. Res. Natl. Bur. Stand.* **59**, 121.
Weeks, R. F. (1975). Plastic optics. *Opt. News, Opt. Soc. Am. Sept.*, p. 5.
Werner, A. J. (1968). Methods in high precision refractometry of optical glasses. *Appl. Opt.* **7**, 837.

CHAPTER 3

Plastic Optical Components

BRIAN WELHAM

U.S. Precision Lens, Cincinnati, Ohio

I. Introduction	79
II. Materials	80
A. General Description	80
B. Optical Properties	82
C. Coatings	84
III. Design	85
A. Optical Design	85
B. Mechanical Design	89
C. Mold Design	90
IV. Manufacturing Methods	92
A. Injection Molding	92
B. Compression Molding	94
C. Optical Replicating	95
D. Fabricating	96
E. Casting	96
Reference	96

I. INTRODUCTION

The term "plastic" is generally applied to a variety of manmade materials created by combining such basic inorganic materials as oxygen, hydrogen, nitrogen, chlorine, and sulfur in varying amounts with various organic ingredients.

Plastic, as we know it today, first became commercially available in the U.S.A. in 1868. In 1869 eyeglass frames were manufactured using cellulose nitrate (Society of Plastic Engineers, 1961). George Eastman used this type of material, called Celluloid, to produce motion picture film around 1882. It was not until over 40 years after the introduction of cellulose nitrate that any significantly new plastics were introduced. Then a wide variety of materials became available; but it was not until 1936 that the material acrylic (methyl methacrylate) was introduced. Acrylic has become one of the basic materials used today for most plastic refracting optics produced in high volume. A second material with a different dispersion was required before basic color correction could be designed into plastic optical systems. The synthetic resin polystyrene was first isolated in 1831 but was not commercially available until 1938, two years later than acrylic. Acrylic 491572 and styrene 590308 enable plastic systems to be color corrected and are the base materials for the majority of refracting plastic optics used in industry today. An additional material, polycarbonate resin (Lexan) 584301, which offers greater strength and higher service temperatures, was introduced around 1957. This material is more difficult to mold than styrene but is becoming more popular because of its attractive mechanical and thermal characteristics.

Billions of pounds of synthetic plastics and resin materials are used annually. Since the optics industry consumes only a fraction of this low cost material, little effort has been directed toward developing additional new materials with different indices and dispersions for optical plastic users. Other materials in general use include allyldiglycol carbonate (CR39), which is a thermosetting plastic used primarily for ophthalmics. A wide variety of optically clear epoxies are available. These are being used on a limited scale in such applications as replicating various geometrical configurations onto glass substrates. Relative to glass, plastic is a recent entry into the field of optics. Most of the development work enabling plastic to be used for refracting optics has been financed by private industry with the goal of fulfilling specific product requirements. These products are marketed in a highly competitive consumer market; therefore, the design and manufacturing technology associated with plastic optics has been kept proprietary by the companies concerned. The result is that a minimum amount of information has been published. Consequently, raw data are not readily available to the optical engineer. An effort has been made in this publication to introduce plastic terminology to the optical engineer and apply the standard glass terminology to plastics.

II. MATERIALS

A. GENERAL DESCRIPTION

When confronted with designing an optical system in plastic, the optical designer first requests the equivalent of the established "glass map." An examination of available optical plastics results in a limited choice of acceptable manufacturing materials. This restriction is considered by many designers as a severe limitation in designing high-performance optical systems. However, many optical systems do not require extremely high performance but must be produced in high volume at low cost. The low material cost of plastic optics used in conjunction with aspheres and irregular shapes provides additional freedoms not available to the traditional optical designer. Exploration of these freedoms requires a basic understanding of the plastic optics manufacturing process.

3. PLASTIC OPTICAL COMPONENTS

TABLE I
Optical Characteristics of Commonly Used Optical Plastics

Material		Methyl methacrylate (acrylic)	Polystyrene (styrene)	Polycarbonate	Methyl methacrylate styrene copolymer (NAS)
Refractive index	N_D	1.491(7)	1.590(3)	1.583(9)	1.562(8)
	N_F	1.497(8)	1.604(1)	1.597(8)	1.574(4)
	N_C	1.489(2)	1.584(9)	1.578(4)	1.558(3)
Abbe value (v)		57.2	30.8	30.1	33.5
Transmittance % (3mm thickness)		92	87–92	85–91	90
Haze (%)		<3	<4	<4	<4
Critical angle		42.1°	39.0°	39.1°	39.8°

TABLE II
Mechanical Characteristics of Commonly Used Optical Plastics

Material	Methyl methacrylate (acrylic)	Polystyrene (styrene)	Polycarbonate	Methyl methacrylate styrene copolymer (NAS)
Coefficient of linear thermal expansion (in/in °F x 10^{-5})	3.6	3.5	3.8	3.6
Deflection temperature (°F) 3.6 F/min, 264 psi	198	180	280	
3.6 F/min, 66 psi	214	230	270	212
Specific gravity (density)	1.19	1.06	1.20	1.09
Hardness (0.25-in. sample)	M97	M90	M70	M75
Impact strength-izod notch (ft-lb/in)	0.3–0.5	0.35	12–17	
Water absorption (%) (immersed 24hr at 73°F)	0.3	0.2	0.15	0.15

Tables I and II list some of the optical and mechanical characteristics of the four most commonly used optical plastics. The data in these tables have been gathered from a variety of sources, including the manufacturer, and are subject to change as more information is generated.

(1) Methyl methacrylate (acrylic) is the plastic equivalent to a crown glass. Acrylic is relatively inexpensive, can be easily molded and polished, and has the best combination of transmission, scratch resistance, and light stability. Optical memory, which is a material's ability to return to its original shape after exposure to heat, is excellent. Moisture absorption is minimal.

(2) Polystyrene (styrene) is the most commonly used equivalent to flint glass. It is the lowest cost optical plastic and is the easiest to mold. Less moisture is absorbed by styrene than by acrylic unless totally immersed in

liquid. However, it is much more difficult than acrylic to machine and polish.

(3) Polycarbonate has similar optical characteristics to styrene as well as very high impact strength and higher temperature stability than styrene and acrylic. It is more expensive than the other materials, more difficult to mold, and scratches easily. It is also considered difficult to machine and polish.

(4) Methyl methacrylate styrene copolymer (NAS) offers an alternate with optical characteristics between those of acrylic and styrene. This material molds well, is considered stable, and can be machined and polished.

(5) Allyldiglycol carbonate (CR39) is a thermosetting plastic used for some optical applications. This material is normally cast and cured under highly controlled temperature conditions. CR39 has high temperature resistance but its high shrinkage rate has limited its optical uses beyond ophthalmic lenses. In this application it is normally machined and polished after casting.

Other materials which are not in general use that could be considered for optical applications are:

(1) Styrene acrylonitrile (SAN) could be used as an alternate to styrene. The material molds well and is stable. It does have a tendency to yellow and is often used in reflecting applications.

(2) Methyl pentene polymer (TPX) has similar optical characteristics to acrylic. It has high temperature resistance, good chemical resistance, good electrical properties, and is soft but very tough. However, its very high shrinkage rate has limited its molded optical applications.

B. Optical Properties

1. *Refractive Index*

The refractive index figures in Table I are given to the 4th decimal place. The accuracy of the 4th decimal place quoted is questionable since data obtained from a variety of sources vary considerably. It is assumed that the designer will interpolate and smooth this data to suit his needs. Experience has shown that the variation of the refractive index of plastic materials from lot to lot does not exceed a tolerance of ± 0.0005. Within any given lot, variation of refractive index has been measured to be consistent within ± 0.0002.

2. *Transmittance*

The transmittance curves for some plastics are shown in Fig. 1. Surface losses for plastic can be assumed to be similar to an equivalent index glass.

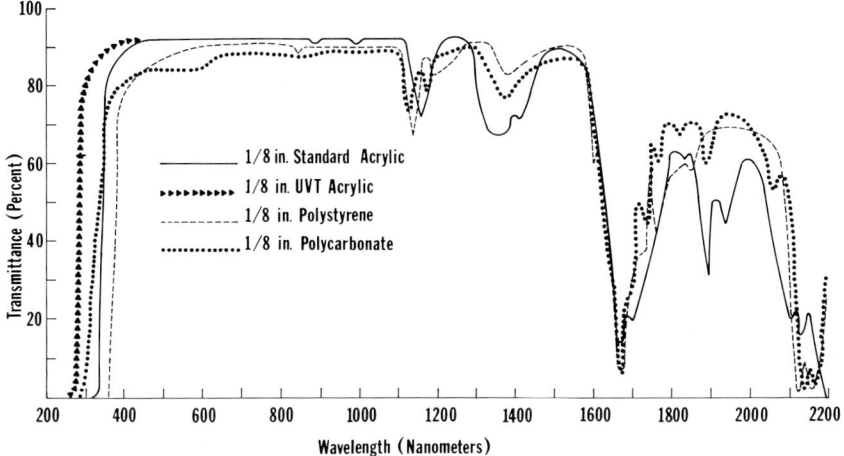

FIG. 1. Transmission curves for commonly used optical plastics.

Acrylic, for example, has a surface reflection loss of 3.9% per uncoated surface. Internal scattering varies slightly, depending on the specific manufacturer of the material measured.

3. *Stress–Strain*

The amount of stress imposed on a plastic optical component is invariably a function of the exact molding conditions applied. Mold design, gate size, shape, and location, injection temperatures and pressures, and cooling rates are a few parameters which contribute to the amount of stress evident in a component. The resultant strain patterns can be limitedly controlled by the molding conditions. Some annealing is performed on critical parts.

4. *Bubbles*

Bubbles, as specified for optical glass, are not evident in plastic optics. Bubbles, however, do occasionally appear when moisture is allowed in the material prior to molding. This can be eliminated by drying the raw material immediately before molding.

5. *Knit and Weld Lines*

Knit and weld lines appear in parts where the material injected into the mold takes separate routes to fill the cavity and arrives for fusion at different temperatures. These anomalies can be controlled but not always eliminated by good mold design, part design, and molding practices.

6. Digs and Scratches

The glass optics' specifications for digs and scratches can be directly applied to plastic optics. The surface quality of the finished part must always be exceeded in the mold optical insert surfaces. However, in general, plastics are comparatively soft and require special handling to maintain the part surface quality after it has been removed from the mold. Since special handling generally incurs higher component cost, careful consideration must be given to the dig and scratch specifications requested. Protective coatings can be considered when the operational environment demands protection of the surface quality. Using MIL–13830A as a reference, 80–50 quality can be produced under normal conditions. Better than 80–50 quality requires special handling and packaging.

C. Coatings

1. Hard Coatings

There are a variety of dip, spin, and spray coatings available which increase the scratch resistance of plastic surfaces. Unfortunately, the application of most of these coatings affects the optical performance of the elements and are costly relative to the uncoated element cost. Consequently, abrasive-resistant coatings are not generally applied to high performance plastic optical surfaces.

2. Antistatic

Static charges generated by the separation of unlike materials that have been in intimate contact are evident on the optical surfaces when a plastic part is ejected from the mold. Since these charges tend to attract undesirable dust particles, an antistatic coating is often applied. Antistatic agents may be applied by wiping the optic with a dampened cloth or by dipping the complete optic in an antistatic solution. These agents show reasonable resistance to dry-wiping, but may be washed off with water. Prior to assembly or coating, parts may have static charges neutralized by blowing the component clean with ionized air.

3. Antireflective

Antireflective coatings are applied to optical elements to increase the transmission and sometimes the contrast of an optical system. Magnesium fluoride (MgF_2) can be vacuum-deposited on plastic. When it is applied to acrylic, the surface reflection is reduced from 3.9% to less than 2%. Because of the lower service temperature of plastics and outgassing caused by plasticizers added to the raw material for molding, the MgF_2 must be applied at less than optimum coating conditions, resulting in relatively poor adhesion. This requires careful handling and cleaning before assembly.

4. *Multilayer*

Multilayer coatings have been developed and applied to refracting surfaces of plastic optics. Since most multilayer coatings are custom designed for specific characteristics pertaining to the end use of the component, a "standard" coating cannot be discussed. However, plastic optical surface reflections in the visible region have been reduced to below 0.2%. The coating process is extremely delicate and should be examined carefully before proceeding. If extreme care is not taken in the design and application of the coating, "crazing" can occur within the high-index layers which transfers to the optical surface resulting in good control of surface reflections but lower transmission of the optic because of internal scattering.

5. *Metallizing*

Silver, aluminum, and gold have been deposited on plastic optics. Since vacuum-deposited metallic film is very thin and has little resistance to abrasion and oxidation, an overcoat is generally required for protection.

III. DESIGN

A. OPTICAL DESIGN

It is common practice for optical designers when considering the design of a new lens to first examine the files and patent records to find suitable prototypes. The image quality of the prototypes is then evaluated after tracing various rays on a computer. The designer then selects a candidate with physical characteristics and aberration corrections similar to the desired product. A recent additional step taken by many designers in the process of selecting the prototype is to calculate "change or sensitivity tables." These tables indicate the selected prototypes' sensitivity to manufacturing variations or deviations from design of such parameters as thickness, centration, curves, index, and tip. A prototype selected for high volume production using plastic materials should be examined with consideration of the following parameters.

1. *Index*

Since as discussed above the 4th decimal place in the refractive index has a high degree of uncertainty caused by such factors as temperature control, measuring equipment, and the ability to produce high quality measuring prisms, the prototype design selected should show relative insensitivity to index variations.

2. Curves

Curve conformance in plastic optics is generally a function of size, thickness, and cross-section variation. Curves generated in the mold are required to be of higher quality than those for the finished component. Finished-part-to-mold conformance can be considered as a thermal heat sink problem. Controlled cooling is essential; and, when possible, cross-sectional area variations should be minimized. Since part thickness directly affects cooling, and therefore cost, it is desirable to design plastic optics with constant sections and minimum thickness consistent with part mechanical stability. These factors directly influence the finished component surface accuracy and quality. Elements up to 150 mm diameter, with good molding characteristics, are currently being manufactured to the following tolerances.

3. Spheres

Radius of Curvature: An acceptable radius of curvature tolerance is 1%, with 2% being more cost effective.

Sphericity: Interferometers and test plates are currently being used to measure surface conformance on production plastic optics. Better than five fringes of power per 10 millimeters of diameter can be achieved in production.

Irregularity: Irregularity, as defined in glass optical code books, is applicable to plastic. As with sphericity, irregularity is introduced during the cooling cycle. Since uneven cooling is a function of cross-sectional area variation, irregularity generally follows accordingly. Surfaces located closest to the largest masses (or heat sinks) take longer to cool and, of course, shrink more. Irregularity will often occur tracing the geometry generated by these mass concentrations. Consequently, geometrical zone irregularity or slope deviation is common. Compensation for slope deviation changes can be and is built into the mold. This mold compensation process takes several iterations, considerable experience, and demands strictly controlled molding conditions.

4. Aspheres

Consider the process of molding a plastic optic. The material is injected into the mold and prevailing conditions are controlled to ensure adherence of the plastic to the mold geometry. If the geometry takes the form of an asphere rather than a sphere, it is inconsequential to the molding process. The use of aspheres, therefore, is only limited by the moldmaker's ability to generate the initial asphere and the molder's ability to mold consistently. Numerically controlled machines are available which can generate a curve dictated by an equation to deviation accuracies of 0.0001″. The optician can then hand correct and polish the metal insert surface to the curvature and quality required. Since the mold optical insert is a master from which millions of parts can be generated, it is economical to invest considerable time and skill to optimize its performance. The primary limitation on asphere generation, production, and quality is the ability to accurately measure, evaluate, and determine deviations from the theoretically desired equation.

A tolerance equivalent for aspheres, to the 1 or 2% spherical radius of curvature tolerance, is an allowable deviation band bordering either side of the nominal aspheric curve. This deviation band should be zero at the apex of the lens then gradually increase to $\pm 0.0002''$ at the desired clear aperture. This tolerance is a suggested guide and should be modified when diameters become large. Irregularity for aspheres can be considered as symmetrical or assymmetrical. Symmetrical irregularity can be specified as zonal slope deviations within the basic tolerance band bordering the aspheric curve.

Assymmetrical irregularity is analogous to spherical curve irregularity and is more a function of component shape and gate location which detrimentally affect cooling. Cylinders, toroids, conics, etc., are manufactured regularly and are treated as aspheres.

5. *Thickness*

A well-designed mold will incorporate thickness adjustment capability. Once the mold has been manufactured, placed in the molding machine, cycled out, and adjusted, optical parts can be consistently molded to a $\pm 0.001''$ thickness tolerance

6. *Centration and Tip*

Most optical elements manufactured contain two refracting surfaces. To enable extraction of the element from the mold, each optical surface is located in one half of the mold. Consequently, there are three basic parameters which dictate the relative curve location within a given element.

(1) Location of the optical insert pocket in each mold half relative to the master guide pins or interlocks which lock the two mold halves together. Master guide pins and hole receptacles can be and are manufactured to tolerances of 0.001″–0.002″ for location. Optical insert pockets are located with $X-Y$ coordinates from the master guide pins within $\pm 0.0005''$.

(2) The optical surfaces are polished in oversized inserts. These inserts are then laser-reflection centered and externally ground to a sliding fit with

the insert pocket. The laser reflection deviation can be specified dependent on the quality required. The insert mounting bearing (for grinding the insert to size) has run out accuracy of 0.00005".

(3) The relative tip of the two component optical surfaces is dictated by the flatness of the two mating halves of the mold, perpendicularity of the optical insert receptacles to these reference surfaces, and the ability of the molding machine to maintain enough pressure to keep the two halves of the mold in contact during the molding cycle.

The two halves of the mold can be ground flat to better than 0.0001". Perpendicularity can be maintained within 0°0'30". The size and available clamping pressure of the molding machine can be selected to ensure that the two halves of the mold remain in contact during molding. In a multi-element optical system, if element orientation is necessary to improve performance, the individual inserts can be marked to ensure the required orientation in assembly.

7. *Plano*

Plano surfaces and prisms are considerably more difficult to mold than elements containing shallow curves. It is desirable to replace plano surfaces with shallow curves wherever possible.

8. *Doublets*

Optical cements are available for cementing plastic optics together. Some plastic doublets have been produced on a custom basis only. Production methods and better cements are required before cemented doublets can be considered as a production item.

9. *Large Optics*

Optics up to 200 mm in diameter and 50 mm thick have been manufactured. Spherical surfaces of 15 fringes at 72 °F can be obtained and aspheres with contour conformance of $\pm 0.0002"$ can be generated. Optics of this size and quality are currently in production. However, because of their size, they are more costly than typical multicavity injection molded optics. Most larger diameter optics, over 500 mm in diameter, are manufactured on a custom basis using glass optics manufacturing techniques.

10. *Color*

The material for molding plastic optics is generally purchased in transparent pellet form. However, the same type of material is available in a wide variety of colors. Judicious mixing and variation in part thickness allows a wide variety of neutral density filters to be molded. Where monochromatic viewing is required, selected colors can be incorporated in the molding material. Colors can be generated for ornamental attractiveness of the end product without destroying the optical function to be performed.

Multielements, irregular shapes, Fresnels, lens arrays, overlapping elements, and deviations from symmetry can all be considered and are only limited by the moldmaker's ability to duplicate a shrink compensated negative of the required configuration in metal.

B. Mechanical Design

To obtain the desired performance of an optical system, it is essential to consider the mechanical mount as an integral part of the system design. With glass optics it is generally accepted that the mechanical mount is a separate component of a different material designed to place the optical elements where they will provide maximum performance at minimum cost. The ability of plastic to conform to irregular shapes provides an added degree of freedom to the mechanical designer. Consideration can be given to integral molding of location bosses, spacers, and attachment pins.

As demonstrated in Fig. 2, the optics are molded integrally with the mechanical mount which contains the location and attachment pins. These pins are used to locate the optic on its mating part and are then heat staked over, attaching the optic permanently to the substrate. Assembly is simple, fast, and economical. Air spaced doublets have been designed and manufactured with integral mounts and spacers allowing the assembler to snap

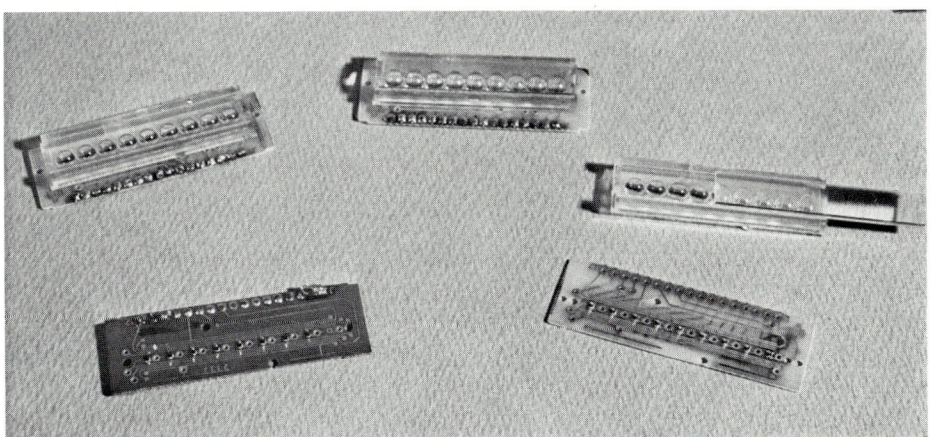

Fig. 2. Molded optics and mechanical mount.

the parts together automatically producing the required airspace and centration.

The thermal characteristics for common optical plastic materials are listed in Table II. The coefficient of linear thermal expansion is approximately 3.5×10^{-5} in./in. °F which is about an order of magnitude larger than common glasses. Compensation for thermal expansion is, therefore, essential when materials are used for critical optical assemblies.

Removal of the material injection gate is a secondary operation if it is required that the gate area be flush with its surround. If possible, the part should be designed with allowance for a slight protrusion in the gate area which, if high volume is required, would allow ultrasonic degating.

Parts with side walls perpendicular to the mold parting surfaces tend to cause drag when being ejected from the mold and often result in parts sticking in the mold. It is desirable, therefore, to design the parts with a minimum of 1° draft angle on vertical walls with the angle being selected relative to the parting line to allow easy removal of the part from the mold.

C. Mold Design

A mold consists essentially of three parts: the A half, the B half, and the ejector mechanism. (See Fig. 3.) The A half of the mold attaches to the fixed vertical platen of the molding machine. The B half, containing the ejector plate, attaches to the retractable platen. As the A and B halves are parted, the ejector plate is forced forward ejecting the molded part and its runners out of the mold cavities.

The two halves of the mold are located relative to each other through precision ground locating pins in the A half mating with their respective holes in the B half. Final location is accomplished with taper-locks located between the A and B plates. All insert cavities in the mold are located relative to these ground pins. The optical inserts are generated and polished individually with allowances for part thickness adjustment. The runner system is designed to balance and control the fill rate and pressure in each of the cavities. Gate size, shape, and location are critical. The gates are generally located to allow material to flow directly into the optic with the mechanical surround being a secondary consideration. Good gate control, in conjunction with good molding procedures, is vital to eliminate weld lines in the final part. Heating and cooling passages are always built into the mold base to allow thermal control during the molding cycle.

Since thermoplastics shrink at a rate of 0.002–0.005 in./in. °F, it is necessary to allow for shrinkage in the mold design and manufacture. When applied to parts with constant cross-sectional areas, these shrinkage allowances are straightforward. When applied to an optical surface located in a part of

Fig. 3. Typical injection molding mold.

varying cross-sectional area, the shrinkage becomes a complex thermal heat sink problem. It is often necessary to aspherize the mold optical insert to provide a sphere on the finished surface and generally several iterative processes are required.

The total number of parts required in conjunction with the requested delivery rate generally determines the tool quality and number of cavities. Adding more cavities does not always satisfy a high delivery rate since a larger molding machine is required and balancing multicavities for consistent high quality parts is difficult. The total number of parts required from a tool will affect the tool materials and built-in maintenance features. Eight-cavity molds have been built which have produced over one million parts.

1. Degating

Since most optics are molded in multicavities and all have runner systems, it is necessary to degate (remove the part from the runner system) the molded part. Parts can be sawed off, milled off, snipped off, or ultrasonically degated. Quality and cost of degating is a function of the end product cosmetics and mounting restrictions. Ultrasonic degating is fast; however, the mold designer has to compensate for this type of degating and the part designer has to allow for a small protrusion in the gate area on the finished part.

2. *Parting Line*

Since the mold consists of two halves, each containing a negative of half of the final product, consideration must be given to the parting line location. It is desirable to keep the parting line in a constant plane located away from the optical surfaces.

3. *Compression Molds*

The precision and control required to generate a Fresnel lens master requires a precise ruling engine. These ruling engines are contained in environmentally controlled chambers (cells) and incorporate ultraprecise mechanisms with the required mechanical motions being numerically controlled by a computer. The cutting tools used are usually diamonds which cut grooves to predetermined angles and depths in a brass master. Depth control of ± 10 millionths of an inch, groove spacings of ± 50 millionths of an inch, and angular tolerances of $\pm 0°0'4''$ can be accomplished. Nickel electroforms are made replicating the brass master and these electroforms are used in the mold. Expected life of an electroform is over 3,000 parts if it is not damaged during handling.

IV. MANUFACTURING METHODS

The raw material used for manufacturing plastic optics can be purchased in many forms including powder, pellets, and sheets of varying size and thicknesses. This selection of raw materials has created several methods for transforming the raw material into a finished product. These methods include injection molding, compression molding, casting, and fabrication. Both injection and compression molding and casting require a mold which is essentially a negative of the required part. The fabrication methods include replicating a thin plastic layer onto a glass substrate and using standard glass manufacturing techniques.

A. Injection Molding

Injection molding is the most popular method of manufacturing plastic optics. Multiple cavities and short cycle times (as low as 20 sec) allow high volume production at low component cost. Injection molding machines are readily available as standard products. An injection molding machine, manu-

Fig. 4. Injection molding machine.

factured for molding precision mechanical parts, can be adapted for optics manufacture. A photograph of a typical molding machine can be seen in Fig. 4. A schematic diagram of the machine can be seen in Fig. 5. The machine basically consists of two platens (one fixed and one retractable), a clamping unit for the movable platen, and an injection unit which inserts plasticized material through the fixed platen.

Raw material in pellet form is fed into the hopper. The material is either predried or a dryer is included in the hopper unit. Material is gravity fed into the injection barrel containing the injection screw. Around the periphery of the injection barrel are located heater bands. These bands heat the material, and the screw turns until the material is in molten form and is approximately 400–500 °F. The screw retracts to develop a shot, then rams forward to inject the material into the mold. The size of the injection barrel and screw determine the shot size available for filling the mold. The second half of the mold is attached to the retractable platen and is held in contact with its mating half during the molding cycle by a clamping unit, either mechanical or hydraulic. Clamping pressure and shot size determine the size of machine

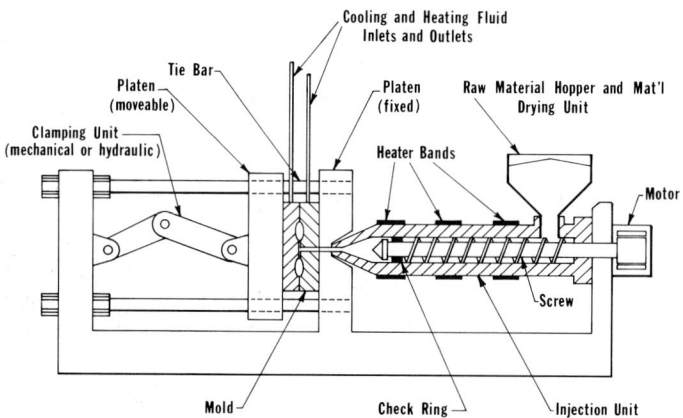

FIG. 5. Injection molding machine schematic.

required for a particular mold. To mold plastic optics a clamping pressure of 4–7 tons/in.² of projected molding area (including runners) is required.

To mold good optics the molding environment must be clean and dry and the material selected must be kept clean and dry. The clamp pressure must be precisely determined, controlled, and held constant for consistent quality. The injection profile, hold pressure, and time must be carefully determined. Instrumentation providing molding parameter feedback is highly desirable.

B. Compression Molding

Most compression molding of plastic optics is dedicated to the manufacture of Fresnel lenses and some lenticular arrays. This molding process lends itself more readily to near constant section optics and provides excellent resolution. A typical compression molding machine is shown in Fig. 6. As with injection molding, the mold is in two halves. One half is attached to the fixed platen and the other half is attached to a retractable platen. Unlike injection molding, the platens are normally arranged in the vertical direction. The material, in powder or sheet form, is placed in the mold prior to closing. Pretreatment of the material for moisture removal, cleanliness, etc., is essential.

The mold is then closed and temperature is applied followed by pressure, both responding to a controlled profile for a predetermined time. The times are relatively long—5–20 min. The material undergoes a chemical change which permanently hardens it in the shape of the mold. The mold is opened, the part removed, and excess material is trimmed. Parts up to 36 in. in diameter can be molded and multicavities are common.

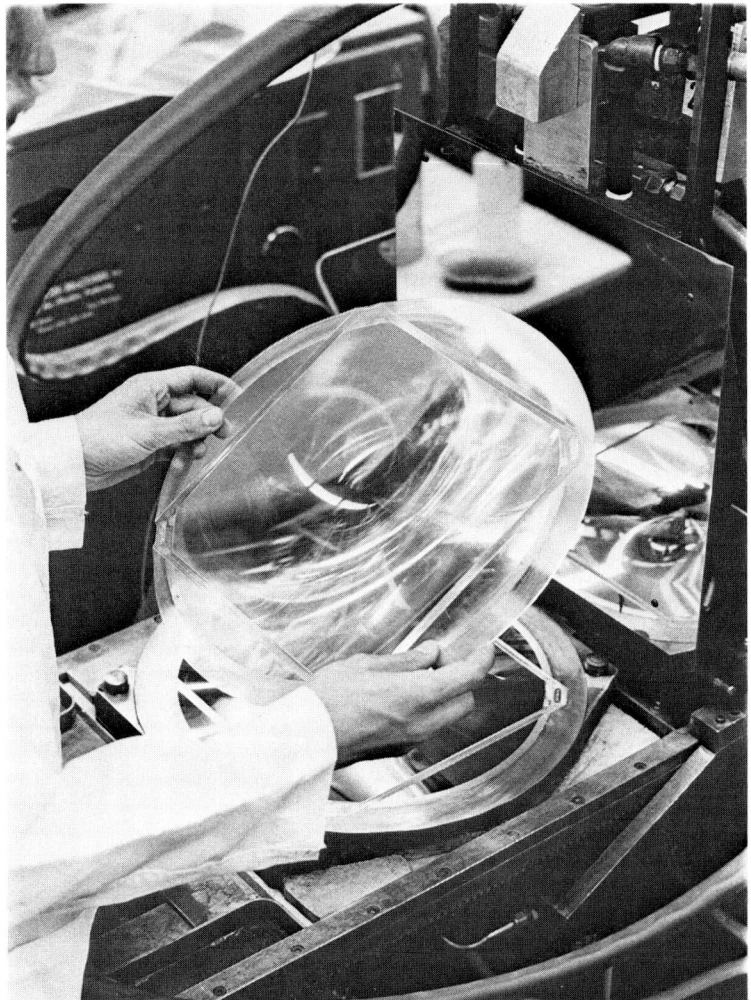

Fig. 6. Compression molding machine.

C. Optical Replicating

The availability of optical quality, low shrinkage epoxies, has led to the use of these materials for replicating a thin layer of epoxy in the form of an asphere or sphere onto a glass substrate. A negative of the required asphere is usually generated on glass, metal, or ceramic. Optical epoxy is deposited on the substrate and a release agent is placed on the master. The two parts are placed in contact and the epoxy is allowed to cure under controlled conditions. The master is then removed leaving the replicated epoxy curve deposited on the glass substrate. One of the main problems with this method of manufacturing is that because of the long curing cycle, many masters are required to provide production quantities. However, high quality aspheres have been generated using these techniques.

D. Fabricating

It is sometimes necessary to produce large (500 mm and up) diameter optics. Because of the economics of purchasing and working glass in this size, plastic is often considered as an economical alternative. In these sizes, plastic is manufactured into large optics using variations of classical glass optical manufacturing techniques.

E. Casting

Thermosetting plastics, such as CR39, have high shrinkage rates. These plastics are normally cast (usually on a glass mold) and then cured in an environmentally controlled area. They are then ground, machined, and polished to achieve their final configuration. Most plastic ophthalmic lenses are manufactured using this technique. Again, a large number of molds is required to achieve production quantities.

REFERENCE

Society of Plastic Engineers (1961). "Plastics." The Society of the Plastics Industry, Inc., New York.

CHAPTER 4

Optical Materials—Reflective

WILLIAM P. BARNES, Jr.

Optical Systems Division
Itek Corporation, Lexington, Massachusetts

I. Introduction	97
II. Mirror Property Requirements	98
A. Geometric Stability	98
B. Absorption and Scattering	100
III. Property Comparisons	101
A. Composition, Structure, and Processing Methods	101
B. Physical Properties	106
C. Stability Comparisons	107
D. Radiation Effects	109
IV. Lightweight Mirrors	110
A. Substrate Fabrication Methods	110
B. Mechanical Behavior	114
C. Thermal Behavior	118
References	119

I. INTRODUCTION

Curved reflecting surfaces are used as image-forming optical elements, offering advantages of near achromatic performance and feasibility of fabrication in much larger sizes when compared to refractive elements. Flat mirrors are used for scanning or redirecting the field of view of an optical system and for folding refractive or reflective lens systems into more convenient packages. Since the spectral reflectance of a surface can be adjusted by the choice of various coatings, we will focus our attention here on the substrate materials suitable for production of a precise and stable surface geometry and an excellent surface specularity.

The geometrical tolerances of reflectors are established by the optical system design. The actual surface geometry is initially determined by optical fabrication operations, but as we shall see in more detail below, may be subsequently affected by quite small strain changes in the substrate behind the reflective surface. The normal requirement is that the original geometry be maintained (or returned to) when the mirror is subjected to sometimes severe variations in its mechanical and thermal environment. Examples are the operation of large terrestrial telescopes at different elevation angles, launch survival and "g" release requirements of satellite optical payloads, and the cryogenic operating temperatures of some infrared optical systems. Less exotic but equally important may be the centrifugal accelerations of point-of-sale and other high speed scanners.

In many ordinary optical applications, imperfections in surface specularity result only in a minor loss of energy scattered away from the normal optical path. The small scatter levels of a normal well-polished mirror ($\sim 0.2\%$) become of concern, however, in imaging a dim object in the presence of a bright near neighbor or in applications of high energy density laser systems.

II. MIRROR PROPERTY REQUIREMENTS

A. Geometric Stability

If we assume that we start with a mirror of perfect geometry, then spherical changes of $\frac{1}{4}$ wavelength in surface displacement will cause a shift in focus and nonspherical changes of the same amount will noticeably affect the aberrations of an imaging system. The displacement of any point in a body may be determined (at least conceptually) by direct integration of the strain distribution. We should thus be able to establish a range of correspondence between tolerable strain levels and tolerable displacements for various mirror sizes (integration distance) whatever may be the initial source of strain change. With these numbers as a guide, we can then examine the potential effects of mechanical, thermal, and temporal strain changes in a substrate on optical performance.

1. *Mechanical Loads*

   a. Elastic strain. Consider a solid disk mirror, simply and uniformly supported at its edge, axis vertical. How different will the surface geometry be when this mirror is used in a satellite? We can get a good approximation with elementary plate bending theory, which yields the surface deflection change ω as

$$\omega = \frac{3\rho g(1-v^2)}{16Eh^2} \left[\frac{5+v}{1+v} a^4 - \frac{2(3+v)}{1+v} a^2 r^2 + r^4 \right] \tag{1}$$

where ρ is the density, E Young's modulus, and v Poisson's ratio of the disk material; h is the thickness and a the outer radius of the disk; and g is the gravitational acceleration. For a given material and assuming the supported edge fixed in relation to other portions of the imaging system, the terms in the brackets represent:

(a) a spacing change proportional to $[(5+v)/(1+v)]a^4/h^2$;

(b) a change in spherical power as measured by total surface displacement, proportional to $[-2(3 + v)/(1 + v)]a^4/h^2$; and

(c) a change in third order spherical aberration proportional to a^4/h^2.

For fused silica, with $\rho = 2200$ kg/m^3, $E = 7.32(10^{10})$N/m^2, and $v = 0.17$, the numerical values of the spherical and spherical aberration components are

$$\omega_{\text{spher}} = -2.91(10^{-7})a^4/h^2 \qquad (2)$$

$$\omega_{\text{aber}} = 5.36(10^{-8})a^4/h^2 \qquad (3)$$

For $a = 1.2$ m (as currently proposed for the NASA space telescope), $h = 0.4$ m (the old 6:1 rule of thumb) we have a total deflection of 3.1 μm and a residual spherical aberration of 0.7 μm. The support arrangement for terrestrial testing thus needs to be well engineered if it is to verify zero g performance to fractional wavelength tolerances in the visible.

The maximum bending stress induced using the above parameters is 3.2 MN/m^2, equivalent to a strain level of 4.4×10^{-5}. This illustrates that strain levels of 10^{-5} may noticeably affect the performance of a large mirror.

b. Nonelastic strain. In usual engineering practice, any irreversible (plastic) strain behavior is ignored up to the yield stress, which is arbitrarily defined as the stress required to produce a permanent strain of 2×10^{-3}. Since strains of 10^{-5} (and perhaps less) are of optical concern, it is of considerable value to know the "microyield" stress of a mirror material. The microyield stress (MYS), sometimes called the precision elastic limit, is usually reported as the stress required to produce a permanent strain of 10^{-6}. Stresses for other values of permanent strain are occasionally reported also.

2. Thermal Effects

A change in temperature ΔT induces a local nonelastic strain equal to $\alpha \Delta T$; this defines the thermal expansion coefficient α. If these strains are constant or linear functions of position, no stresses are induced, and surface displacements will depend directly on the thermal strain distribution. Nonlinear distributions and "bimetallic" discontinuities will cause stresses to develop, somewhat limiting the thermal strain effect and usually making exact solutions much more complex.

Since we often have incomplete control over the temperature environment, a near zero value of α is obviously desirable. A major portion of the effort in developing mirrors and mirror materials in the past decade or two has been devoted to "zero-expansion" materials. This has brought us to the state at which spatial variations in α from point to point are often comparable to the maximum absolute value of α at any point. In these cases, the changes caused by a completely uniform change in temperature level become of concern. We should note, of course, that a completely uniform change in $\alpha \Delta T$ throughout an all-reflective telescope structure would be of no consequence, since it would impose only a very small change of scale on the system, the relative geometry remaining constant.

3. *Temporal Effects*

Although changes in mechanical loading or supports and in temperature are the major causes of surface geometry changes, other effects, usually time dependent under essentially constant ambient conditions, may often be observed. Some of these have been found to result from residual stress relaxation, phase changes in materials, and sensitivity of some materials to atmospheric humidity.

B. Absorption and Scattering

Since the principal function of a mirror is to redirect light to a predetermined direction, any increase in absorption or scattering at the surface is undesirable to some extent. Although the early telescope mirrors (of Newton, Herschel, Rosse) were all made of speculum metal, Newton himself recognized the greater ease of highly polishing glass and experimented with a quick-silvered (second surface) mirror. A second surface mirror, of course, carries all the disadvantages of any refractor and more extensive use of glass reflectors had to await the development of first chemical (Common, Brashear) and then physical (Strong) processes for producing a uniform highly reflecting film on the first surface.

The preference for glass over metal from the point of view of achievable specularity persists to this day with one exception. The preference is a result of the fact that the physical structure of a glass has no long range order—atomic positions are correlated to each other over only a few interatomic distances—while most metals are polycrystalline. In metals, the action of optical polishing may vary with crystal orientation, leading to surface discontinuities at crystallite boundaries. These boundaries also have a tendency to accumulate whatever impurities exist and their physical structure itself may introduce further nonuniformities in polishing action. The one metallic exception is the nearly amorphous layer which can be obtained by the autocatalytic deposition of nickel with varying amounts of phosphorus. Scattering levels lower than those of conventional pitch-polished glass have been achieved on such nickel–phosphorus (electroless nickel) coatings. The very small crystallite structure of some low-expansion glass ceramics (Pyroceram, CER-VIT) has also permitted the achievement of good specularity on these materials.

III. PROPERTY COMPARISONS

For convenience of discussion, the more useful mirror substrate materials may be segregated into four classes:

(a) low-expansion glasses,
(b) very low-expansion glasses and glass-ceramics,
(c) metals, and
(d) other materials.

The properties of importance in selecting a mirror material are listed in Table I for some representative members of these classes. Some understanding of the composition, production methods, and forms of supply may be useful in the selection of the proper material for a given application.

A. Composition, Structure, and Processing Methods

1. *Low-Expansion Glasses*

These include Pyrex (Corning Glass Works), Duran 50 (Schott), and E-6 (Ohara) glasses, all of which are borosilicate compositions produced by conventional pot melting of oxides and in which B_2O_3 replaces all of the CaO and MgO, and part of the Na_2O of soda–lime–silica window glass. Thermal expansion coefficients for Pyrex and Duran 50 are 3.2×10^{-6} K^{-1} (300–670 K) and 2.4×10^{-6} K^{-1} for E-6. The softening and melting temperatures of these materials are much higher than those of soda–lime glass, thus they are somewhat more difficult to process. Their cost, however, remains significantly lower than that of the very low-expansion materials.

2. *Very Low-Expansion Glasses and Glass-Ceramics*

These include two true glasses—high purity fused silica and a 7% titania–silica glass (Corning ULE™ titanium silicate, Code 7971) in which the temperature at which the thermal strain has zero slope is shifted from 140 K (for fused silica) to 300 K.

High purity fused silica is further subdivided into several grades which differ in production method (fusion of natural quartz or synthesized by thermal decomposition and oxidation of $SiCl_4$), water impurity content, and optical transmission quality. These differences are of little consequence for mirror applications except as they may affect bubble content, residual strain level, and homogeneity of thermal expansion—all of which should be specified for critical mirror applications.

Several grades of both fused natural quartz and synthetic fused silica are available from Heraeus-Amersil, Inc., while Corning Glass Works produces both pure and titanium-doped silica synthetically from the appropriate

TABLE

Property (units)	Duran 50	Fused silica	ULE (7971)	CER-VIT (C-101)	Beryllium (I-70)
Young's modulus (10^{10} N/m²)	6.17	7.32	6.77	9.18	30.4
(10^6 lb/in.²)	8.9	10.6	9.8	13.3	44
Poisson's ratio	0.20	0.167	0.176	0.252	0.025
Density (10^3 kg/m³)	2.23	2.202	2.20	2.50	1.85
(lb/in.³)	0.0806	0.0796	0.0795	0.0903	0.066
Ultimate tensile stress (10^7 N/m²)	7.8	5.0	5.0	5.7	24
(10^3 lb/in.²)	11.3	7.2[a]	7.2[a]	8.3[a]	35
Microyield stress (10^7 N/m²)	>UTS	>UTS	>UTS	>UTS	1.7
(10^3 lb/in.²)					2.5
Hardness	—	500 (Knoop)	460 (Knoop)	540 (Knoop)	—
Stress optical coefficient (10^{-12} m²/N)	(Pyrex) 3.88	3.47	3.99	3.08	—
Thermal expansion coefficient (10^{-6} K^{-1})	3.2	0.56	0 ± 0.03	0 ± 0.03	11.2
α homogeneity (total excursion) (10^{-9} K^{-1})	—	5	15	15	40
Specific heat (J/kg K)	835	741	766	840	1.82×10^3
(Btu/lb F)	0.20	0.177	0.183	0.20	0.436
Thermal conductivity (W/m K)	1.02	1.37	1.31	1.70	220
(Btu/hr ft F)	0.59	0.80	0.76	1.0	127
Thermal diffusivity (10^{-6} m²/sec)	0.548	0.840	0.777	0.809	65
Volumetric specific heat (10^6 J/m³ K)	1.86	1.63	1.68	2.10	3.37
Volume resistivity (ohm m)	—	10^{11}	10^{10}	10^{13}	4.2×10
Surface smoothness (Å (rms))	5	5	5	5	60–80[c]
Maximum service temperature					
Continuous (K)	—	1170	1070	980	—
Short term (K)	—	1370	1270	1090	—
Melting point (K)	—	—	—	—	1560

[a] Modulus of rupture, abraded.
[b] Annealed.
[c] Sputtered.

4. OPTICAL MATERIALS—REFLECTIVE

PROPERTIES OF MIRROR SUBSTRATE MATERIALS

Aluminum 6061-T6	Copper (OFHC)	Molybdenum (TZM)	Electroless nickel (8–10% P)	Glassy carbon (PC 2000)	Silicon carbide (CVD)	Graphite–epoxy GY-70/x-30 "isotropic"
6.90	11.7	31.8	14.7	2.8	38	9.3
10.0	17	46	21	4.0	55	13.5
—	—	—	—	0.12	—	—
2.71	8.94	10.20	7.8–8.1	1.47	3.10	1.78
0.0979	0.3230	0.3685	—	0.0531	0.1120	0.0643
31	22[b]	86	73	14	17	21
45	32[b]	125	106	20	24	31
12–14	—	16	—	—	—	—
18–20	—	24	—	—	—	—
95 (Brinell)	40 (Rockwell F)	200 (Vickers)	500–800 (Knoop)	200 (Vickers)	9–9.5 (MOH)	—
—	—	—	—	—	—	—
23.0	16.7	5.0	13–14.5[e]	3.95	3.5	+0.02
—	—	—	—	—	—	100
960	385	272	—	750	1.42×10^3	—
0.23	0.092	0.065	—	0.18	0.34	—
171	392	146	4.4–5.6	4.2	71	35 (in plane)
99	226	84.5	2.5–3.2	2.4	40	20
66	114	53	—	3.8	9.1	—
2.6	3.44	2.77	—	1.10	4.4	—
8×10^{-8}	1.71×10^{-8}	5.2×10^{-8}	30–60	3×10^{-5}	—	—
200 + (?)	40	10	5	10	10	—
—	—	—	470	—	3600[d]	—
—	—	—	(Crystallizes)	—	—	—
850–920	1350	2880	—	—	3100	—

[d] Nonoxidizing.
[e] Some proprietary coatings more closely match the expansion of beryllium to yield low stress coatings.

tetrachlorides. Other sources include Dynasil Corporation, the General Electric Lamp Glass Division, Thermal American Fused Quartz, Inc., Quartz et Silice (France), and the Thermal Syndicate Ltd. (England).

Very low-expansion glass-ceramics have been developed from Corning's Pyroceram technology and have been produced as CER-VIT™ (Owens-Illinois), Zerodur™ (Schott), and Crystron-Zero™ (Hoya). All are melted and poured as lithia–alumina–silica glasses, which, on subsequent heat treatment, develop a microcrystalline structure in a glass matrix. The crystallite structure is similar to that of β quartz (Simmons, 1967) yielding a negative coefficient of thermal expansion of the crystallite and a macroscopic expansion coefficient which can be held (in premium grade material) to $0 \pm 1 \times 10^{-7}$ K^{-1} (0–38 °C). By special selection, material with $\alpha = 0 \pm 3 \times 10^{-8}$ K^{-1} may be available in both CER-VIT and ULE.

3. *Metals*

Metals differ most markedly from glasses in their thermal transport properties and this may recommend their use in, for example, high energy density pulsed laser optics. Beryllium stands out additionally because of its light weight and high modulus of elasticity, making it preferable for high speed scanning mirrors. The one low-expansion metal (36% nickel–iron or Invar) is not discussed here because it offers no advantage in weight, stiffness or strength, its temporal stability is suspect unless carefully machined and heat treated, and its thermal conductivity (~ 10 W/m K) is much lower than the base metals included in Table I. The properties of electroless nickel are included because of its utility as a thick (tenths of millimeters) coating suitable for both optical shaping (figuring) and polishing to good specularity.

Aluminum, copper, and molybdenum are cubic in crystal structure, while beryllium is hexagonal and highly anistropic in thermal expansion. Wrought forms of the cubic metals are preferred for mirror applications, since castings may suffer from microporosity. Aluminum, copper, and molybdenum are generally available from metal distributors. The properties listed are for 6061-T6 aluminum, oxygen free high conductivity copper (annealed), and a titanium–zirconium–molybdenum alloy (TZM).

Excessive grain growth and shrinkage make the casting of beryllium a difficult task. After production of a chemically pure but mechanically imperfect vacuum casting, the engineering material beryllium is fabricated by powder metallurgy techniques. The metal is chipped and ground fine, blended, commonly canned in steel under vacuum to prevent reaction with the atmosphere (or pressed under vacuum in graphite dies), and sintered at pressures of 1000–2000 psi and temperatures of 1050–1150 °C. The two U.S. producers of beryllium are Brush-Wellman and Kawecki-Berylco Corporations.

Within the last 15 years, a significant effort to evaluate and improve beryllium for use as a mirror substrate material has been pursued. The characteristics that have inhibited and continue to influence the successful engineering of beryllium mirrors have been (Barnes, 1966):

(1) The elemental crystal structure is hexagonal and some measure of anisotropy persists through most methods of production. Current "optical" grades are specially processed by impact attritioning to reduce the anisotropy of thermal expansion in the transverse plane to 0.1–0.2% (Stonehouse et al., 1973). Expansion differences between the longitudinal (pressing) direction and the transverse directions remain 1–2% in directionally pressed material. Isostatic pressing and pressureless sintering techniques have also been used to improve isotropy.

(2) The crystal is susceptible to basal plane slip at low stress levels.

(3) The residual voids of powder pressing, beryllium oxide inclusions, and the beryllium oxide film on each powder particle all interfere with successful polishing to a highly specular surface. Improvements have been achieved by use of vacuum deposited or sputtered layers of high purity beryllium.

4. *Other Materials*

We have included some materials under current development because of special properties which may prove of value in the future.

Glassy carbon is made by pyrolysis of thin (~3 mm maximum) layers of a highly cross-linked organic precursor. It offers a glassy, polishable structure of light weight, with a thermal conductivity comparable to electroless nickel. It may also be produced as a coating (Rossi, 1974) on substrates which can withstand the pyrolysis temperatures of 1100 K and higher.

Silicon carbide, particularly the chemically vapor deposited (CVD) form which offers lower porosity than the powder metallurgy product may become of value in short wavelength, high energy density laser applications. One might also consider, for example, the use of a glassy carbon coating on a silicon carbide substrate for applications at elevated temperatures.

Progress in composite material engineering has produced a graphite-fiber/epoxy structural laminate material with weight and stiffness properties comparable to beryllium and thermal expansion characteristics nearly equal to the very low-expansion glasses and glass-ceramics. It remains necessary to combine this material with a polishable face sheet but applications in which weight and stiffness are at a premium are expected to require continued development of such composite mirror substrates.

The GY-70/X-30 composite of Table I contains eight layers with fiber directions of 0°–45°–90°–135° symmetrically arranged about the midplane of the final sheet. The in-plane properties are nearly isotropic. The GY-70 graphite fibers (Celanese Corporation) have a tensile modulus of 52×10^{10} N/m² (nearly twice that of beryllium) and a negative coefficient of thermal expansion. These properties are obtained by pyrolysis of drawn polyacrylonitrile (PAN) fibers at temperatures in the neighborhood of 3300 K. A relatively low modulus X-30 epoxy resin (Fiberite Corporation) is used in this composite to reduce the temperature at which trans-laminar microcracking occurs to about 150 K. These microcracks between fibers result from directional differences in the macroscopic expansion between plys of the layup. They do not strongly affect the strength and stiffness, but do cause a change in effective thermal expansion of the laminate. Additional development, including the use of woven fiber plys, is proceeding rapidly.

As with most plastics, the dimensions of the X-30 resin will change when its moisture content is varied, and the shape of an unprotected graphite–epoxy mirror substrate will vary with changes in ambient humidity levels. All of the evidence currently available indicates that this effect is completely reversible to the levels needed for good optical performance, however.

B. Physical Properties

The properties listed in Table I are for materials at a normal laboratory ambient temperature of 293 K and are suitable for preliminary intercom-

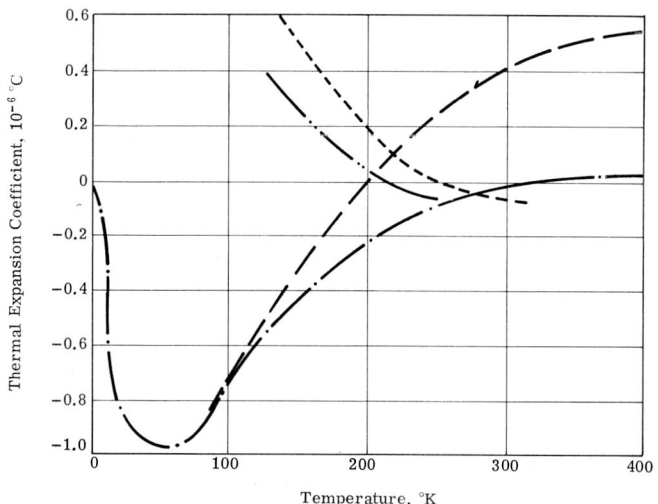

FIG. 1. Thermal expansion coefficients of low-expansion materials. ——: SiO₂, vitreous (AIP Handbook); —·—: ULE titanium silicate 7971, $\alpha = 0.015 \times 10^{-6}/°C$ (5–35 °C) (unpublished data, Corning Glass Works); ---: CER-VIT C-101, $\alpha = 0.05 \times 10^{-6}/°C$ (0–38 °C) (unpublished data, Owens-Illinois); —··—: CER-VIT C-101 (catalog data).

parisons of materials. For most of the materials, strength and stiffness values are not strong functions of temperature. In applications which depart significantly from 293 K operation, the variation of thermal expansion coefficient with temperature becomes important, and the typical behavior of α for the very low-expansion materials is shown in Fig. 1. Similar data on the metals is readily available, for example, in the American Institute of Physics Handbook, while a fuller discussion of the low temperature behavior of graphite–epoxies is beyond the scope of this work.

The thermal expansion of the metals is a strong function of temperature also, while their thermal conductivity depends on composition and impurities, as well as temperature.

C. Stability Comparisons

We consider here primarily those surface displacements which directly affect the aberrations of an optical system, i.e., surface figure irregularities rather than mere changes in spherical power.* Although a rough relation between strain levels and induced irregularities exists—that strains larger than 10^{-6} may affect the performance of high quality visible light systems—the precise effect on surface figure depends on not only the maximum strain magnitude but also the strain distribution in the substrate. Direct measurements of optical effects are needed to provide a complete experimental picture and the available data are somewhat limited.

1. *Thermal Expansion Homogeneity*

Estimates of the expected homogeneity of some materials are shown in the $\Delta\alpha$ row of Table I. These data summarize the results of many thermal tests of mirror specimens at both Perkin-Elmer Corporation and Itek Corporation, and direct measurements by Corning Glass Works (Hagy and Smith, 1969) and Owens-Illinois (Stewart and Davis, 1970).

Expansion inhomogeneity causes the development of a stress as the blank is cooled slowly through the strain point, thus the residual birefringence of a well-annealed blank is a measure of the inhomogeneity. If we assume that the stress is $\frac{1}{2}$ that which would develop if an inhomogeneous volume were rigidly restrained, this stress σ is

$$\sigma = E(\Delta\alpha)(\Delta T)/2$$

and the birefringence will be σ times the stress–optical coefficient. Using the values of Table I for fused silica (and a strain point of 1300 K) we find

---

* It should be noted, however, that a spherical change in a flat mirror used at non-normal incidence will introduce astigmatism.

that the birefringence corresponding to a $\Delta\alpha$ of 5×10^{-9} is 6.35×10^{-7} or 6.35 nm/cm. This value is consistent with the actual specification for well-annealed material of "less than 10 nm/cm."

2. Thermal Cycling

An extensive sequence of thermal cycling tests (Jerke and Platt, 1972) have shown that any thermal hysteresis in the low expansion materials in the 195–418 K range is less than $1/100\lambda$ rms ($\lambda = 633$ nm).

When beryllium mirrors are not given etching, stress relief, and low temperature stabilization treatments, some permanent figure change may be noted after any low temperature exposure. Cycling to temperatures of 100 K has produced deformations of as much as 1 wavelength rms (633 nm), but these instabilities have been completely reversible in mirrors that were previously cycled (preferably two to three times) to the same or lower temperatures.

3. Mechanical Load Cycling

Glasses, and particularly fused silica, are among the most purely elastic materials known. Cyclic stressing will not result in permanent changes in surface shape until it reaches levels which will initiate crack growth.

CER-VIT is also elastic, but in some tests has exhibited a time dependent increase in strain at constant load and in strain recovery. We have observed that a 14 cm CER-VIT C-101 disk stressed to 5000 psi for 65 days required 5 days to fully recover its initial surface shape. A similar experiment with fused silica led to fracture after a few days.

Cycling stressing of beryllium beyond the microyield stress may, of course, produce irreversible surface changes of optical importance.

4. Temporal Effects

The very low-expansion glass and glass-ceramic materials are not in thermodynamic equilibrium since phases with a fully developed crystal structure have lower free energies (and different volumes). Any trend toward equilibrium is extremely slow, however, and no effect on surface displacements has been observed in the absence of impurities, inclusions, or environmental factors which would accelerate a phase transition or devitrification.

Properly etched and stress relieved specimens of beryllium have also exhibited good stability in test durations of 14–25 months (Jerke and Platt, 1972).

We have previously noted that plastic composites are sensitive to moisture content. The effects of atmospheric humidity, which are believed to be completely reversible, may be manifested as temporal changes as a result of seasonal humidity changes.

D. Radiation Effects

In general, fused silica is more resistant than other glasses to all forms of naturally occurring radiation. It is also quite resistant to isotope and low level reactor exposure. For this reason, and because ULE and CER-VIT are relatively recent developments, there does not exist any extensive literature on radiation damage.

The first incentive for production of beryllium was for nuclear reactor uses as a moderator and as a low neutron cross-section structural material. Its resistance to damage by all forms of radiation is substantially better than the low-expansion mirror materials.

1. Electromagnetic

Other than the results of the conversion of some fraction to thermal energy, the only optical effect of electromagnetic radiation that has been observed is the development of color centers by ultraviolet or shorter wavelength radiation. The observations extend to 10^{10} rad of 1-MeV γ radiation and the effect is of consequence only for transmitting applications. The visible light transmission of fused silica is slightly affected at 10^{10} rad, while that of ULE is reduced 25% for 10^5 rad of 1-MeV γ photons.

2. Neutron

No data on the effects of neutron irradiation on ULE and CER-VIT have been found.

The Corning literature reports no change in transmission, but a 2% increase in density for a neutron dose of $10^{20}/\text{cm}^2$ in fused silica. A nonuniform change of density would, of course, affect the optical figure. This dosage level, however, seems possible only in the interior of a nuclear reactor or on exposure to a nuclear detonation.

3. Electron

In a study of replica grating stability (Gunter, 1969), 2- to 4-in. specimens of BSC-2, Pyrex, fused silica, and CER-VIT were exposed to 1-MeV electrons at the dosage rates needed to simulate one full year of peak Van Allen belt radiation in a few days. At this level, changes of 1λ ($\lambda = 546$ nm) in Pyrex and $\sim 5\lambda$ in CER-VIT were reported.

4. General Considerations

Since the glasslike materials are in metastable states (albeit with geological decay times under normal conditions), we should expect that large dosages of high energy radiation may accelerate the transition to the lower free energy crystalline forms. The effects of the accompanying density change are completely analogous to the volume change effect of thermal expansion and the potential effects on optical performance of mirrors are completely similar. We have identified only three cases for which the energy and dosage levels of concern might possibly be reached. They are

(1) Within the shield of a nuclear reactor.
(2) On exposure to a nuclear detonation.
(3) On exposure continuously for at least 1 year at the highest electron fluxes found in the Van Allen belts (electron radiation only).

IV. LIGHTWEIGHT MIRRORS

Efforts in developing lightweight mirrors have been devoted primarily to constructions of fused silica, ULE, CER-VIT, and beryllium.

A. SUBSTRATE FABRICATION METHODS

1. *Fused Silica*

Prior to 1967 fused silica was the only low-expansion material that had demonstrated promise as a lightweight mirror structural material. In the late 1950's and early 1960's (and, in fact, on and off since George Ritchey's work with glass structures in the late 1920's), many artisans, engineers, and scientists had made small specimens of pet ideas for machining or assembling struts, plates, and tubes into a lightweight structure. Until the later 1960's, the most successful of these ideas was the "eggcrate" structure of fused silica developed by Corning Glass Works.

The eggcrate fused silica lightweight mirror structure is assembled from machined fused silica parts and consists usually of two circular plates between which is fitted a core structure of slotted, interlocking ribs (similar to the dividers in old fashioned eggcrates). This rather fragile plate and core assembly is fired to the softening point of the silica and the force of gravity promotes fusion at the surfaces where the plates and ribs join. Little, if any, fusion is achieved between the ribs themselves at the areas where they interlock.

A typical silica eggcrate structure of recent years is shown in Fig. 2a. Also shown, in Fig. 2b, is a less efficient (from the point of view of weight reduction) silica structure consisting of a drilled core and "back plate" of opaque fused silica to which is fused a clear fused silica faceplate (the lower plate in the figure). Lightweighting of fused silica may also be done by diamond machining of holes through a solid boule of material to form an ultralightweight core structure. Front and back plates would then be fused to this one-piece core in the conventional manner.

4. OPTICAL MATERIALS—REFLECTIVE

FIG. 2a,b. Lightweight mirror constructions. (a) Silica eggcrate (Corning); (b) silica, partially drilled core (Heraeus-Schott).

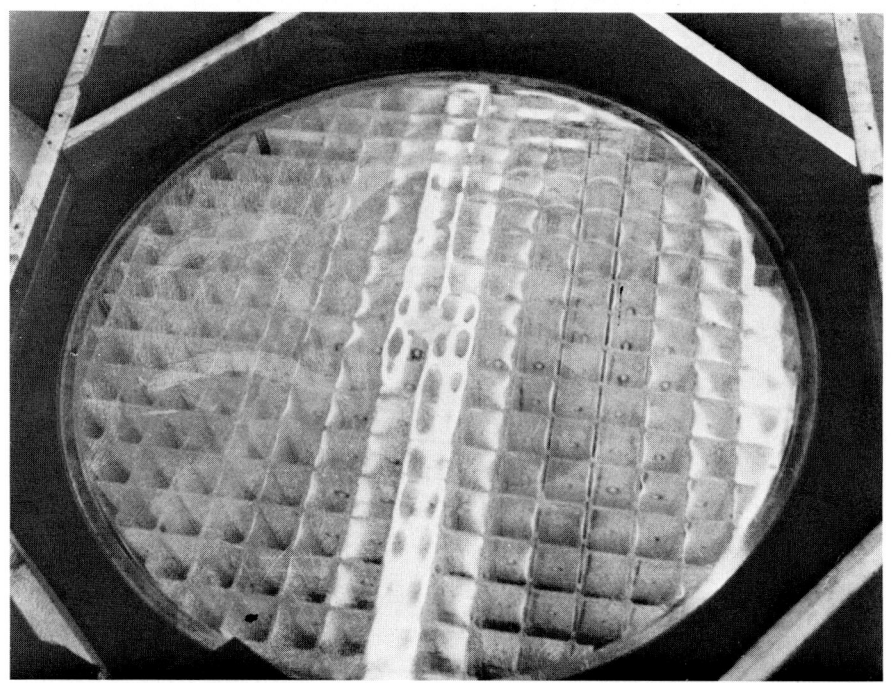

Fig. 2c,d. Lightweight mirror constructions. (c) ULE, fully fused (Corning); (d) CER-VIT, machined (Owens-Illinois).

2. ULE Titanium Silicate

The fusion of pure silica requires heating the entire structure to prevent fracturing because of differential thermal expansion. The thermal expansion of ULE, however, is sufficiently low that it may be welded to itself at one point in a structure while the major portion of the structure remains at room temperature, much as in the fusion welding of steel. This makes practical the preassembly of a fully fused core structure, which is subsequently joined to front and back plates as in the eggcrate structure. The additional joining obtained at the rib intersections contributes measurably to the overall stiffness of a blank of given size and weight. A typical blank is shown in Fig. 2c. Machining of ULE for lightweighting of core structures could in principle be done in the same manner as for fused silica, but the successful development of economical welding processes for ULE has rendered the pursuit of machining technology less pressing.

3. CER-VIT

CER-VIT, in the glass stage, is cast by the same techniques used for high temperature glasses. Care must be used in mold design and intricate shapes are difficult to achieve because of the shrinkage on solidification and cooling and the further shrinkage in the nucleation stages during heat treatment. Direct casting of lightweight mirror structures is thus difficult, and machining, using metal-bonded diamond tools, directly from solid disks is preferred for this material.

The final finishing of the internal surfaces of lightweight CER-VIT mirrors is accomplished by chemical milling with hydrofluoric acid,* removing any microcracks that may be left by the machining operations. The major structural difference of the machining approach is the necessity of perforating the back plate to provide access for removal of "core" material, as may be seen in Fig. 2d.

For CER-VIT, fusion bonding and heat treating are mutually exclusive operations. Thus self-bonded CER-VIT structures are not obtainable.

4. Beryllium

The fastening of beryllium remains a challenging area of engineering technology; in fact, many areas of use of beryllium as an engineering material are only beginning to mature and many challenges remain.

Welding is difficult because of grain growth, atmospheric reaction, and toxic vapor production. The most useful techniques at present include mechanical fasteners, adhesive bonding, and furnace brazing with aluminum or silver based filler materials. The fabrication of mirror substrates using these or other joining techniques, e.g., diffusion bonding, remains in the very

---

* Such etching is well recommended subsequent to machining any glassy material.

early stages of development. Weight reduction is thus now restricted, as in CER-VIT, to configurations that can be produced by machining and chemical milling operations. The removal of residual machining stresses by etching and heat treatment are extremely important in ensuring the dimensional stability of a beryllium mirror.

B. Mechanical Behavior

Some first order deflection analyses for ribbed lightweight structures have been published (Barnes, 1969, 1970). These analyses have been compared with holographic deflection measurements on four lightweight mirrors approximately 1 m in diameter. These mirrors were loaded centrally (by hydraulic piston) and supported at three points near the edge. A comparison of deflection is summarized in Table II. A typical holographic interferogram, from which the experimental deflections were measured, is shown in Fig. 3.

From Table II we see that the ULE structure most closely fits the mathematical model. The result that the blank is stiffer than theoretically predicted is unexpected. However, in consideration of the variability in the data input to the theory (even though dimensions were taken from several points on the blank, not from drawings) the result does not appear exceptional. For the CER-VIT blank the neglected effect of the holes in the rear of the structure is probably the principal reason that the measured deflection is 18% higher than calculated.

The values for the Heraeus-Schott drilled core silica blank and the Corning silica eggcrate illustrate their respective penalties of inefficient use of the weight in the core structure and reduced stiffness because of the open joints in the core.

The local deflection behavior of these lightweight structures is also of importance since the local response to polishing lap loads may determine whether a given mirror can be finished to the optical tolerances required in a reasonable length of time. One can, of course, reduce the violence of the polishing operation almost to the vanishing point, but only at the expense of increased polishing time.

TABLE II

Comparison of Lightweight Mirror Deflections
(Central Load, Three-Point Support)

Mirror blank	Measured deflection / Calculated deflection
Welded ULE (Corning)	0.91
Machined CER-VIT (Owens-Illinois)	1.18
Drilled core silica (Heraeus-Schott)	1.26
Silica eggcrate (Corning)	1.39

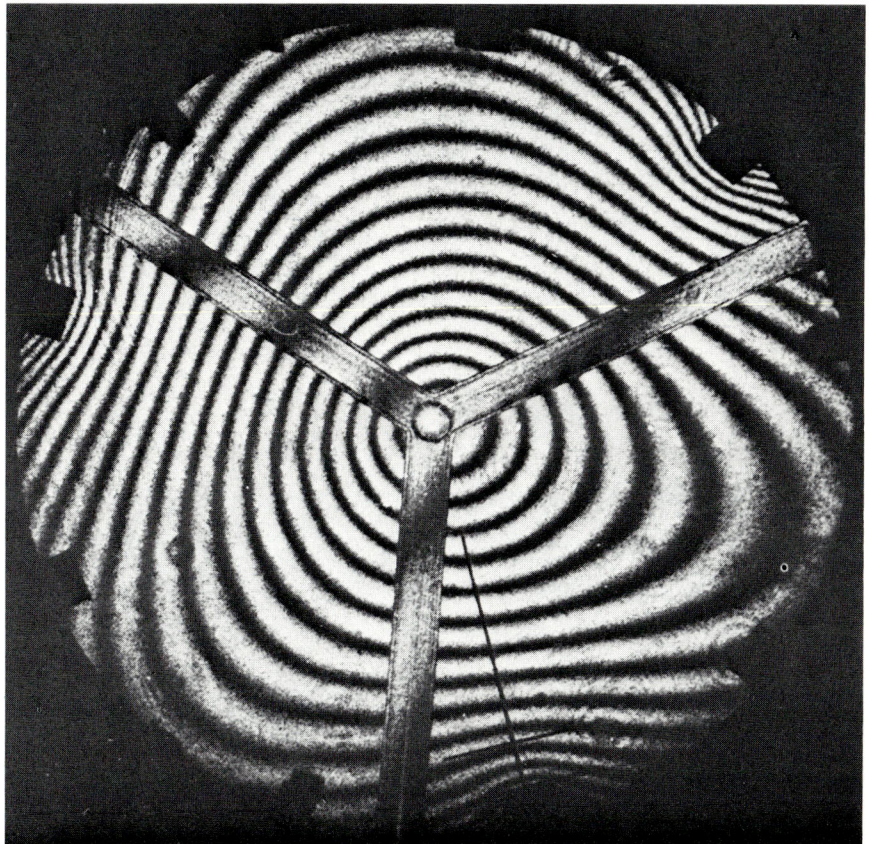

FIG. 3. Holographic interferogram of lightweight mirror deflection.

In order to examine this local deflection under uniform pressure load and to obtain an experimental comparison of some effects of rib configuration on both overall and local deflections, we constructed two 75-cm Plexiglas structural models, one with triangular cells and one with hexagonal cells (Barnes, 1972).

We found a significant difference in both overall and local deflection behavior of these two models. The overall central deflections for two loading cases are shown in Table III. (Both mirrors were tested with a 30° change of orientation with respect to the support points and no effect of this rotation on overall deflection was found.) We see that the deflection of the triangular cell model is significantly higher, particularly in the case of the three-point support, for which it is sometimes argued that the rib continuity of the triangular cell configuration provides a more advantageous "shear flow." Figure 4 shows the interferograms for the ring-supported case for both models. Note that the triangular core symmetry is apparent in the deflection

TABLE III
Comparison of Triangular and Hexagonal Cell Mirror Models

Cell	Loading	Central deflection under 20-lb load (micrometers)
Triangular cell, plastic	Central load, three-point edge support	8.1
Hexagonal cell, plastic	Central load, three-point edge support	6.4
Triangular cell, plastic	Central load, ring support at 83% radius	5.2
Hexagonal cell, plastic	Central load, ring support at 83% radius	4.7

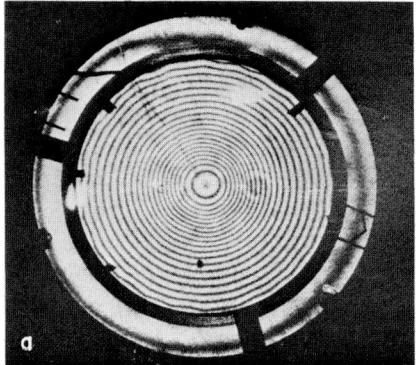

FIG. 4. Holographic interferograms, ring-supported models. (a) Hexagonal cells; (b) triangular cells.

of this model, while the circular symmetry of the load and support seems well preserved by the hexagonal core structure.

As a check on the local behavior under uniform pressure, we plugged the outside vent holes originally included in these models and pressurized them internally with air. Typical interferograms are shown in Fig. 5. Some nonuniform stretching of the ribs is apparent. This deviation from the usually postulated clamped edge behavior is more serious in the triangular cell case (three fringes, as compared to less than one fringe). As a result, the comparable local deflection irregularities under uniform pressure load are in the ratio of 7:4.

These comparative measurements of both the overall and local deflection behavior lead to a definite preference for the hexagonal core structure. From

4. OPTICAL MATERIALS—REFLECTIVE

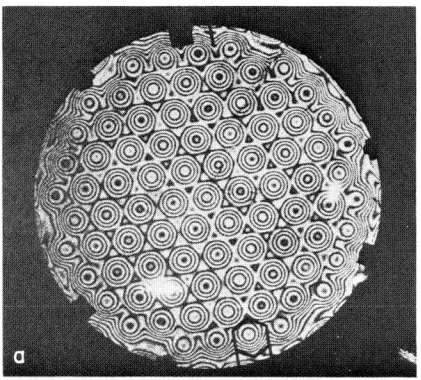 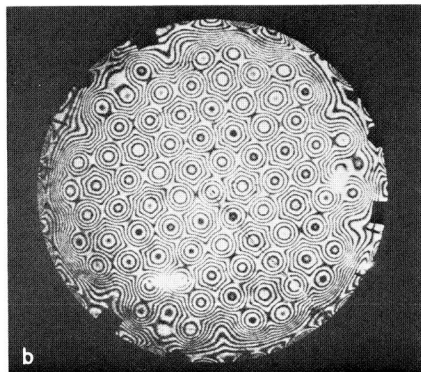

FIG. 5. Holographic interferograms, internally pressurized models. (a) Hexagonal cells; (b) triangular cells.

considerations of comparative symmetry, the square cell construction should lie in the middle—in personal judgment a bit closer to the hexagonal case. The hexagons remain preferred, but the penalties of square cell construction should be less severe than those for triangles.

The first-order analyses yield expressions for the maximum deflections for several cases summarized in Table IV. These expressions are useful for

TABLE IV
CIRCULAR MIRROR DEFLECTION PARAMETERS ($v = 0.2$)

	Axial acceleration = g		Central load = P	
	Full edge support	3 point support	Full edge support	3 point support
Maximum bending deflection	$6.77 \times 10^{-2} \dfrac{\rho g h a^4}{D_F}$ [a,b]	$14.33 \times 10^{-2} \dfrac{\rho g h a^4}{D_F}$	$5.30 \times 10^{-2} \dfrac{P a^2}{D_F}$	$7.71 \times 10^{-2} \dfrac{P a^2}{D_F}$
Maximum shear deflection	$0.60 \dfrac{\rho g h a^2}{D_S}$	$1.60 \dfrac{\rho g h a^2}{D_S}$	$1.15 \dfrac{P}{D_S}$	$1.46 \dfrac{P}{D_S}$

[a] *Nomenclature*: ρ is the effective mirror density (kg/m^3) and is equal to $\rho_{\text{solid}}(2\alpha + 2t)$ for rib plate construction; g is the acceleration imposed by earth's gravitation (9.8 m/sec^2); h is the disk overall thickness (m); a is the disk outer radius (m); D_F is the flexural rigidity $= Eh^2/11.52$ for solid disk and is equal to $(D_F)_{\text{solid}}[6t + \alpha]$, N-M for rib plate construction; E is Young's modulus of the mirror material (N/m^2); α is the ratio of rib thickness to rib spacing; t is the ratio of plate thickness to total thickness h; and D_S is the shear rigidity $= 2Eh/3$ for solid and equal to $Eh\alpha(1-t)$ for lightweight construction.

[b] Note: Because of approximations using the assumption $\alpha, t \ll 1$, the expressions for D_F and D_S do not converge to the values for a solid as $\alpha \to 1$ and $t \to \tfrac{1}{2}$.

preliminary design analysis. More intensive study of deflection details and of more complex structures may be undertaken with currently available finite element programs for structural analysis (Richard and Malvick, 1972). In using such programs, care should be taken to ensure that the element subdivision is sufficiently fine to faithfully display all the important behavioral details of the real structure.

C. Thermal Behavior

The low-expansion predominantly SiO_2 materials are nearly blackbodies at temperatures below 450 K. The transfer of heat by radiation within the cells of a lightweight mirror is comparable in magnitude to the transfer of heat by conduction through the ribs. To demonstrate this, let us calculate typical effective heat transfer coefficients for both radiation and conduction. For radiation, assuming a view factor of 1.0, we may write an average heat transfer coefficient h_r as

$$h_r = 4\sigma\varepsilon T^3 \qquad (4)$$

where σ is Boltzmann's constant $= 5.71 \times 10^{-12}$ W/cm² K⁴ and ε the normal total emissivity $= 0.93$. For $T = 300$ K, this yields

$$h_r = 0.573 \times 10^{-3} \quad \text{W/cm}^2 \text{ K} \qquad (5)$$

The analogous heat transfer coefficient for conduction h_c is simply the thermal conductivity divided by the total thickness of the material. For fused silica we have

$$h_c = (13.7 \times 10^{-3})/t \quad \text{W/cm}^2 \text{ K} \qquad (6)$$

and, for thickness t of about 24 cm ($\sim$10 in.), the radiation and conduction paths are equivalent.

Since the cell has no heat capacity, the overall thermal response time of a lightweight mirror will be substantially less than that of a solid mirror. For ground-based installations, one might consider reducing the response time further by forced ventilation of the core.

Because of the much higher thermal conductivity of beryllium, lightweight structures may be subject to more severe local effects since the only significant heat flux path will be through the structure itself. No other general conclusions can be drawn and the thermal–optical behavior of particular beryllium lightweight constructions should be examined in detail for each specific application.

For detailed analysis of the thermal response behavior, digital computation is again the most powerful tool available. In the thermal case, one may first determine the temperature distribution within the mirror and then convert the temperature distributions into surface and body force distributions using a mathematical analogy such as the one described by Timoshenko (1934). Some closed form solutions may, however, be derived from the formulations given in Boley and Weiner (1960).

One case of particular interest is that of a slowly heated disk, described in Boley and Weiner (1960, Article 9.9). For a heat flux q into one face, with the second surface insulated ($q = 0$), the curvature change is

$$\Delta(1/R) = -\alpha q/2k \qquad (7)$$

where R is the initial spherical radius and k is the thermal conductivity. For the steady state case of uniform heat flux to, through, and away from the disk, the curvature change is exactly twice that of Eq. (7), equal in magnitude to $\alpha q/k$. The formalism of Boley and Weiner is well recommended for the development of other particular solutions and as an aid in finite element analyses.

REFERENCES

Barnes, W. P., Jr. (1966). *Appl. Opt.* **5**(12), 1883–1886.
Barnes, W. P., Jr. (1969). *Appl. Opt.* **8**(6), 1191–1196.
Barnes, W. P., Jr. (1970). "Optical Instruments and Techniques. Proc. ICO-8, Reading, 1969" (J. Home Dickson, ed.). Oriel, Newcastle upon Tyne.
Barnes, W. P., Jr. (1972). *Appl. Opt.* **11**(12), 2748–2751.
Boley, A., and Weiner, J. H. (1960). "Theory of Thermal Stresses." Wiley, New York.
Gunter, R. G., Jr. (1969). *NASA Rep.* **N69-17992** (Jan.).
Hagy, H. E., and Smith, A. F. (1969). *J. Can. Ceram. Soc.* **38**(1), 63–68.
Jerke, J. M., and Platt, R. J., Jr. (1972). *NASA Tech. Note*, **NASA TN D-6626** (Jan.).
Richard, R. M., and Malvick, A. J. (1972). *J. Opt. Soc. Am.* **62**, 1339A.
Rossi, R. C. (1974). U.S. Patent 3,854,979 (Dec.). Aerospace Corp., El Segundo, California.
Simmons, G. A. (1967). *Opt. Spectra* **1**(2), 25–29.
Stewart, D. R., and Davis, W. L. (1970). *Appl. Opt.* **9**(4), 938–941.
Stonehouse, A. G., *et al.* (1973). Tech. Rep. No. 491 (Mar.) Brush Wellman, Cleveland, Ohio.
Timoshenko, S. (1934). "Theory of Elasticity," Ch. 7, Art. 65. McGraw-Hill, New York.

CHAPTER 5

Photographic Detectors

R. SHAW

Xerox Corporation, Webster, New York

I. Properties of Photographic Grains	121
A. Latent Image Formation	121
B. Quantum Efficiency	122
C. Size Distribution	124
II. Statistical Properties of Images	124
A. The Characteristic Curve	124
B. Image Noise	126
C. Detective Quantum Efficiency	128
D. Noise-Equivalent Input	129
E. Factors Determining DQE	130
III. Spatial Frequency Analysis of Images	132
A. The Modulation Transfer Function	132
B. The Wiener (Noise) Spectrum	136
C. Signal-to-Noise Ratio and Information	138
IV. Problems of Detection and Information Storage	141
A. Signal Detection Parameters	141
B. Principles of Information Storage	144
C. Signal Amplification	146
V. Photographic Detection for Unconventional Exposures	148
A. DQE for Electrons	148
B. DQE for X-Rays	150
References	152

I. PROPERTIES OF PHOTOGRAPHIC GRAINS

A. LATENT IMAGE FORMATION

The silver halide grains in a photographic emulsion layer interact with exposure quanta to form the latent image and subsequently act as the individual units of the developed image. Thus the grains act both as quantum receptors and recorders and a large amount of published literature is concerned with understanding these procedures and their mechanisms. James (1977) is a standard reference and Hamilton (1972) has given a concise review aimed at those working in applied optics, of these and other essential properties of photographic grains.

Briefly, the crystal lattice of silver and halide ions contains imperfections and impurities which act as trapping sites for the photoelectrons, which are raised from the valence to the conduction band by the absorption of exposure quanta of sufficient energy. The valence and conduction bands are separated by around 2.5 eV, which implies a quantum of energy associated with a wavelength of about 495 nm or less, although this may be extended to longer wavelengths by the presence of adsorbed spectral sensitizers, with a practical limit at around 1200 nm to prevent latent image formation due to naturally occurring thermal fluctuations. When a photoelectron is trapped at a site a silver ion is then attracted and they combine to form a silver atom. This process is repeated, with at least three or four silver atoms being required to achieve a stable and developable latent image. Thus following absorption of a sufficient number of quanta the grain has the property that it may be preferentially reduced to silver when brought into contact with the developer, which acts as a reducing agent in a very high gain process whereby about 10^9 silver atoms are produced for each stable silver atom in the latent image.

Further absorbed quanta are wasted once a latent image has been formed and hence the photographic grain operates in a binary state with the state being changed from no-image to image at a quantum threshold. Understanding the implications of this binary behavior in terms of both macro-and micro-image properties and signal-to-noise transfer during photographic recording is crucial to the proper utilization of photographic materials in scientific applications in general and detection problems specifically.

B. Quantum Efficiency

The quantum response of grains to exposure is expressed in various ways associated with latent image mechanisms, for example, in terms of the numbers of essential latent image events or the overall quantum yield (silver atoms in the developed grain to number of exposure quanta). However, to those interested in quantifying photographic response a "black-box" method is often most suitable. Since for light exposures the developed photographic grain is a many-input/one-output phenomenon, quantitative expression of the numbers associated with this required quantum input can completely specify quantum efficiency aspects of grains. In this way the quantum sensitivity distribution of the grains refers to the proportion (α_{Q_R}) which require Q_R quanta to become developable. If Q_R denotes the average value of this required number, then the responsive quantum efficiency (RQE) is defined by $100(\bar{Q}_R)^{-1}$. For example, $\bar{Q}_R = 20$ would correspond to RQE = 5%.

Deductive methods of measuring the quantum sensitivity distribution have been reported by several authors (Webb, 1948; Haase, 1960; Frieser and Klein, 1960; Farnell and Chanter, 1961; Marriage, 1961; Spencer, 1971; Shaw, 1973a), and a recent summary has been given by Spencer (1976).

From the black-box viewpoint a simple model of quantum sensitivity can be used (Shaw, 1973a). This model ignores the complicated mechanisms of latent image formation, but establishes the important macro- and micro-image dependencies on quantum sensitivity, and it is these that are relevant to the photographic engineer.

The model assume that latent image formation can be completely described by only two essential grain properties. First, it is assumed that all grains require the same fixed number T of exposure quanta for latent image formation (where T might be anticipated to be a small number in the region of 2, 3, or 4). Second, it is assumed that all absorbed quanta have a random probability ($p = 1/m$) of making a contribution to the number T. Thus from a statistical viewpoint all grains are assumed to have identical properties. Following these assumptions, elementary statistical theory leads to a negative binomial distribution of the form

$$\alpha_{Q_R} = \frac{(Q_R - 1)!}{(Q_R - T)!(T - 1)!} \left\{\frac{1}{m}\right\}^T \left\{1 - \frac{1}{m}\right\}^{Q_R - T} \quad (1)$$

This distribution has the property that $\bar{Q}_R = Tm$ and hence the responsive quantum efficiency is defined by $RQE = 100(Tm)^{-1}\%$. Figure 1 shows a sensitivity distribution generated by assuming $T = 3$, $m = 4$, implying that each grain requires three essential events for latent image formation, and that one in four of all absorbed quanta contribute at random to this number. The implications of the parameters T and m on the characteristic curve and signal-to-noise transfer during image recording will be discussed later.

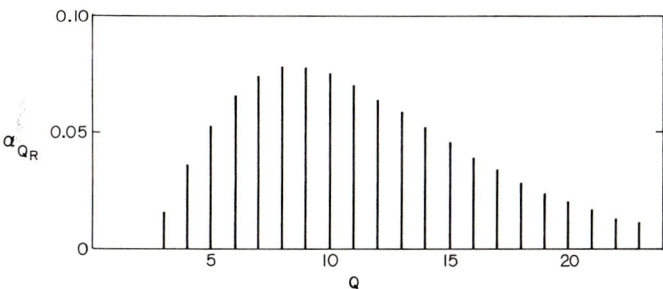

FIG. 1. A quantum sensitivity distribution generated (Shaw, 1973a) for a population of grains assumed to require three quanta for latent image formation, assuming that on average only one in four absorbed quanta contribute to this number.

C. Size Distribution

Conventional silver halide emulsions usually contain grains with a wide range of sizes of shapes. The projection grain area is an important parameter both during latent image formation where it influences the capture cross section for exposure quanta and in the output sense where it determines the grain contribution to image density. The projection areas of a grain before and after development are not necessarily identical. Figure 2 shows typical size distributions, which often approximate to log–normal distributions. Mean areas vary from 0.002 μm^2 for the finest-grained high resolution films to in excess of 2 μm^2 for x-ray emulsions.

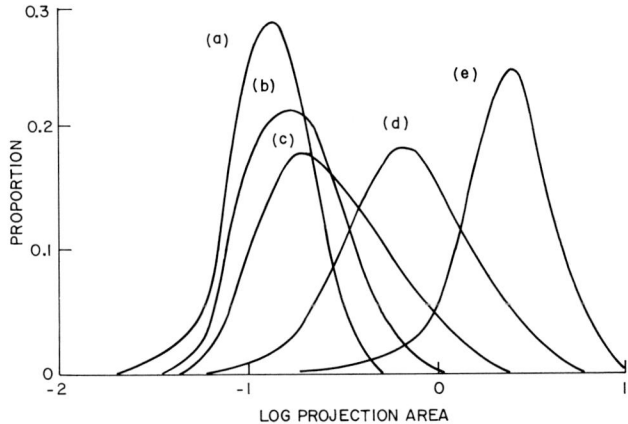

FIG. 2. Grain size distributions for typical silver halide emulsions: (a) positive-type film; (b) fine-grain roll film; (c) portrait film; (d) medium-speed orthochromatic film; (e) x-ray film. The area is in terms of μm^2.

II. STATISTICAL PROPERTIES OF IMAGES

A. The Characteristic Curve

1. Exposure Statistics

The relationship between grain developability and exposure can be quantified assuming Poisson statistics for the exposure quanta. The probability p of an image contribution from grains requiring Q_R quanta and having access to Q on average follows as

$$p = 1 - e^{-Q} \sum_{r=0}^{Q_R - 1} \frac{Q^r}{r!} \tag{2}$$

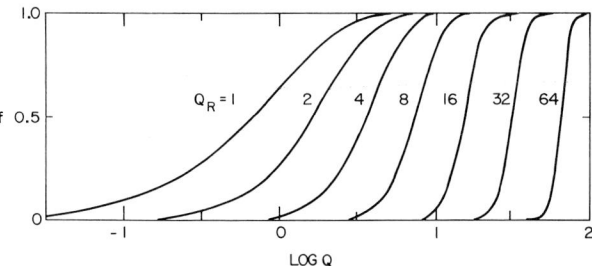

FIG. 3. The fraction f of grains made developable as a function of the logarithm of the average number of quantum exposure Q. The constant quantum requirements Q_R are shown on the curves.

For monosized grains the fraction of grains f made developable will thus be defined by

$$f = \sum_{Q_R} \alpha_{Q_R} p = \sum_{Q_R} \alpha_{Q_R} \left\{ 1 - e^{-Q} \sum_{r=0}^{Q_R - 1} \frac{Q^r}{r!} \right\} \quad (3)$$

Figure 3 shows this fraction computed as a function of the logarithm of the quantum exposure for constant values of Q_R. The curves become steeper for higher quantum requirements, approaching a steplike function in the limit.

For a distribution of quantum requirements the curves must be suitably averaged in the sense of Eq. (3) to obtain the overall curve. When using the model sensitivity distribution described above this averaging is replaced by a very simple transformation of the exposure scale. The shape of the curve is determined entirely by the $T = Q_R$ monosensitivity curve, but the entire curve is moved bodily by $\log_{10} m$ units towards higher log exposure levels compared to the $m = 1$ case. In this way it is possible to compare grain-fraction curves measured on practical emulsions with model curves and with the limits imposed by grain sensitivity. Such a comparison is shown in Fig. 4. An unsensitized emulsion is shown along with the same emulsion

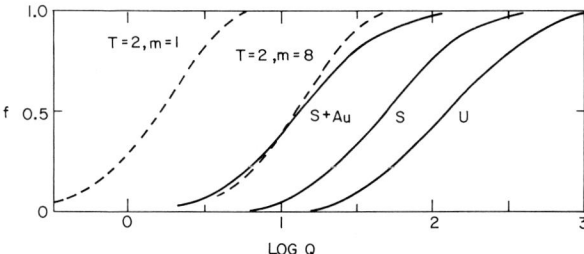

FIG. 4. The fraction f of grains made developable as a function of the logarithm of the average number of exposure quanta Q, for an unsensitized emulsion, a sulfur-sensitized emulsion, and a sulfur-plus-gold sensitized emulsion (Spencer, 1976). Also shown are the model curves for $T = 2$, with $m = 1$ and 8, respectively (Shaw, 1973a).

sulfur-sensitized and sulfur-plus-gold sensitized, the latter representing the state of the art in emulsion sensitivity. These three curves are nearly identical in shape, suggesting that in model terms sensitization has the effect of a variation only in m and not in T. The log exposure shift from U to S is 0.4 units, while that from S to S + Au is 0.6 units, i.e., an overall shift of one unit, corresponding to a ten fold improvement in the utilization of absorbed quanta. The nearest match to these experimental curves by a model curve with a single T value is provided by the $T = 2$ curve, and the $m = 8$ curves then gives a good fit with the practical S + Au curve. The $m = 1$ curve thus represents the ultimate sensitivity in this case.

2. Image Density

The image density can be calculated in terms of the sensitivity and sizes of the grains by an extension of the well-known Nutting (1913) formula. Based on a simple geometrical representation of the developed emulsion layer and its subsequent stopping power for light during densitometry, this states that the density D is related to the average grain area a and the average number (n) of grains per image area A by:

$$D = \log e(na/A) \tag{4}$$

The ratio of the number of image grains to the total number N available for image formation will also be equal to the fraction f of Eq. (3). Hence the density can also be related to the quantum exposure level. When there is a distribution of grain sizes the summations of Eq. (3) have to be carried out for each size class and appropriately averaged. The grain fraction curves of Fig. 3 and Fig. 4 can readily be converted into densities, since if the saturation density is denoted by

$$D = \log e(Na/A) \tag{5}$$

it follows that

$$D/D_s = n/N = f \tag{6}$$

Reviews and examples of the influence of both grain size and sensitivity distributions on the characteristic curve are given in Dainty and Shaw (1974).

B. Image Noise

Selwyn (1935) and Siedentopf (1937) found interesting noise relationships when measuring the mean-square density fluctuation $\overline{\Delta D_A^2}$ with relatively large circular apertures (area A). Selwyn found that for a wide range of aperture sizes the product

$$S = 2\{A\,\overline{\Delta D_A^2}\}^{1/2} \tag{7}$$

was approximately independent of A, and S has subsequently become known as the Selwyn granularity. Siedentopf related the mean-square fluctuation in density to the mean density level by the equation:

$$\overline{\Delta D_A^2} = \log e \, \frac{\bar{a}D}{A} \left\{ 1 + \frac{\overline{\Delta a^2}}{(\bar{a})^2} \right\} \tag{8}$$

Where $\bar{a}$ now denotes the mean grain projection area and $\overline{\Delta a^2}$ the mean-square distribution of areas. The Siedentopf equation explains the constancy of Selwyn granularity, since at a given density level it is seen to depend only on the (fixed) statistics of the grain sizes.

The Siedentopf equation also predicts that increase in the spread of grain sizes will be accompanied by increase in the noise. In the special case when all the grains have identical size, it reduces to

$$A \, \overline{\Delta D_A^2} = \log e \, (aD) \tag{9}$$

Verification of the predications of this equation with practical measurements made on emulsions with special monosized grains has been carried out by Ericson and Marchant (1972). The product $G = A \, \overline{\Delta D_A^2}$ will be seen to have special significance when the full Fourier spectrum of the noise is introduced. Assuming 1 μm² grains, a value for G of 0.22 μm² would be predicted at an image density level of 0.5. Figure 5 shows the product as a function of density for a range of six types of aerial films (Shaw, 1965), with values ranging from 0.4 to 2 μm² at this density level.

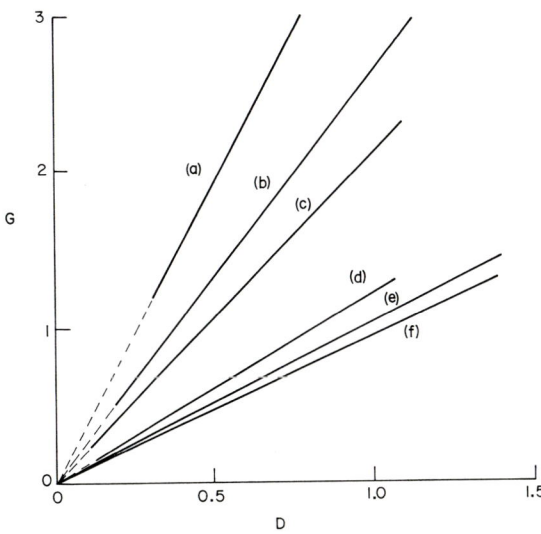

FIG. 5. The product $G = A \, \overline{\Delta D_A^2}$ as a function of average density level for a range of six types of aerial films (Shaw, 1965), from (a) a high-speed film to (f) a high-resolution film.

C. Detective Quantum Efficiency

The responsive quantum efficiency (RQE) is inadequate to determine the signal-to-noise ratio (SNR) which is associated with a quantum conversion of signal from one stage to another. In particular RQE has no theoretical upper limits and cannot provide satisfactory answers to questions sent as those of ultimate detector performance. Following Rose (1946), investigations of the performance of diverse detector types such as TV camera tubes, photoemissive and photoconductive devices, infrared detectors, and the photographic process, led to the concept of detective quantum efficiency (DQE) as a common and universal method of comparing radiation detectors in general. DQE defines the relationship between detector input and output *fluctuations*, rather than between *mean levels*.

The concept of DQE can be understood as follows. Suppose that a detector produces a mean image density level D when exposed to Q quanta per unit area. The gain g associated with the conversion of exposure quanta to image density can be defined as dD/dQ. Again it is assumed that the exposure quanta have associated Poisson statistics and hence the input fluctuation will be $\overline{\Delta Q^2} = Q$. The fluctuation in the output associated with unit image area will be $\overline{\Delta D^2}$ and for an ideal detector which introduces no spurious noise it follows that

$$g^2 = \overline{\Delta D^2}/\overline{\Delta Q^2} \tag{10}$$

In general, for practical detectors the output noise will be greater than that defined by Eq. (10) and the ratio

$$\varepsilon = \frac{g^2}{\overline{\Delta D^2}/\overline{\Delta Q^2}} = \frac{g^2 Q}{\overline{\Delta D^2}} \tag{11}$$

provides the basic definition of DQE. In terms of the more general image density fluctuation associated with area A, assuming that $G = A \overline{\Delta D_A^2}$ is a constant independent of A, then since $\overline{A \Delta D_A^2} = \overline{\Delta D^2}$, where the latter denotes the product for unit area,

$$\varepsilon = g^2 Q/G \tag{12}$$

When the gain is expressed in terms of the slope (γ) of the characteristic curve (D–$\log Q$ or D–$\log E$) it follows that

$$\varepsilon = [(\log e)^2 \gamma^2]/QG \tag{13}$$

Rose (1974) has summarized much of his earlier work on detector types and mechanisms in SNR and DQE terms in a recent book. Jones (1959) has given one of the first extensive reviews of DQEs, with example for various detector types. Other reviews include those due to Zweig (1964) and Dainty

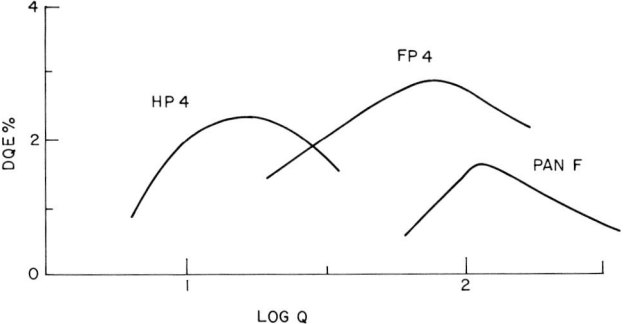

FIG. 6. Measured DQE values for three conventional negative processes (Shaw and Shipman, 1969).

and Shaw (1974). Jones (1958a) and Fellgett (1958) made the first DQE measurements for the photographic process, and subsequently DQE values up to a maximum of around 3% have been reported (Shaw and Shipman, 1969), and even higher values are claimed to be possible using special hypersensitization techniques in astronomical detection applications (Scott and Smith, 1974; Latham, 1977). DQE values for three conventional negative processes (Shaw and Shipman, 1969) are shown in Fig. 6. Other DQE measurements, including the influence of a range of photographic parameters, have been reported in the literature (Bird et al., 1969; Vendrovski et al., 1972; Burton et al., 1973; Sayhun, 1975).

D. NOISE-EQUIVALENT INPUT

The concept of DQE leads to the possibility of expressing the output SNR of a detector as the noise-equivalent input, and specifically in terms of the noise-equivalent number of exposure quanta (NEQ). For radiation-limited signals the SNR associated with Q input quanta may be defined as

$$\text{SNR}_{in}^2 = (Q/\sqrt{Q})^2 = Q \tag{14}$$

For a practical detector the output SNR will be equivalent to the lesser number Q' of quanta, which thus defines the number an ideal detector would have needed to yield the same SNR. Q' is termed the NEQ, and the ratio of output to input SNRs provides another way of expressing DQE:

$$\varepsilon = \text{SNR}_{out}^2/\text{SNR}_{in}^2 = Q'/Q \tag{15}$$

where Q' is in turn defined [from Eq. (13)] by

$$Q' = (\log e)^2 \gamma^2 / G \tag{16}$$

Since the product QG does not vary widely from one photographic process to another (bigger grains imply more noise but need less exposure quanta) DQE tends not to vary strongly with the grain size of the emulsion. However the variation of the NEQ per unit image area from emulsion to emulsion is dominated by the term G in the denominator of Eq. (16). Thus NEQ values range from less than 0.01 μm^{-2} for coarse-grained emulsions to in excess of 1 μm^{-2} for high resolution materials.

E. Factors Determining DQE

Zweig (1961a) made the first attempt to model the photographic parameters which reduce DQE from 100% to the much lower values measured in practice. He restricted his analysis to a checkerboard array of grains each requiring exactly Q_R exposure quanta to be rendered developable. In this case he found that DQE could be expressed as

$$\varepsilon = Qt^2/s(1-s) \tag{17}$$

with the function s and t defined by

$$s = e^{-Q}\sum_{r=0}^{Q_R-1}\frac{Q^r}{r!}; \quad t = e^{-Q}\frac{Q^{Q_R-1}}{(Q_R-1)!} \tag{18}$$

Figure 7 shows the DQE curves for a series of constant quantum requirements. As Q_R becomes large, maximum DQE becomes $2/\pi$ or 64%, independent of Q_R, at an exposure level of $Q = Q_R - \frac{1}{2}$ quanta per grain.

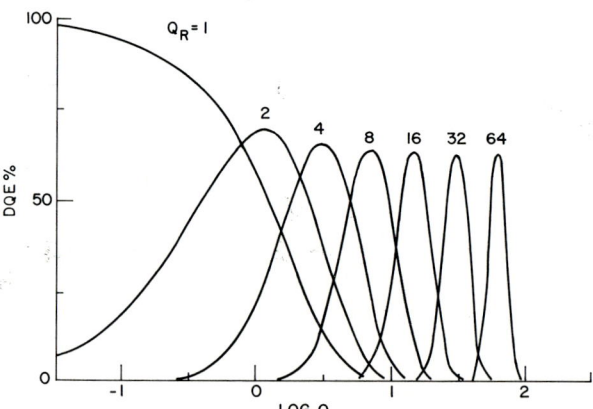

FIG. 7. DQE curves for grains with constant quantum requirements as shown, according to the Zweig checkerboard model (Zweig, 1961).

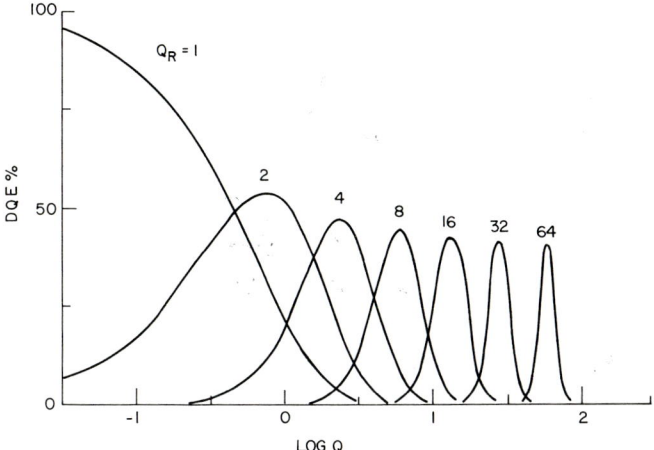

FIG. 8. DQE curves as for Fig. 7, but assuming a random grain array (Shaw, 1968).

Extensions to the Zweig model have been made to include the influence of sensitivity and size distributions, random grain arrays, and other more realistic photographic parameters (Shaw, 1967, 1968, 1969, 1972a,b, 1973b). For a random array of monosized and monosensitive grains the DQE expression becomes

$$\varepsilon = Qt^2/(1 - s) \qquad (19)$$

Figure 8 shows the DQE curves for the random array which correspond to those of Fig. 7 for the checkerboard array. The maximum value is now approximately 40% for higher values of Q_R.

The influence of the spread of quantum sensitivities on DQE may be investigated (Dainty and Shaw, 1974; Shaw, 1967, 1968) by extending the definitions of the functions s and t to:

$$s = \sum_{Q_R} \alpha_{Q_R} \left\{ e^{-Q} \sum_0^{Q_R - 1} \frac{Q^r}{r!} \right\}, \qquad t = \sum_{Q_R} \alpha_{Q_R} \left\{ e^{-Q} \frac{Q^Q R^{-1}}{(Q_R - 1)!} \right\} \qquad (20)$$

It is found that the spread in quantum sensitivity makes the most severe contribution to loss of DQE for practical photographic emulsions. This can be confined by use of the two-parameter model for α_{Q_R} as discussed previously. As with the characteristic curve the model is equivalent to a uniform filter of transmission $1/m$ placed over monosensitive ($Q_R = T$) grains. Thus the filter acts as merely a change of exposure scale and reasoning along these lines leads to

$$\varepsilon_{T,m}(mQ) = (1/m)\varepsilon_T(Q) \qquad (21)$$

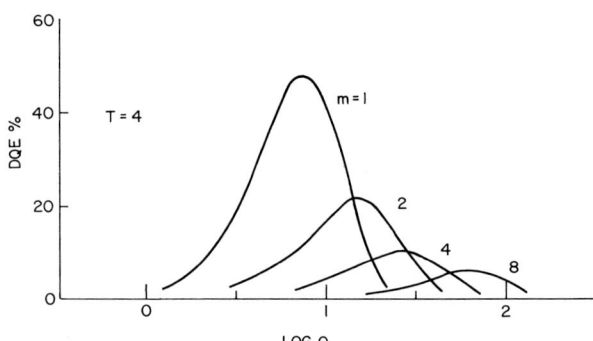

FIG. 9. Model DQE curves for $T = 4$, with $m = 1, 2, 4,$ and 8.

Thus the DQE can be deduced from the monosensitivity curves of Fig. 8, and likewise the maximum DQE levels will follow from these curves. For example, when $T = 4$, $\varepsilon_{\max} = 48\%$, and hence for $m = 2, 4,$ and 8 this will be reduced to 24%, 12%, and 6%, respectively. Figure 9 shows the full DQE curves for these values of T and m. Each curve is scaled according to $1/m$ and shifted log 2 units on the log exposure scale as m changes by a factor of 2.

III. SPATIAL FREQUENCY ANALYSIS OF IMAGES

A. THE MODULATION TRANSFER FUNCTION

The Fourier approach to optical images pioneered by Duffieux (1946) was applied to photographic images by Schade (1955), Ingelstam et al. (1956), and others, allowing signal transfer characteristics to be expressed on the microscale within a rational analytical framework. This has led to the widespread use of the modulation transfer function (MTF) as a spatial frequency descriptor of photographic reproduction, with the conveniences associated with cascading transfer functions for linear systems.

If the point spread function (SF) of an imaging process is denoted by $f(x, y)$, where x and y are space coordinates, then the MTF in spatial frequency (u,v) coordinates is defined by

$$\text{MTF}(u,v) = \iint f(x,y) e^{-2\pi i(ux+vy)} \, dx \, dy \bigg/ \iint f(x,y) \, dx \, dy \qquad (22)$$

and by definition $\text{MTF}(0,0) = 1$. The SF and MTF are thus Fourier transform pairs with suitable scale adjustments. Imaging processes are often isotropic in the statistical sense and the use of single space (r) and spatial frequency (w) coordinates is then adequate. Strictly speaking Eq. (22) defines the so-called optical transfer function (OTF), however for rotationally symmetric spread functions the OTF and MTF are identical since there is no imaginary term. In general

$$\text{OTF}(w) = R(w) + iI(w); \quad \text{MTF}(w) = [R^2(w) + I^2(w)]^{1/2}$$

A sine-wave exposure is imaged with amplitude reduced according to the value of the MTF at that spatial frequency.

Photographic nonlinearities pose problems when attempting to apply linear systems analysis. Kelly (1960) defined these problems and ways of circumventing them by dividing the photographic process into a series of separate stages. He described the photographic process as:

(a) an optical diffusion stage within the emulsion layer which operates on the exposure;

(b) a large-scale exposure-density conversion as the diffused exposure interacts with the silver halide grains; and

(c) a chemical diffusion stage associated with the development process.

It is the latter, which operates as a local space processer especially at sharp boundaries, which can introduce the most serious nonlinearity. Following Kelly, Lorber (1970) extended the model to deal with coherent light exposures as in holographic information recording. For this he modified the optical diffusion stage for coherent light and added a fourth stage to account for granularity and bleaching of the emulsion as in phase holography.

Kelly (1971) published some of the first measurements of chemical transfer functions showing that these could enhance spatial frequencies as high as 100 cycles/mm. In an engineering approach, Schade (1964) treated these local space-processing effects as a negative feedback loop during image formation, with magnitude depending on both the average exposure level and difference in adjacent exposure levels, as well as depending on the spatial frequency. Subsequent authors have since developed elegant models to describe the non linearities of chemical diffusion in transfer function terms (e.g., Kriss *et al.*, 1974; Silevitch *et al.*, 1977).

For scientific photography it can be useful to combine the macro- and micro-signal transfer characteristics (Dainty and Shaw, (1974) and express them in terms of the contrast transfer function (CTF), where $\text{CTF}(w) = \gamma \text{MTF}(w)$. If as the mean exposure level varies a series of low-contrast MTFs are measured about each exposure level, then a response solid can be built up as illustrated in Fig. 10, thus establishing the practical working region of the process.

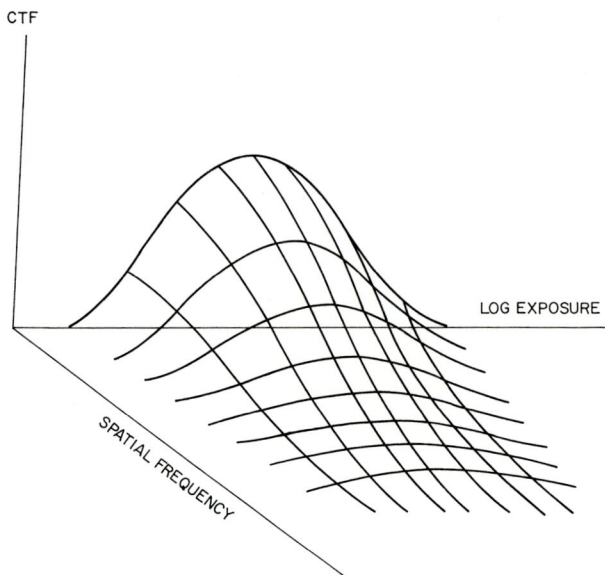

Fig. 10. Illustration of three-dimensional photographic contrast-transfer characteristics for low-contrast input exposures.

In general, the transfer function approach has been found to work well when applied to photographic processes, in spite of the various difficulties due to nonlinearities. Figure 11 shows an example where the theoretical cascading of transfer functions representing lens and negative and positive processes is seen to be in good agreement with practical measurement

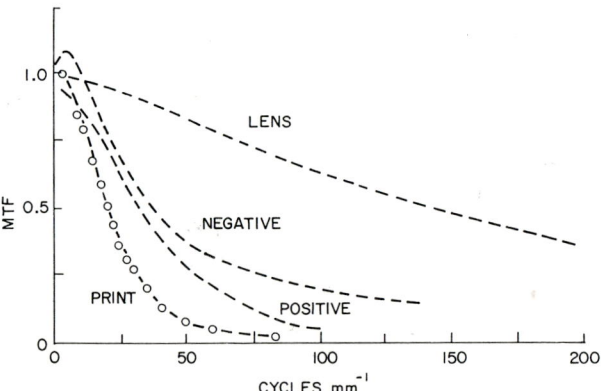

Fig. 11. The cascading of the modulation transfer functions of the components of a photographic system (Higgins, 1971): ○, values measured in the print; ———, product of the component MTFs.

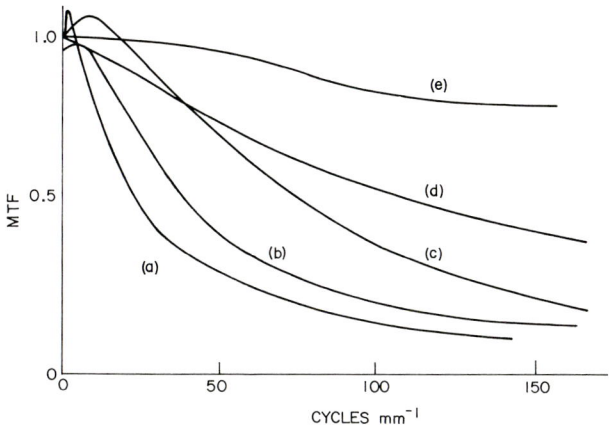

FIG. 12. Some representative MTFs for a range of film types, from (a) fast panchromatic to (e) a very fine-grain emulsion.

(Higgins, 1971). Representative MTFs of various photographic processes are shown in Fig. 12.

Two important types of MTF modelling have been carried out. The first of these is essentially a black-box approach, of interest to the systems engineer, which consists of fitting the experimental MTF (or more usually the spread function) with two or three characteristic parameters (Frieser, 1960; Gilmore, 1967). In this context Johnson (1970) had success in fitting a wide variety of electro-optical imaging devices with a model MTF according to

$$\text{MTF}(w) = \exp(-\{w/w_c\}^n) \tag{23}$$

Johnson termed w_c and n the frequency constant and the device index, respectively. A log–log plot of $-\ln \text{MTF}(w)$ against w results in a straight line, with the index n being the slope of this line. The frequency constant is the spatial frequency at which the MTF has fallen to e^{-1}. In this way he arrived at ways of determining the parameters for two elements in combination and for predicting the limiting resolution.

The second important type of model concerns the mechanistic approach, leading to an understanding of the properties of the emulsion layer which influence the MTF (De Belder et al., 1965; Wolfe et al., 1968; DePalma and Gasper, 1972). DePalma and Gasper (1972) found that realistic simulations are possible in terms of six parameters: the thickness of the emulsion layer, the refractive indices of the gelatin and base, the probability that a photon which collides with a grain will be absorbed, the probability distribution for the angular scattering of a photon which collides with a grain but is not absorbed, and the distribution function for the distance travelled by a

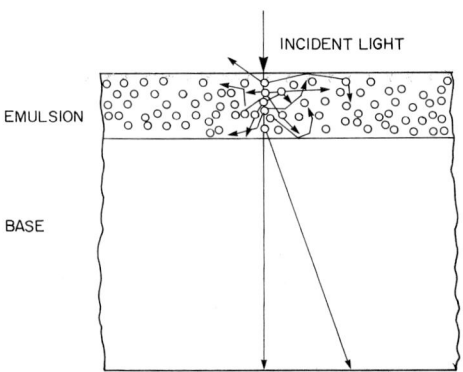

Fig. 13. Illustration of the diffusion of light in a typical photographic layer (DePalma and Gasper, 1972).

photon between collisions with grains. Due to the complexity of these mechanisms a Monte Carlo computer simulation method was used, with many incident rays of light being traced through the emulsion layer in the probabilistic sense (see Fig. 13). In the limit this determines the spread function and hence the MTF. Good agreement was found between MTFs determined by this simulation technique and those measured experimentally.

B. The Wiener (Noise) Spectrum

The noise, as discussed previously, based on the product $G = A \overline{\Delta D_A^2}$ corresponding to large-scale scanning apertures of area A, represents in Fourier terms the very low spatial frequency components of the noise. Following Jones (1955), Zweig (1956), and others, a full spatial frequency analysis of the noise is possible. The mathematical framework for this had been established by Wiener (1930), who had shown that the autocorrelation function (ACF) and Wiener spectrum (WS) of the noise are Fourier transform pairs (just as the SF and MTF are the Fourier transform pairs in signal transfer analysis). For a two-dimensional fluctuation, say of image density in the space domain $\Delta D(x, y)$, the autocorrelation function is defined as

$$C(s,t) = \lim_{X,Y \to \infty} \frac{1}{2X} \frac{1}{2Y} \int_{-X}^{+X} \int_{-Y}^{+Y} \Delta D(x, y) \Delta D(x + s, y + t) \, dx \, dy \quad (24)$$

where s,t denote the respective correlation intervals and the fluctuation is defined over some image area $(\pm X, \pm Y)$. It follows that $C(0,0)$ is simply the mean-square density fluctuation. The corresponding definition of WS is:

$$WS(u,v) = \lim_{X,Y \to \infty} \left\langle \frac{1}{2X} \frac{1}{2Y} \left| \int_{-X}^{+X} \int_{-Y}^{+Y} \Delta D(x, y) e^{-2\pi i(ux + vy)} \, dx \, dy \right|^2 \right\rangle \quad (25)$$

Since $C(s,t)$ and $WS(u,v)$ are Fourier transform pairs, it follows that the total "power" in the Wiener spectrum is, again, equal to the mean-square density fluctuation. Thus the Wiener spectrum can be interpreted simply as the Fourier components of the density fluctuations.

The influence of a microdensitometer scanning aperture on the measured WS, denoted by $WS'(u,v)$, can be stated as

$$WS'(u,v) = WS(u,v)\,MTF_A^2(u,v) \tag{26}$$

Where MTF_A denotes the MTF of the aperture. This and other practical WS relationships have been reviewed by Dainty and Shaw (1974). From Eq. (28) it follows that

$$A\,\overline{\Delta D_A^2} = WS(0,0) \tag{27}$$

thus providing an interpretation of the Selwyn granularity coefficient, i.e.,

$$S^2 = 2\,WS(0,0) \tag{28}$$

In general, both the ACF and WS for photographic detectors exhibit rotational symmetry due to the isotropic nature of the grain statistics. Hence single space and spatial frequency variables are adequate to specify these functions and the mean-square density fluctuation can be expressed as

$$\overline{\Delta D_A^2} = \int_{w=0}^{\infty} WS(w)\,MTF_A^2(w)\,2\pi w\,dw \tag{29}$$

The transfer of noise during imaging, as, for example, when printing a negative process, is defined by the relationship (Doerner, 1962)

$$WS_{tot}(w) = WS_{neg}(w)\,CTF_{pos}^2(w) + WS_{pos}(w) \tag{30}$$

Any enlargement during printing must be taken into account. For example, with linear magnification x of the negative

$$WS_{tot}(w) = x^2\,WS_{neg}\{w/x\}\,CTF_{pos}^2(w) + WS_{pos}(w) \tag{31}$$

For conventional photographic processes the Wiener spectrum is generally found to be fairly flat, at least in the spatial frequency range up to 100 cycles/mm. Thus the scale value of $WS(0,0)$ tends to be the important practical variable in establishing the magnitude of the noise in any practical imaging problem. This magnitude of the WS can be determined from models involving the particle statistics, as, for example, by the Siedentopf relationship. Combining Eq. (8) and (27) leads to

$$WS(0) = \log e\,\bar{a}\,D\{1 + \overline{\Delta a^2}/(\bar{a})^2\} \tag{32}$$

Substitution of typical grain size distributions in Eq. (8) leads to predictions for $WS(0,0)$ typically in the range of 0.1 μm^2 for emulsions with small grains at low image density levels, to 10 μm^2 for emulsions with largest practical grain sizes at maximum density. These predictions are generally in good agreement with measured values, as demonstrated by the values shown in Fig. 5.

Jones (1955) showed the manner in which the full shape of the WS might be related to the statistics of the image particles. For example, for grains of constant size (circular, area $a = \pi r^2$) the autocorrelation function will be that of a circular disk whose Fourier transform leads to the well-known Bessel function, and hence

$$\text{WS}(w) = \log e\, (aD)\{2J_1(y)/y\}^2 \tag{33}$$

where $y = \pi rw$. Frieser et al. (1959) and Frieser (1959) considered the case where, as in electron exposures, a clump of several grains is developed for each exposure event. If this number of grains is Poisson-distributed with mean value m, then

$$\text{WS}(w) = \log e\, maD\left[\left\{\frac{2J_1(Y)}{Y}\right\}^2 + \frac{1}{m}\left\{\frac{2J_1(y)}{y}\right\}^2\right] \tag{34}$$

where $Y = \pi RW$, and R is now the radius associated with an average clump of grains. Whereas no general expressions have been derived for the Wiener spectrum where there is both grain clumping and the grain have a distribution of sizes, the scale value follows by simple fluctuation analysis (Shaw, 1977) as

$$\text{WS}(0) = \log e\, maD\left\{1 + \frac{1}{m} + \frac{\overline{\Delta a^2}}{m(\overline{a})^2}\right\} \tag{35}$$

Berwart (1969) has carried out detailed comparisons of models and measurements for both experimental and commercial emulsion types, and noise analysis and modelling relating to a range of practical emulsion parameters has been described in the literature (Benton and Kronauer, 1971; Trabka, 1971; Lawton et al., 1972; Castro et al., 1973; Trabka and Doerner, 1976; Benton, 1977), and includes studies of noise associated with color layers (Trabka and Lawton, 1974) and dye clouds (Trabka, 1977).

C. Signal-to-Noise Ratio and Information

By use of a full spatial frequency analysis of both signal and noise it is possible to extend the expressions for SNR and DQE to include the detector MTF and Wiener spectrum (Shaw, 1963a). Specifically, in spatial frequency terms Eqs. (13) and (16) become

$$Q'(w) = (\log e)^2[\gamma^2 \text{MTF}^2(w)/\text{WS}(w)] \tag{36}$$

$$\varepsilon(w) = (\log e)^2[\gamma^2 \text{MTF}^2(w)/Q\text{WS}(w)] \tag{37}$$

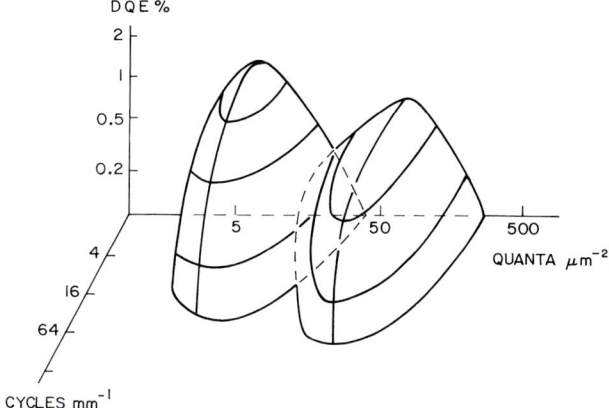

FIG. 14. DQE curves as a function of both spatial frequency and exposure level for two negative processes (Vendrovski et al., 1972).

Thus, the noise-equivalent input and the detective quantum efficiency of a photographic process may both be expressed as three-dimensional solids by plotting them as functions of exposure level and spatial frequency. Vendrovski et al. (1972), have expressed the DQE of conventional emulsions in this way, as shown in Fig. 14. Johnson (1977) has reviewed this three-dimensional solid as the information-throughput-manifold (ITM) and given examples for various photoelectronic imaging devices. This notation is due to the close relationship between DQE and information which becomes apparent when both are expressed in terms of their spatial frequency parameters.

The classical Shannon (1948) formula for the information capacity C of a one-dimensional power-limited channel may be expressed as

$$C = \log_2\{1 + WS_s(f)/WS_n(f)\}\,\Delta f \tag{38}$$

where f denotes temporal frequency, and WS_s and WS_n are, respectively, the Wiener spectra of the signal and noise. Fellgett and Linfoot (1955) applied this to two-dimensional spatial images, rather than one-dimensional time series, and arrived at the equivalent expression.

$$C = \frac{1}{2}\int_{-\infty}^{+\infty}\int_{-\infty}^{+\infty} \log_2\left\{1 + \frac{WS_s(u,v)}{WS_n(u,v)}\right\}\,du\,dv \tag{39}$$

Assuming that both the signal and noise have statistical properties which are isotropic and can be expressed by a single spatial frequency variable, Eq. (39) reduces to

$$C = \pi \int_0^\infty \log_2\left\{1 + \frac{WS_s(w)}{WS_n(w)}\right\} w\,dw \tag{40}$$

In these equations both signal and noise must be expressed in equivalent terms, for example, in image density units. Thus the signal spectrum is that after transfer through the imaging characteristics. If the power spectrum of the signal in terms of log exposure fluctuations is denoted by $P(w)$, then

$$WS_s(w) = (\log e)^2 \gamma^2 \mathrm{MTF}^2(w) P(w) \tag{41}$$

Equation (37) allows this to be re-expressed in DQE terms as

$$WS_s(w) = Q\, WS_n(w) \varepsilon(w) P(w), \tag{42}$$

and hence

$$C = \pi \int_0^\infty \log_2\{1 + Q\varepsilon(w)P(w)\}\, w\, dw \tag{43}$$

We may deduce from Eq. (43) that since the only variable of the imaging process is DQE, the information properties and DQE of a process are equivalent properties. Variations and approximations based on Eqs. (39)–(43) have been discussed in the literature (Linfoot, 1961; Jones, 1961; Shaw, 1962, 1963b).

Since the general DQE expression contains all the relevant imaging parameters that define information, comparisons between imaging process may be made in these simpler terms. In particular, when as is often the case the image noise spectrum is substantially flat over the frequency range within which the MTF is finite, then DQE may be factorized as

$$\varepsilon(w) = \varepsilon(0)\,\mathrm{MTF}^2(w) \tag{44}$$

where

$$\varepsilon(0) = (\log e)^2 \gamma^2 / Q\, WS(0)$$

Thus we may compare two imaging processes A and B in the DQE sense by the ratio

$$\varepsilon_A(0)\,\mathrm{MTF}_A^2(w) / \varepsilon_B(0)\,\mathrm{MTF}_B^2(w)$$

and in the informational sense by the ratio

$$\varepsilon_A(0) \int_0^\infty \mathrm{MTF}_A^2(w)\, w\, dw \Big/ \varepsilon_B(0) \int_0^\infty \mathrm{MTF}_B^2(w)\, dw$$

In this way informational efficiency I_{eff} and informational quality I_{qual} may be defined (Dainty and Shaw, 1974) over any spatial frequency range $0-w_0$ as

$$I_{\mathrm{eff}} = \varepsilon(0) \int_0^{w_0} \mathrm{MTF}^2(w)\, w\, dw \Big/ \int_0^{w_0} w\, dw \tag{45}$$

$$I_{\mathrm{qual}} = Q'(0) \int_0^{w_0} \mathrm{MTF}^2(w)\, w\, dw \Big/ \int_0^{w_0} w\, dw \tag{46}$$

Where $I_{\text{qual}} = QI_{\text{eff}}$. Closely related information-based integrals have been discussed by Linfoot (1964) and Kriss et al. (1977), among others.

Analysis in DQE terms of two-stage information transfer systems, such as the negative/positive processes of conventional photography, leads to the conclusion (Dainty and Shaw, 1974; Shaw, 1977) that the first (available-light) stage requires a detector with high DQE, while the second (print) stage requires high noise-equivalent input, irrespective of DQE. In this way information can be preserved and the final image will have a high SNR. Specifically for two stages, A and B, their combination can be expressed by

$$\varepsilon_{A+B}(w) = \frac{\varepsilon_A(w)}{1 + [WS_B(w)/WS_A(w)\gamma_B^2 \, MTF_B^2(w)]} \tag{47}$$

Thus the DQE of the first stage can be preserved if

$$WS_A(w)\gamma_B^2(w) \, MTF_B^2(w) \gg WS_B(w)$$

In other words, the noise of the first stage, as seen through the transfer characteristics of the second stage, must be large compared to the noise of second stage.

IV. PROBLEMS OF DETECTION AND INFORMATION STORAGE

A. Signal Detection Parameters

For exposures to visible light the photographic process has features which are generally undesirable for detection problems. DQE is a strong function of exposure and peaks at low density levels, implying that SNR transfer is a nonlinear function of mean signal level and signal contrast. Thus the SNR_{out} will not linearly increase as SNR_{in} increases, but will likewise peak before declining at higher densities. Furthermore, this peak does not coincide with the peak in DQE, as illustrated in Fig. 15. This has given rise to the apparent contradiction that when using the photographic process as a detector there must be a choice between exposing for maximum SNR_{out} or for the optimum rate of SNR transfer. However Fellgett (1961) pointed out that the former is only appropriate when the available light is of sufficient amount to exploit the full storage capabilities of the emulsion, while the latter is appropriate when the amount of available light is the limiting factor. He also reasoned that when magnification is a disposable parameter it can be chosen to yield any specified SNR_{out} while still using any reasonable exposure level at maximum DQE.

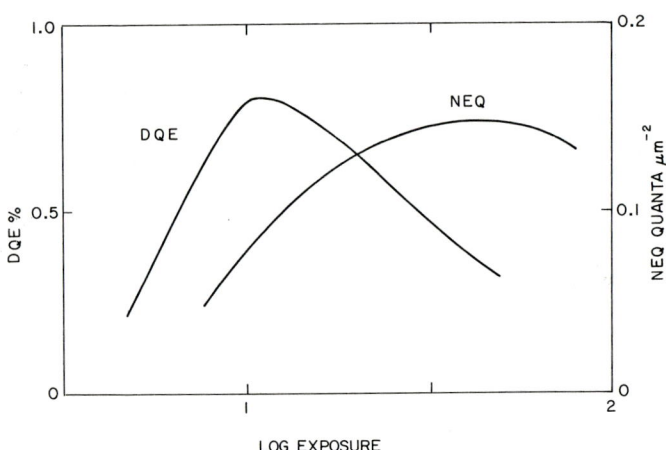

FIG. 15. Low spatial frequency values of DQE and NEQ for a typical negative process showing the peak values at different exposure levels (Shaw, 1965).

Following these agreements it is possible to define the way in which the signal statistics should be matched to the detector in the DQE sense by making appropriate tradeoffs between exposure time and detector area. In this way it can be demonstrated (Shaw, 1972c) that if the photographic process is used at the DQE level ε when some other higher level ε_{max} is available, either for a different emulsion–development combination or for the same combination at a different exposure level, then for a fixed exposure time the SNR loss may be expressed as the ratio $(\varepsilon_{max}/\varepsilon)^{1/2}$. It is stressed that this conclusion applies only to radiation-limited signals and not to problems where the dominant noise is due to background signal rather than radiation noise. In the limit where such background dominates, the detector DQE will play only a small role in helping to distinquish the signal from its own background. Other discussions of the photographic detection techniques in SNR and DQE terms, and especially related to problems in astronomy, have been given by Marchant (1964), Marchant and Millikan (1965), and Fellgett (1970). Hoag and Miller (1969) gave a review of these and other problems in the application of photographic materials in astronomy, with extensive bibliography.

Detection concepts and parameters based on SNR analysis and closed related to DQE have been discussed by Jones (1958b, 1959, 1960) and Zweig (1961b). They defined parameters which correspond to the ability of the recording process to detect increments of exposure energy and signal contrast. Jones defined detectivity D_e as the output SNR per unit input quantum

$$D_e = SNR_{out}/Q \qquad (48)$$

Ericson, R. H., and Marchant, J. C. (1972). *Photogr. Sci. Eng.* **16**, 253.
Farnell, G. C., and Chanter, J. B. (1961). *J. Photogr. Sci.* **9**, 73.
Fellgett, P. B. (1958). *Mon. Not. R. Astron Soc.* **118**, 224.
Fellgett, P. B. (1961). *J. Photogr. Sci.* **9**, 201.
Fellgett, P. B. (1970). In "Optical Instrument and Techniques" (J. Home Dickson, ed.), pp. 475–506. Oriel Press, London.
Fellgett, P. B., and Linfoot, E. H. (1955). *Philos. Trans. R. Soc. London, Ser. A* **247**, 369.
Frieser, H. (1959). *Photogr. Sci. Eng.* **3**, 164.
Frieser, H. (1960). *Photogr. Sci. Eng.* **4**, 324.
Frieser, H., and Klein, E. (1960). *Photogr. Sci. Eng.* **4**, 264.
Frieser, H., Klein, E., and Zeitler, E. (1959). *Z. Angew. Phys.* **11**, 90.
Gilmore, H. F. (1967). *J. Opt. Soc. Am.* **57**, 75.
Haase, G. (1960). *Naturwissenschaften* **47**, 320.
Hamilton, J. F. (1972). *Appl. Opt.* **11**, 13.
Hamilton, J. F., and Marchant, J. C. (1967). *J. Opt. Soc. Am.* **57**, 232.
Higgins, G. C. (1971). *Photogr. Sci. Eng.* **15**, 106.
Hoag, A. A., and Miller, W. C. (1969). *Appl. Opt.* **8**, 2417.
Ingelstam, E., Djurle, E., and Sjögren, B. (1956). *J. Opt. Soc. Am.* **46**, 707.
James, T. H., ed. (1977). "The Theory of the Photographic Process," 4th Ed. Macmillan, New York.
Johnson, C. B. (1970). *Photogr. Sci. Eng.* **14**, 413.
Johnson, C. B. (1977). In "Image Analysis and Evaluation" (R. Shaw, ed.), pp. 103–113. Soc. Photogr. Sci. Eng., Washington, D.C.
Jones, R. C. (1955). *J. Opt. Soc. Am.* **45**, 799.
Jones, R. C. (1958a). *Photogr. Sci. Eng.* **2**, 57.
Jones, R. C. (1958b). *Photogr. Sci. Eng.* **2**, 191.
Jones, R. C. (1959). *Adv. Electron. Electron Phys.* **11**, 87.
Jones, R. C. (1960). *J. Opt. Soc. Am.* **50**, 1058.
Jones, R. C. (1961). *J. Opt. Soc. Am.* **51**, 1159.
Kelly, D. H. (1960). *J. Opt. Soc. Am.* **50**, 269.
Kelly, D. H. (1971). *J. Opt. Soc. Am.* **51**, 1159.
Kriss, M. A., Nelson, C. N., and Eisen, F. C. (1974). *Photogr. Sci. Eng.* **18**, 131.
Kriss, M. A., O'Toole, J., and Kinard, J. C. (1977). In "Image Analysis and Evaluation" (R. Shaw, ed.), pp. 122–133. Soc. Photogr. Sci. Eng., Washington, D.C.
Lamberts, R. L. (1966). *Photogr. Sci. Eng.* **10**, 213.
Lamberts, R. L., and Higgins, G. C. (1966). *Photogr. Sci. Eng.* **10**, 219.
Latham, D. W. (1977). In "Image Analysis and Evaluation" (R. Shaw, ed.), pp. 452–454. Soc. Photogr. Sci. Eng., Washington, D.C.
Lawton, W. H., Trabka, E. A., and Wilder, D. R. (1972). *J. Opt. Soc. Am.* **65**, 659.
Levi, L. (1958). *J. Opt. Soc. Am.* **48**, 9.
Linfoot, E. H. (1961). *J. Photogr. Sci.* **9**, 188.
Linfoot, E. H. (1964). "Fourier Methods in Optical Image Evaluation." Focal Press, London.
Lorber, H. W. (1970). *IBM J. Res. Dev.* **14**, 515.
McGee, J. D., and Wheeler, B. E. (1961). *J. Photogr. Sci.* **9**, 106.
Marchant, J. C. (1964). *J. Opt. Soc. Am.* **54**, 798.
Marchant, J. C., and Millikan, A. G. (1965). *J. Opt. Soc. Am.* **55**, 907.
Marriage, A. (1961). *J. Photogr. Sci.* **9**, 93.
Miller, W. C. (1964). *Publ. Astron. Soc. Pac.* **16**, 328.
Millikan, A. G. (1974). *Am. Sci.* **62**, 324.
Nutting, P. G. (1913). *Philos. Mag.* **26**, 423.
Rose, A. (1946). *J. Soc. Motion Pict. Eng.* **47**, 273.

Rose, A. (1974). "Vision: Human and Electronic." Plenum, New York.
Sayhun, M. R. V. (1975). *Photogr. Sci. Eng.* **19**, 38.
Schade, O. H. (1955). *J. Soc. Motion Pict. Telev. Eng.* **64**, 593.
Schade, O. H. (1964). *J. Soc. Motion Pict. Telev. Eng.* **73**, 81.
Scott, R. L., and Smith, A. G. (1974). *Astron. J.* **79**, 656.
Selwyn, E. W. H. (1935). *Photogr. J.* **75**, 571.
Shannon, C. E. (1948). *Bell Syst. Tech. J.* **27**, 379.
Shaw, R. (1962). *Photogr. Sci. Eng.* **6**, 281.
Shaw, R. (1963a). *J. Photogr. Sci.* **11**, 199.
Shaw, R. (1963b). *J. Photogr. Sci.* **11**, 313.
Shaw, R. (1965). *J. Photogr. Sci.* **13**, 308.
Shaw, R. (1967). *J. Photogr. Sci.* **15**, 78.
Shaw, R. (1968). *J. Photogr. Sci.* **16**, 170.
Shaw, R. (1969). *J. Photogr. Sci.* **17**, 141.
Shaw, R. (1972a). *Photogr. Sci. Eng.* **16**, 192.
Shaw R. (1972b). *Photogr. Sci. Eng.* **16**, 395.
Shaw R. (1972c). *J. Photogr. Sci.* **20**, 174.
Shaw, R. (1973a). *J. Photogr. Sci.* **21**, 25.
Shaw, R. (1973b). *Opt. Acta* **20**, 749.
Shaw, R. (1975). *Proc. Soc. Photo-Opt. Instrum. Eng.* **70**, 359.
Shaw, R. (1977). *In* "Image Science Mathematics" (C. O. Wilde and E. Barrett, eds.), pp. 1–9. Western Period., Hollywood, California.
Shaw, R. (1978). *Rep. Prog. Phys.* **41**, 1103.
Shaw, R., and Shipman, A. (1969). *J. Photogr. Sci.* **17**, 205.
Siedentopf, H. (1937). *Phys. Z.* **38**, 454.
Silevitch, M. B., Gonsalves, R. A., and Ehn, D. C. (1977). *Photogr. Sci. Eng.* **21**, 7.
Smith, G. H. (1972). Ph. D. Thesis, Univ. of Arizona. Tucson.
Spencer, H. E. (1971). *Photogr. Sci. Eng.* **15**, 468.
Spencer, H. E. (1976). *J. Photogr. Sci.* **24**, 34.
Thourson, T. L. (1975). *Proc. Soc. Photo-Opt. Instrum. Eng.* **56**, 225.
Trabka, E. A. (1971). *J. Opt. Soc. Am.* **61**, 800.
Trabka, E. A. (1977). *Photogr. Sci. Eng.* **21**, 183.
Trabka, E. A., and Doerner, E. C. (1976). *J. Appl. Photogr. Eng.* **2**, 1.
Trabka, E. A., and Lawton, W. H. (1974). *J. Photogr. Sci.* **22**, 131.
Valentine, R. C. (1966). *Adv. Opt. Electron Microsc.* **1**, 180.
Vendrovski, K. V., Veitzman, A. I., and Ptashenchuk, V. M. (1972). *Zh. Nauchn. Prikl. Fotogr. Kinematogr.* **17**, 426.
Wagner, R. F. (1977). *Photogr. Sci. Eng.* **21**, 252.
Webb, J. H. (1948). *J. Opt. Soc. Am.* **38**, 312.
Wiener, N. (1930). *Acta. Math.* **55**, 117.
Wilcock, W. L., Emberson, D. L., and Weekley, B. (1960). *Nature (London)* **185**, 370.
Wolfe, R. N., Marchand, E. W., and DePalma, J. J. (1968). *J. Opt. Soc. Am.* **58**, 1245.
Zweig, H. J. (1956). *J. Opt. Soc. Am.* **46**, 805.
Zweig, H. J. (1961a). *J. Opt. Soc. Am.* **51**, 310.
Zweig, H. J. (1961b). *Photogr. Sci. Eng.* **5**, 142.
Zweig, H. J. (1964). *Photogr. Sci. Eng.* **8**, 305.

CHAPTER 6

Propagation of Laser Beams

H. KOGELNIK
Electronics Research Laboratory
Bell Laboratories, Holmdel, New Jersey

I. Paraxial Ray Propagation	156
A. Ray Transfer Matrix	157
B. Ray Transfer Matrix of Composite Systems	158
C. Refraction at an Interface	159
D. Translation through Free Space	160
E. Translation through a Uniform Dielectric	161
F. Translation through Compound Dielectric Layers	161
G. Refraction at a Tilted Interface	162
H. Translation through a Brewster Window	164
I. Refraction at a Curved Dielectric Interface	165
J. Ray Matrix of a Thin Lens	165
K. Ray Passage through Multiple Thin Lenses	166
L. The Thick Lens	167
M. Combinations of Lenses and Translations	168
N. The Telescope	171
O. Principal Planes and Focal Length	172
P. Periodic Sequences	173
Q. Lenslike Medium	174
II. Propagation of Gaussian Laser Beams	175
A. The Wave Equation	175
B. The Gaussian Laser Beam	176
C. Beam Transformation by a Lens	178
D. The $ABCD$ Law	179
E. Mode Matching	179
F. Circle Diagrams	183
III. Fresnel Diffraction	185
A. Rectangular Symmetry	186
B. Circular Symmetry	187
C. Fraunhofer Diffraction	187
D. Far Fields for Special Cases	188
E. Effects of a Lens on Fresnel Fields	188
References	190

The following collection of formulas tries to provide the tools required for the analysis of the propagation of laser beams. It is divided into three parts, each of which corresponds to a different level of analysis. The first and simplest level is the analysis of laser beam propagation in terms of paraxial rays. A key element of this analysis is the ray matrix and we have listed the ray matrices for a considerable number of optical components and systems. With the help of the *ABCD* law, these matrices can also be used for the second level of analysis in terms of laser beams with a Gaussian intensity profile, the "Gaussian beams." This second part follows very closely the review paper of Kogelnik and Li (1966) which can be consulted for more tutorial detail. The basis for the third and most sophisticated level of analysis of laser beam propagation is the theory of Fresnel diffraction where we provide the basic formulas and some standard examples.

I. PARAXIAL RAY PROPAGATION

The study of the propagation of paraxial rays through optical systems reveals important clues for the propagation of laser beams (Kogelnik and Li, 1966). In particular, the *ray transfer matrices* which are used to describe paraxial ray propagation can be used very directly for a detailed description of the propagation of laser beams. We will describe here the ray concepts needed for the understanding of laser beams and discuss the properties of a list of elementary optical systems. More complicated optical systems can be decomposed into several elementary systems for analysis.

Paraxial rays are those rays that remain close to the axis of the optical system of interest and which are almost parallel to this axis. As indicated in Fig. 1, a paraxial ray in a given cross section ($z =$ constant) of the system is characterized by its distance x from the optic axis (i.e., the z axis) and by its angle α with respect to it. For rays propagating to the right α is defined as positive if it has an inclination as shown in the figure. A convenient measure of ray direction is the optical direction-cosine x', defined as

$$x' = n \sin \alpha \approx n\alpha \qquad (1)$$

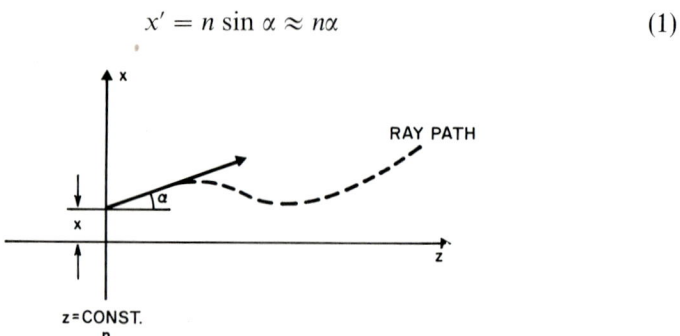

FIG. 1. Ray path in an optical system and reference axes. The z axis is the optic axis, x the ray distance from the optic axis, and α the ray angle.

where n is the refractive index in the cross section $z =$ constant. In the paraxial approximation the sine function is replaced by its argument as indicated.

The ray position x and the direction-cosine $x' = n\alpha$ are combined into a ray vector $\mathbf{X}$ defined as

$$\mathbf{X} = \begin{pmatrix} x \\ x' \end{pmatrix} \tag{2}$$

For systems of cylindrical symmetry the behavior of paraxial rays in the y dimension perpendicular to the x–z plane is the same as that in the x dimension. Therefore, it is often sufficient to restrict the analysis to the x dimension. However, there are special cases of practical interest, such as astigmatic elements, where we have to treat the two dimensions separately.

A. Ray Transfer Matrix

The ray path through a given optical structure, such as that sketched in Fig. 2, may be relatively complicated. However, in the paraxial ray approximation, the ray position x_2 and direction-cosine x_2' in the output plane of the system are always linearly related to the ray position x_1 and direction-cosine x_1' in the input plane (Kogelnik and Li, 1966; Brower, 1964; O'Neill, 1963; Gerrard and Burch, 1975)

$$x_2 = Ax_1 + Bx_1', \qquad x_2' = Cx_1 + Dx_1' \tag{3}$$

The constants A, B, C, and D are characteristics of the system, which are generally related by

$$AD - BC = 1 \tag{4}$$

The linear relation (3) is conveniently written in matrix form (Kogelnik and Li, 1966; Brower, 1964; O'Neill, 1963; Gerrard and Burch, 1975)

$$\begin{pmatrix} x_2 \\ x_2' \end{pmatrix} = \begin{pmatrix} A & B \\ C & D \end{pmatrix} \begin{pmatrix} x_1 \\ x_1' \end{pmatrix} \tag{5}$$

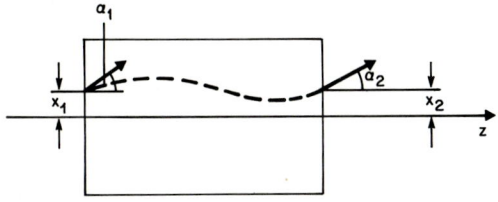

INPUT PLANE   OUTPUT PLANE

Fig. 2. Ray position and ray angles at the input and output of an optical system.

The matrix **M**

$$\mathbf{M} = \begin{pmatrix} A & B \\ C & D \end{pmatrix} \tag{6}$$

is called the ray transfer matrix. In terms of the ray vectors $\mathbf{X}_1$ and $\mathbf{X}_2$ in the input and output planes, the matrix relation (5) can be written in the compact form

$$\mathbf{X}_2 = \mathbf{M}\mathbf{X}_1 \tag{7}$$

Equation (4) implies that the determinant of all ray matrices det **M** is equal to unity, which can be used to check more complicated calculations.

When the system is turned around so that its former input plane becomes the output plane and vice versa, its ray transfer matrix $\tilde{\mathbf{M}}$ can be expressed in terms of the coefficients A, B, C, D of the original system as

$$\tilde{\mathbf{M}} = \begin{pmatrix} D & B \\ C & A \end{pmatrix} \tag{8}$$

If the optical system is symmetrical, we have

$$A = D \tag{9}$$

B. Ray Transfer Matrix of Composite Systems

When a system consists of several parts for which the ray matrices are known, the overall system matrix can be calculated by matrix multiplication. The example shown in Fig. 3 shows three parts with the matrices $\mathbf{M}_1$, $\mathbf{M}_2$,

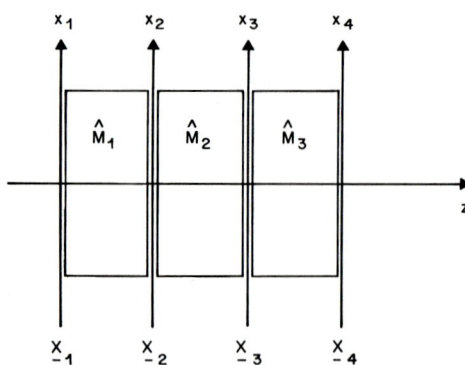

Fig. 3. Composite optical system consisting of three subsystems. Underscores indicate vectors.

and $\mathbf{M}_3$, respectively. The overall matrix $\mathbf{M}$ relating the input and output ray vectors $\mathbf{X}_1$ and $\mathbf{X}_4$ by

$$\mathbf{X}_4 = \mathbf{M}\mathbf{X}_1 \tag{10}$$

is calculated by multiplying in three matrices in the form

$$\mathbf{M} = \mathbf{M}_3 \cdot \mathbf{M}_2 \cdot \mathbf{M}_1 \tag{11}$$

The rule for obtaining the elements of the product matrix of two matrices $\mathbf{M}_1$ and $\mathbf{M}_2$ with elements $A_1, B_1, C_1, D_1,$ and A_2, B_2, C_2, D_2, respectively, is

$$\mathbf{M}_2 \cdot \mathbf{M}_1 = \begin{pmatrix} A_2 & B_2 \\ C_2 & D_2 \end{pmatrix} \begin{pmatrix} A_1 & B_1 \\ C_1 & D_1 \end{pmatrix} = \begin{pmatrix} A_1 A_2 + C_1 B_2 & B_1 A_2 + D_1 B_2 \\ A_1 C_2 + C_1 D_2 & B_1 C_2 + D_1 D_2 \end{pmatrix} \tag{12}$$

C. Refraction at an Interface

The first elementary example of our list is the plane interface between two dielectric media of refractive indices n_1 and n_2 as sketched in Fig. 4. The interface is chosen perpendicular to the z axis. A light ray incident from the left at the angle α_1 is refracted at the interface and continues at the right of the interface with an angle α_2 according to Snell's law

$$n_1 \sin \alpha_1 = n_2 \sin \alpha_2 \tag{13}$$

Snell's law implies that the direction-cosines in the two media are the same, i.e.,

$$x_1' = x_2' \tag{14}$$

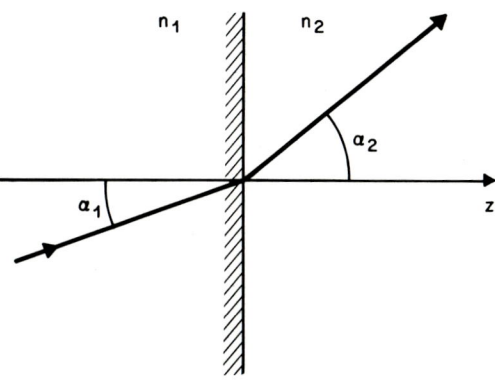

Fig. 4. Refraction of a ray at the plane interface between two dielectric media of refractive indices n_1 and n_2.

When the input and output planes are chosen immediately to the left and right of the interface the input and output ray positions x_1 and x_2 are equal,

$$x_1 = x_2 \tag{15}$$

and the corresponding ray vectors are also equal

$$\mathbf{X}_1 = \mathbf{X}_2 \tag{16}$$

The ray transfer matrix for a dielectric interface is simply the unit matrix

$$\mathbf{M} = \begin{pmatrix} 1 & 0 \\ 0 & 1 \end{pmatrix} \tag{17}$$

D. Translation through Free Space

Another basic example is the propagation of light through free space with refractive index $n = 1$. Here the ray path is a straight line and the direction-cosine remains constant ($\alpha_1 = \alpha_2$, $x_1' = x_2'$). However, the ray position changes, as sketched in Fig. 5. For a distance d between the input and output planes we have

$$x_2 = x_1 + d \tan \alpha_1 \approx x_1 + d\alpha = x_1 + dx_1' \tag{18}$$

The ray matrix for translation through free space has the form

$$\mathbf{M} = \begin{vmatrix} 1 & d \\ 0 & 1 \end{vmatrix} \tag{19}$$

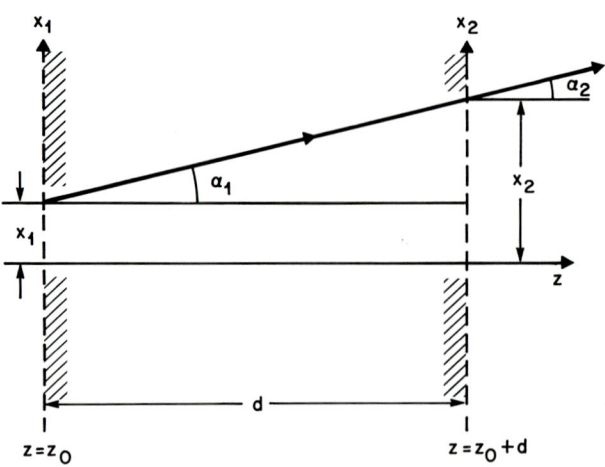

Fig. 5. Translation of a ray through free space over a distance d.

E. Translation through a Uniform Dielectric

When the light propagates in a uniform dielectric medium of index n the ray path is again a straight line as in the previous case. Propagation through a glass plate is an example. As $x' = n\alpha$, we have for the output ray position x_2 the relation

$$x_2 = x_1 + d \sin \alpha_1 \approx x_1 + d\alpha_1 = x_1 + (d/n)x_1' \tag{20}$$

The ray matrix for translation through a dielectric layer of thickness d has the form

$$\mathbf{M} = \begin{pmatrix} 1 & d/n \\ 0 & 1 \end{pmatrix} \tag{21}$$

Here we notice that the presence of the dielectric causes the appearance of a "reduced distance" d/n in the characterization of paraxial rays. Objects viewed through a glass plate appear closer than they actually are.

F. Translation through Compound Dielectric Layers

Figure 6 illustrates the propagation of a ray through two adjacent layers (or compound plates) of refractive indices n_1 and n_2 and thicknesses d_1 and d_2, respectively. The ray matrix for this compound system is obtained by matrix multiplication and has the form

$$\mathbf{M} = \begin{pmatrix} 1 & d_2/n_2 \\ 0 & 1 \end{pmatrix} \begin{pmatrix} 1 & d_1/n_1 \\ 0 & 1 \end{pmatrix} = \begin{pmatrix} 1 & d_1/n_1 + d_2/n_2 \\ 0 & 1 \end{pmatrix} \tag{22}$$

The reduced distance, which appears as the upper right matrix element B, is now

$$B = d_1/n_1 + d_2/n_2 \tag{23}$$

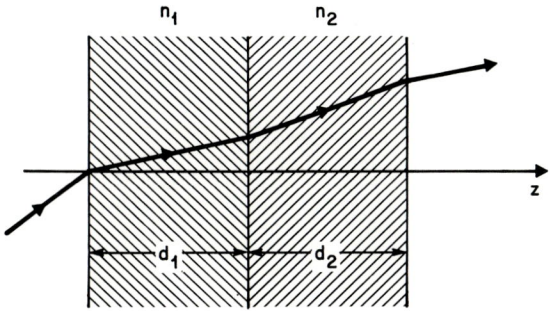

FIG. 6. Propagation of a ray through two dielectric plates of indices n_1 and n_2.

Similarly one gets for the compound of a multiple of dielectric layers or plates a reduced distance

$$B = \sum_i \frac{d_i}{n_i} \tag{24}$$

equal to the sum of the reduced distances d_i/n_i of all individual layers. When the refractive index $n(z)$ varies continuously the sum in this expression should be replaced by an integral

$$B = \int_0^d \frac{dz}{n(z)} \tag{25}$$

G. Refraction at a Tilted Interface

In several practical cases one inserts optical elements with a tilt relative to the axis of the light beam. Examples are the Brewster-angle windows or

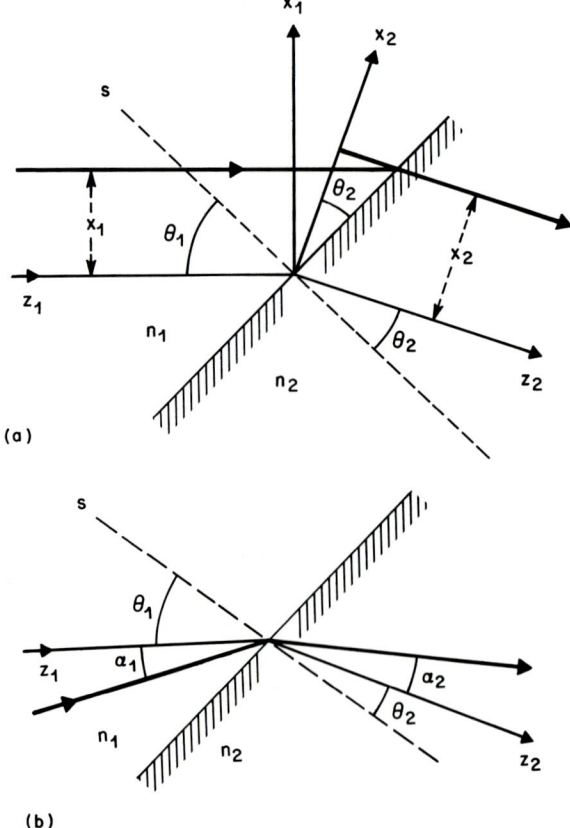

FIG. 7. Refraction of a ray packet at a tilted interface: (a) shows foreshortening of the ray positions; (b) shows input and output ray angles.

etalon plates inserted into laser cavities. One is then dealing with refraction at a tilted dielectric interface (Born and Wolf, 1959; Hanna, 1969; Kogelnik et al., 1972) such as sketched in Fig. 7 where the interface is assumed to be perpendicular to the plane of the drawing. The central ray of a ray packet or beam, incident at an angle θ_1, is refracted according to Snell's law

$$n_1 \sin \theta_1 = n_2 \sin \theta_2 \tag{26}$$

For simplicity this central ray is chosen as the z_1 and z_2 axis on the two sides of the interface as indicated in the figures. The ray vectors of the paraxial rays deviating from this beam axis are characterized relative to the (x_1, y_1, z_1) and (x_2, y_2, z_2) coordinate systems, where the y_1 and y_2 axes are perpendicular to the plane of the drawing. Because of the special geometry of this case, paraxial rays behave differently in the x and y dimensions and we have to consider these two cases separately.

As the z_1 and z_2 axes are tilted with respect to each other the ray positions x_1 and x_2 are foreshortened as illustrated in Fig. 7a where s is the normal to the interface. We have, to first order,

$$x_1/x_2 = \cos \theta_1 / \cos \theta_2 \tag{27}$$

Paraxial rays incident with an angle deviating a small amount α_1 from θ_1 lead to refracted rays deviating by α_2 from θ_2 as shown in Fig. 7b. We have

$$n_1 \sin(\theta_1 + \alpha_1) = n_2 \sin(\theta_2 + \alpha_2), \tag{28}$$

and, to first order in α_1,

$$x_1' \cos \theta_1 = x_2' \cos \theta_2 \tag{29}$$

For the x dimension the ray matrix $\mathbf{M}_x$ for refraction at the tilted interface has the form

$$\mathbf{M}_x = \begin{pmatrix} \cos \theta_2 / \cos \theta_1 & 0 \\ 0 & \cos \theta_1 / \cos \theta_2 \end{pmatrix} \tag{30}$$

For the y dimension the ray matrix $\mathbf{M}_y$ remains the same as that for an untilted interface

$$\mathbf{M}_y = \begin{pmatrix} 1 & 0 \\ 0 & 1 \end{pmatrix} \tag{31}$$

When the interface is set at the Brewster angle θ_{B1} relative to the beam axis reflection of light with an electric field polarized in the (x,z) plane is minimized. The Brewster angle follows from

$$\tan \theta_{B1} = n_2/n_1 = 1/\tan \theta_{B2} \tag{32}$$

where

$$\theta_{B1} + \theta_{B2} = \pi/2 \tag{33}$$

For this case the ray matrix $\mathbf{M}_x$ simplifies to

$$\mathbf{M}_x = \begin{pmatrix} n_1/n_2 & 0 \\ 0 & n_2/n_1 \end{pmatrix} \quad (34)$$

H. Translation through a Brewster Window

For a beam propagating through a dielectric plate set at Brewster's angle with respect to the beam axis, we have to consider one refraction at the tilted input interface, a translation of distance d through the dielectric medium, and another refraction at the tilted output interface (Hanna, 1969; Kogelnik et al., 1972). The choice of the input and output coordinate systems is indicated in Fig. 8. The window thickness is t and the refractive index of the window n. The translation distance d is related to t by

$$d = t/\cos\theta_{B2} \quad (35)$$

For the x dimension the ray matrix $\mathbf{M}_x$ for translation through the window is given by

$$\mathbf{M}_x = \begin{pmatrix} n & 0 \\ 0 & 1/n \end{pmatrix} \begin{pmatrix} 1 & d/n \\ 0 & 1 \end{pmatrix} \begin{pmatrix} 1/n & 0 \\ 0 & n \end{pmatrix} = \begin{pmatrix} 1 & dn \\ 0 & 1 \end{pmatrix} \quad (36)$$

The corresponding ray matrix $\mathbf{M}_y$ for the y dimension is

$$\mathbf{M}_y = \begin{pmatrix} 1 & d/n \\ 0 & 1 \end{pmatrix} \quad (37)$$

Note that the reduced thickness of the Brewster window is different for the x and y dimensions ($B_x = dn$, $B_y = d/n$).

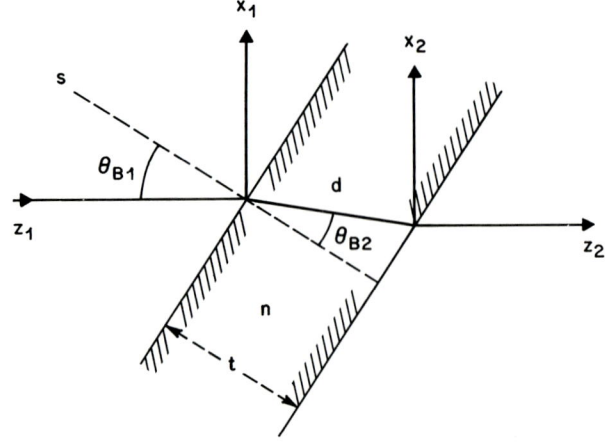

FIG. 8. Translation of a ray through a Brewster window of thickness t and index n. θ_{B1} is Brewster's angle.

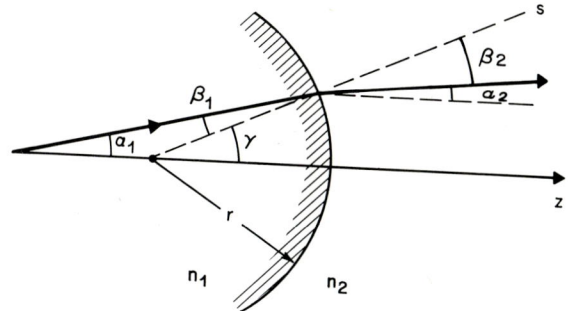

FIG. 9. Refraction at a curved dielectric interface of curvature radius r. The refractive indices of the two media are n_1 and n_2.

I. Refraction at a Curved Dielectric Interface

The curved surface of a dielectric medium such as glass provides the key to the functioning of a lens. Figure 9 shows a curved interface of radius r, assumed positive as drawn, between two media of refractive index n_1 and n_2, respectively. The label s denotes the surface normal. The angles β_1 and β_2 between ray and surface normal are governed by Snell's law

$$n_1 \sin\beta_1 = n_2 \sin\beta_2 \tag{38}$$

Furthermore we have the relation

$$\gamma = \alpha_1 + \beta_1 = \alpha_2 + \beta_2 \tag{39}$$

Within the paraxial ray approximation one gets for the direction-cosine at the output

$$x_2' = [(n_2 - n_1)/r]x_1 + x_1' \tag{40}$$

which reflects the fact, expected for a curved surface, that the direction of the refracted ray depends on the ray position x_1 where the ray strikes the surface.

The ray matrix for refraction at the curved surface has the form

$$\mathbf{M} = \begin{pmatrix} 1 & 0 \\ (n_2 - n_1)/r & 1 \end{pmatrix} \tag{41}$$

The surface acts as a lens of focal length

$$-f = r/(n_2 - n_1) \tag{42}$$

J. Ray Matrix of a Thin Lens

The transfer of rays through a thin lens can be calculated from the imaging law

$$1/p + 1/q = 1/f \tag{43}$$

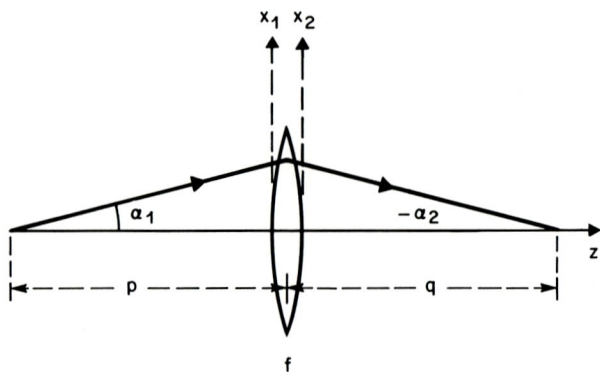

FIG. 10. Refraction of a ray by a thin lens of focal length f.

where we refer to the geometry sketched in Fig. 10. The output and input reference planes are chosen to be immediately to the left and the right of the lens which is assumed to be surrounded by free space. From Eq. (43) follows a relation for the direction-cosines of paraxial rays of the form

$$x_2' = -(x_1/f) + x_1' \tag{44}$$

For the special case of collimated incident rays we have $x_1' = 0$ and all rays are seen to be focused at the focal point at a distance f to the right of the lens. The ray matrix of a thin lens is

$$\mathbf{M} = \begin{pmatrix} 1 & 0 \\ -1/f & 1 \end{pmatrix} \tag{45}$$

The ray matrix of a thin lens can also be calculated with the help of Eq. (41). Assuming a doubly concave lens of refractive index n and curvatures of radius r_1 and r_2, respectively, one finds by matrix multiplication

$$\mathbf{M} = \begin{bmatrix} 1 & 0 \\ (1-n)/r_2 & 1 \end{bmatrix} \begin{bmatrix} 1 & 0 \\ (n-1)/(-r_1) & 1 \end{bmatrix} = \begin{bmatrix} 1 & 0 \\ -(n-1)(1/r_1 + 1/r_2) & 1 \end{bmatrix} \tag{46}$$

Comparison of Eqs. (45) and (46) yields the lens maker's formula for a thin lens

$$1/f = (n-1)(1/r_1 + 1/r_2) \tag{47}$$

The focal length f of a spherical mirror with a radius of curvature r is given by

$$f = r/2 \tag{48}$$

K. Ray Passage through Multiple Thin Lenses

When two thin lenses of focal lengths f_1 and f_2 are placed adjacent to each other as shown in Fig. 11 their focusing powers add. The ray matrix of the

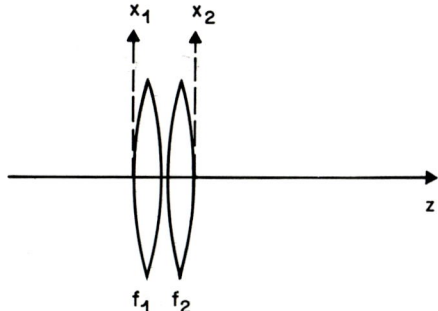

FIG. 11. Pair of adjacent thin lenses of focal lengths f_1 and f_2.

lens compound is

$$\mathbf{M} = \begin{bmatrix} 1 & 0 \\ -1/f_2 & 1 \end{bmatrix} \begin{bmatrix} 1 & 0 \\ -1/f_1 & 1 \end{bmatrix} = \begin{bmatrix} 1 & 0 \\ -(1/f_1 + 1/f_2) & 1 \end{bmatrix} \quad (49)$$

The focal length f of the lens compound is given by

$$1/f = 1/f_1 + 1/f_2 \quad (50)$$

Similarly one has for the focal length of a compound of several lenses of focal length f_i

$$1/f = \sum_i 1/f_i \quad (51)$$

L. THE THICK LENS

When a ray passes through a thick lens or a dielectric rod with curved end surfaces, it experiences (a) refraction at the curved entrance surface, (b) translation through the dielectric, and (c) refraction at the curved exit surface. The lens goemetry is shown in Fig. 12, where r_1 and r_2 are the radii of curvature of the entrance and exit surfaces, d is the length of the rod, and n the refractive index of the lens medium. The ray matrix of the thick lens is obtained by matrix multiplication

$$\mathbf{M} = \begin{bmatrix} 1 & 0 \\ \dfrac{1-n}{r_2} & 1 \end{bmatrix} \begin{bmatrix} 1 & d/n \\ 0 & 1 \end{bmatrix} \begin{bmatrix} 1 & 0 \\ \dfrac{1-n}{r_1} & 1 \end{bmatrix}$$

$$= \begin{bmatrix} 1 + \dfrac{d(1-n)}{nr_1} & \dfrac{d}{n} \\ (1-n)\left(\dfrac{1}{r_1} + \dfrac{1}{r_2}\right) + \dfrac{d(1-n)^2}{nr_1 r_2} & 1 + \dfrac{d(1-n)}{nr_2} \end{bmatrix} \quad (52)$$

Note that $\det \mathbf{M} = 1$ as required.

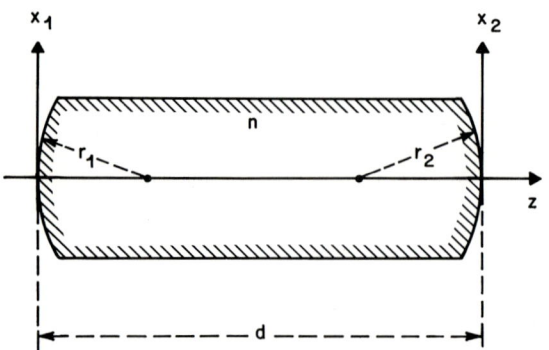

FIG. 12. Dielectric rod with curved end faces (thick lens).

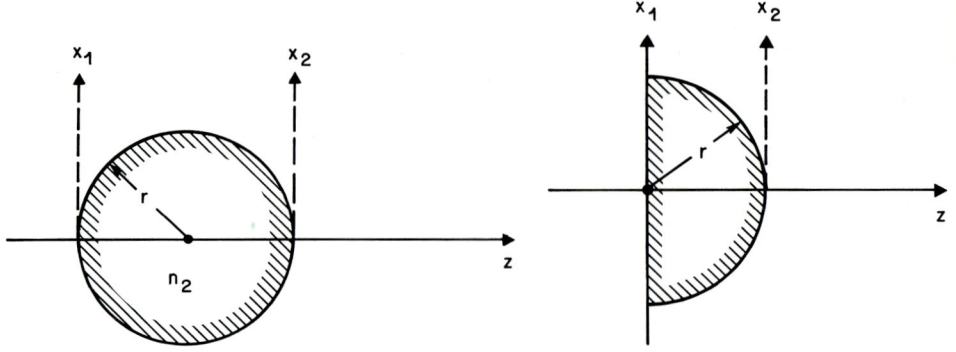

FIG. 13. Dielectric sphere of radius r.

FIG. 14. Dielectric half sphere.

A dielectric sphere of radius r and refractive index n is a special case of a thick lens with $r_1 = r_2 = r$ and $d = 2r$. This is sketched in Fig. 13. The ray matrix of the sphere is

$$\mathbf{M} = \begin{bmatrix} 2/n - 1 & d/n \\ [2(1 - n)]/nr & 2/n - 1 \end{bmatrix} \tag{53}$$

A half sphere as shown in Fig. 14 is another special case with $r_1 = \infty$, $r_2 = r$, and $d = r$. Its ray matrix is

$$\mathbf{M} = \begin{bmatrix} 1 & d/n \\ (1 - n)/r & 1/n \end{bmatrix} \tag{54}$$

M. COMBINATIONS OF LENSES AND TRANSLATIONS

In the analysis of laser beam propagation through practical optical systems one encounters various combinations of lenses and translations through free space. The ray matrices of these systems can be obtained by multiplying the matrices of known subsystems. We list below the ray matrices of some elementary combinations that can serve as a starting point for such an analysis.

Consider first a translation of rays over a distance d, followed by a lens of focal length f, as illustrated in Fig. 15a. The ray matrix of this combination is:

$$\mathbf{M} = \begin{bmatrix} 1 & 0 \\ -1/f & 1 \end{bmatrix} \begin{bmatrix} 1 & d \\ 0 & 1 \end{bmatrix} = \begin{bmatrix} 1 & d \\ -1/f & 1 - d/f \end{bmatrix} \tag{55}$$

When the lens is in the input plane of the combination and the translation follows, we have the situation of Fig. 15b and a ray matrix

$$\mathbf{M} = \begin{bmatrix} 1 & d \\ 0 & 1 \end{bmatrix} \begin{bmatrix} 1 & 0 \\ -1/f & 1 \end{bmatrix} = \begin{bmatrix} 1 - d/f & d \\ -1/f & 1 \end{bmatrix} \tag{56}$$

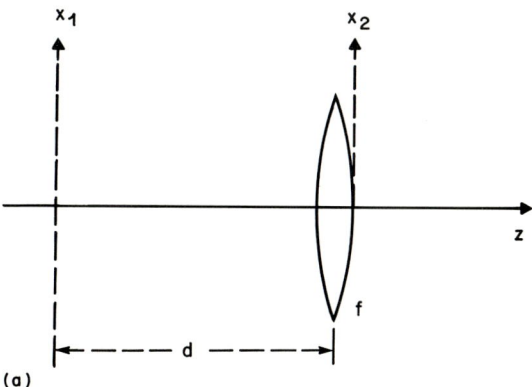

(a)

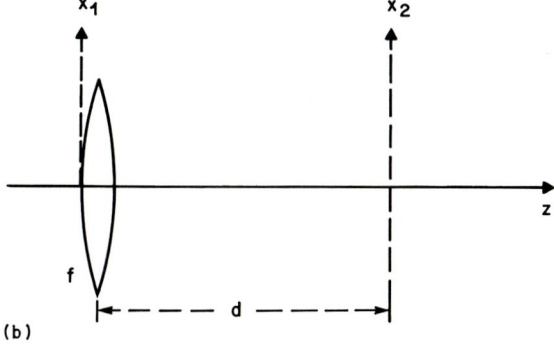

(b)

FIG. 15. Combinations of thin lenses and translation: (a) translation followed by lens, (b) lens followed by translation.

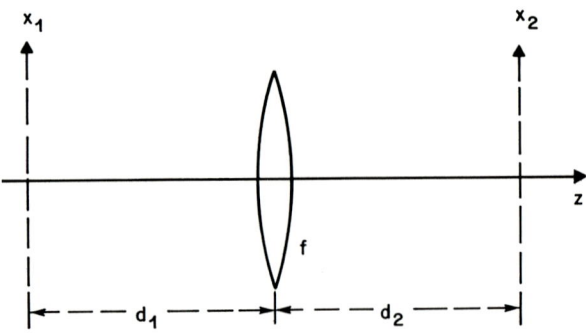

FIG. 16. Translation-lens-translation system.

Figure 16 shows a translation over a distance d_1, followed by a lens, and that followed by a translation over d_2. The ray matrix for this combination is

$$\mathbf{M} = \begin{bmatrix} 1 - d_2/f & d_1 + d_2 - d_1 d_2/f \\ -1/f & 1 - d_1/f \end{bmatrix} \quad (57)$$

In the special case where the input plane is imaged into the output plane we have

$$1/d_1 + 1/d_2 = 1/f \quad (58)$$

and the ray matrix of Eq. (57) simplifies to

$$\mathbf{M} = \begin{bmatrix} -d_2/d_1 & 0 \\ -1/f & -d_1/d_2 \end{bmatrix} \quad (59)$$

where d_2/d_1 is the image magnification. Note that for this case the output ray position is independent of the direction-cosine of the rays at the input (i.e., the object) plane.

The dual complement to the translation-lens-translation system discussed above is the combination of a lens of focal length f_1, followed by a translation over a distance d, and that followed by a lens of focal length f_2, as shown in Fig. 17. The ray matrix of this system is

$$\mathbf{M} = \begin{bmatrix} 1 - \dfrac{d}{f_1} & d \\ -\left(\dfrac{1}{f_1} + \dfrac{1}{f_2} - \dfrac{d}{f_1 f_2}\right) & 1 - \dfrac{d}{f_2} \end{bmatrix} \quad (60)$$

A special case of this combination is the "afocal system" or telescope, where

$$f_1 + f_2 = d \quad (61)$$

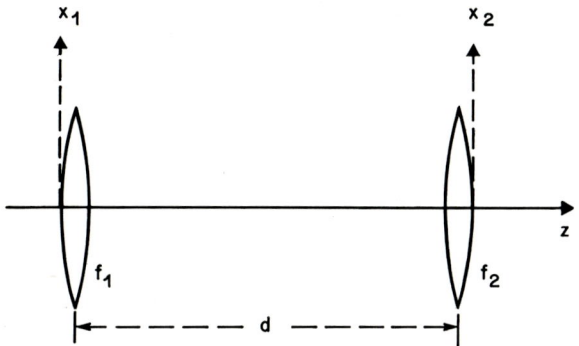

FIG. 17. Lens-translation-lens system.

The ray matrix of the telescope is

$$\mathbf{M} = \begin{bmatrix} -f_2/f_1 & d \\ 0 & -f_1/f_2 \end{bmatrix} \tag{62}$$

The system transforms an incident bundle of parallel rays into another bundle of parallel rays emerging at the output at a different angle, where f_2/f_1 is the angular magnification.

N. THE TELESCOPE

A telescope is a handy tool for the expansion of laser beams. The ray matrix of a telescopic system is given in Eq. (62). In practice one usually adds free-space sections before the first and after the second lens, as shown in Fig. 18. This causes additional ray translations over the distances d_1 and d_2, respectively. The ray matrix of the overall system is

$$\mathbf{M} = \begin{bmatrix} -f_2/f_1 & d - d_1 f_2/f_1 - d_2 f_1/f_2 \\ 0 & -f_1/f_2 \end{bmatrix} \tag{63}$$

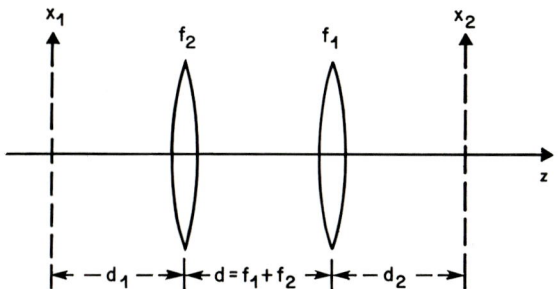

FIG. 18. Telescopic system.

O. Principal Planes and Focal Length

A general optical system is conventionally characterized by its focal length and its principal planes. This allows the tracing of rays through the system according to simple geometrical construction rules (Born and Wolf, 1959). We assume that the system is surrounded by free space of $n = 1$ and characterize the location of the principal planes of their distances h_1 and h_2 from the input and output planes as indicated in Fig. 19. The focal length f is measured from the principal planes. The tracing of two rays is illustrated in the figure. The ray drawn as a solid line passes through the focal point at the input side and emerges parallel to the optic axis at the output. Its intersection with the input principal plane determines the output ray position. The input ray marked by the dash-dotted line intersects the axis in the input principal plane. The corresponding output ray emerges parallel to the input ray and intersects the optic axis at the output principal plane.

When the ray matrix elements (A, B, C, and D) of the system are known, the focal length f and the principal plane positions are determined as

$$f = -1/C \tag{64}$$
$$h_1 = (D - 1)/C \tag{65}$$
$$h_2 = (A - 1)/C \tag{66}$$

For known f, h_1, and h_2, the matrix elements can be expressed as

$$A = 1 - h_2/f \tag{67}$$
$$B = h_1 + h_2 - h_1 h_2/f \tag{68}$$
$$C = -1/f \tag{69}$$
$$D = 1 - h_1/f \tag{70}$$

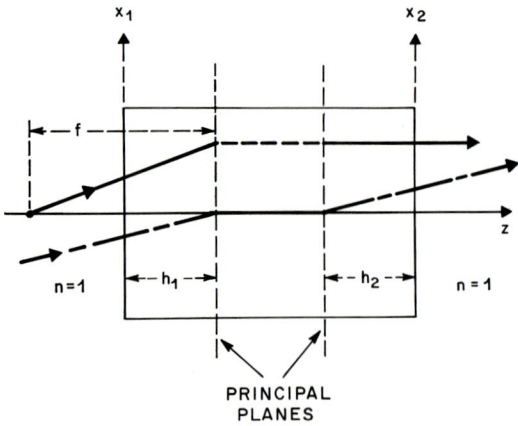

FIG. 19. Principal planes and focal lengths of an optical system.

P. Periodic Sequences

Some optical systems of practical interest are composed of a sequence of identical optical subsystems of known characteristics. Fig. 20 indicates schematically a sequence of n identical subsystems, each characterized by the same ray matrix elements A, B, C, and D. The transfer of rays through this sequence of n subsystems is described by a ray matrix $\mathbf{M}_n$ which is the nth power of ray matrix $\mathbf{M}$ of an individual subsystem, which has the form

$$\mathbf{M}_n = \mathbf{M}^n = \begin{bmatrix} A & B \\ C & D \end{bmatrix}^n$$

$$= \frac{1}{\sin\theta} \begin{bmatrix} A\sin n\theta - \sin(n-1)\theta & B\sin n\theta \\ C\sin n\theta & D\sin n\theta - \sin(n-1)\theta \end{bmatrix} \quad (71)$$

Here we have set

$$\cos\theta = \tfrac{1}{2}(A + D) \quad (72)$$

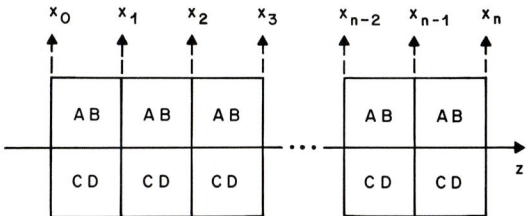

FIG. 20. Periodic sequence of idential optical subsystems.

One classifies periodic sequences as either *stable* or *unstable*. A sequence is stable when the trace $(A + D)$ of the subsystem ray matrix obeys the inequality

$$-1 < \tfrac{1}{2}(A + D) < 1 \quad (73)$$

If the inequality is not obeyed then the sequence is unstable. It is characteristic of a stable sequence that ray bundles passing through it are periodically refocused. In unstable systems a ray bundle becomes more and more dispersed as it propagates through the sequence.

A simple case of a periodic sequence is a sequence of equal lenses of focal length f with a lens spacing d as shown in Fig. 21. If we choose the reference planes just to the right of each lens as indicated, we have for the ray matrix $\mathbf{M}_n$ of a sequence of n lenses the following expression

$$\mathbf{M}_n = \frac{1}{\sin\theta} \begin{bmatrix} \sin n\theta - \sin(n-1)\theta & d\sin n\theta \\ -1/f \sin n\theta & (1 - d/f)\sin n\theta - \sin(n-1)\theta \end{bmatrix} \quad (74)$$

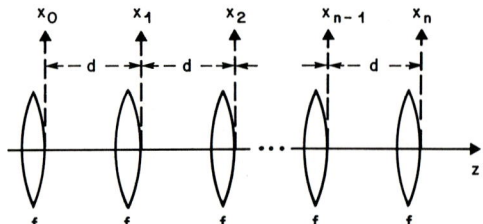

FIG. 21. Periodic sequence of thin lenses.

where

$$\cos \theta = 1 - (d/2f) \tag{75}$$

The sequence is stable if

$$0 < d/4f < 1 \tag{76}$$

Q. LENSLIKE MEDIUM

A lenslike medium is a guiding medium with a graded refractive index $n(x,y,z)$ which changes quadratically with the distance $(x^2 + y^2)$ from the optic axis,

$$n = n_o(z) - \tfrac{1}{2}n_2(z)(x^2 + y^2) \tag{77}$$

This is schematically indicated in Fig. 22. An important example of such a medium is an optical fiber with a graded refractive index profile, such as the "selfoc" fiber. To trace rays in a lenslike medium one uses a differential ray equation which, for paraxial rays, has the form

$$n_o \frac{d^2 x}{dz^2} = -n_2 x \tag{78}$$

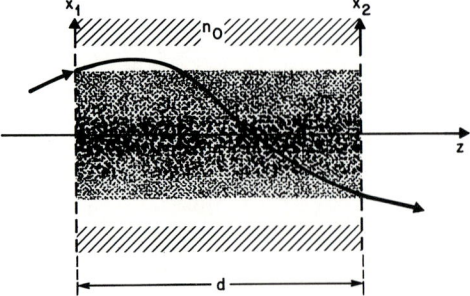

FIG. 22. Lenslike medium with optical density indicated by the density of dots.

This determines the ray position $x(z)$ from the z axis. A corresponding relation holds for $y(z)$.

For a lenslike medium of length d which is uniform along the axis $[n_2(z) = \text{const}]$ we have sinusoidal ray paths and a ray matrix given by

$$\mathbf{M} = \begin{bmatrix} \cos d(n_2/n_o)^{1/2} & (n_o n_2)^{-1/2} \sin d(n_2/n_o)^{1/2} \\ -(n_o n_2)^{1/2} \sin d(n_2/n_o)^{1/2} & \cos d(n_2/n_o)^{1/2} \end{bmatrix} \quad (79)$$

II. PROPAGATION OF GAUSSIAN LASER BEAMS

In this section the wave nature of laser beams is taken into account but diffraction effects due to the finite size of apertures are neglected. The latter will be discussed in Section III. The results derived here are applicable to optical systems with "large apertures," i.e., with apertures that intercept only a negligible portion of the beam power.

A. The Wave Equation

Laser beams are similar in many respects to plane waves; however, their intensity distributions are not uniform but are concentrated near the axis of propagation and their phase fronts are slightly curved. A field component or potential u of the coherent light satisfies the scalar wave equation

$$\nabla^2 u + k^2 u = 0 \quad (80)$$

where $k = 2\pi/\lambda$ is the propagation constant in the medium.

For light traveling in the z direction one writes

$$u = \psi(x,y,z) \exp(-jkz) \quad (81)$$

where ψ is a slowly varying complex function which represents the differences between a laser beam and a plane wave; namely, a nonuniform intensity distribution, expansion of the beam with distance of propagation, curvature of the phase front, and other differences discussed below. The function ψ obeys the parabolic wave equation

$$\frac{\partial^2 \psi}{\partial x^2} + \frac{\partial^2 \psi}{\partial y^2} - 2jk \frac{\partial \psi}{\partial z} = 0 \quad (82)$$

where it has been assumed that ψ varies so slowly with z that its second derivative $\partial^2 \psi/\partial z^2$ can be neglected.

B. THE GAUSSIAN LASER BEAM

As a solution to the parabolic wave equation, one obtains a laser beam with a Gaussian intensity profile. This is also called the fundamental TEM_{00} mode. The field u of the beam is given by

$$u(r,z) = \frac{w_0}{w} \exp\left\{-j(kz - \Phi) - r^2\left(\frac{1}{w^2} + \frac{jk}{2R}\right)\right\} \quad (83)$$

where

$$\Phi = \arctan(\lambda z / w_0^2) \quad (84)$$

$R(z)$ is the radius of curvature of the wave front that intersects the axis at z, and $w(z)$ is a measure of the decrease of the field amplitude u with the distance from the axis. This decrease is Gaussian in form, as indicated in Fig. 23, and w is the distance at which the amplitude is $1/e$ times that on the axis. The parameter w is often called the beam radius or "spot size" and $2w$ the beam diameter.

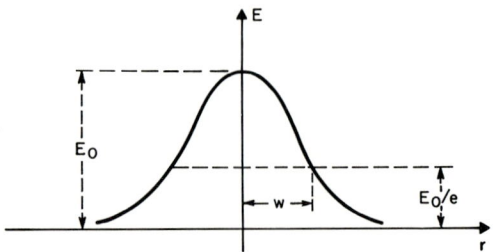

FIG. 23. Amplitude distribution in the cross section of a Gaussian laser beam. The definition of the "spot size" w is indicated.

The variation of the beam parameters w and R is described by

$$w^2(z) = w_0^2[1 + (\lambda z/\pi w_0^2)^2] \quad (85)$$

and

$$R(z) = z[1 + (\pi w_0^2/\lambda z)^2] \quad (86)$$

Figure 24 shows the expansion of the beam. The beam contour $w(z)$ is a hyperbola with asymptotes inclined to the axis at an angle

$$\theta = \lambda/\pi w_0 \quad (87)$$

This is the far-field diffraction angle of the fundamental mode.

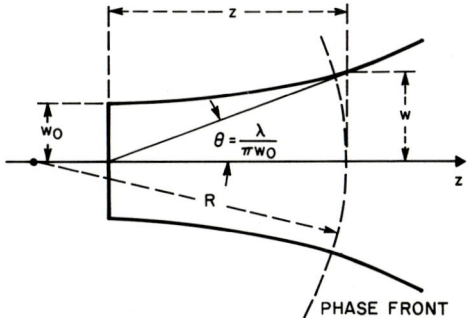

FIG. 24. Expansion of a Gaussian laser beam with a minimum beam diameter $2w_0$ at the waist.

The Gaussian beam contracts to a minimum diameter $2w_0$ at the *beam waist* where the phase front is plane. The distance z is measured from that plane.

The beam parameters w_0 and z can be expressed in terms of w and R by

$$w_0^2 = w^2 \bigg/ \left[1 + \left(\frac{\pi w^2}{\lambda R}\right)^2\right] \tag{88}$$

$$z = R \bigg/ \left[1 + \left(\frac{\lambda R}{\pi w^2}\right)^2\right] \tag{89}$$

It is often convenient to introduce a complex beam parameter q defined by

$$1/q = 1/R - j(\lambda/\pi w^2) \tag{90}$$

In terms of this parameter, the above propagation laws assume the simple form

$$q(z) = j(\pi w_0^2/\lambda) + z \tag{91}$$

An important parameter of a Gaussian laser beam is the confocal parameter b, which is defined as

$$b = 2\pi w_0^2/\lambda \tag{92}$$

b is equal to the minimum radius of phase-front curvature that occurs in a given beam. The distance of this phase front from the beam waist is $b/2$ and the spot size at that location is $w_0\sqrt{2}$.

A coherent light beam with a Gaussian intensity profile is not the only solution of the parabolic wave equation, but this fundamental mode solution is the most important one. The properties of higher order modes are discussed in Kogelnik and Li (1966).

C. Beam Transformation by a Lens

A lens can be used to focus a laser beam to a small spot or to produce a beam of suitable diameter and phase-front curvature for injection into a given optical structure. An ideal lens leaves the transverse field distribution of a beam mode unchanged, i.e., an incoming fundamental Gaussian beam will emerge from the lens as a fundamental beam and a higher order mode remains a mode of the same order after passing through the lens. However, a lens does change the beam parameters $R(z)$ and $w(z)$. The relationship between the parameters of an incoming beam (labeled here with the index 1) and the parameters of the corresponding outgoing beam (index 2) are given below.

An ideal thin lens of focal length f transforms an incoming spherical wave with a radius R_1 immediately to the left of the lens into a spherical wave with the radius R_2 immediately to the right of it, where

$$1/R_2 = 1/R_1 - 1/f \tag{93}$$

Figure 25 illustrates this situation. The radius of curvature is taken to be positive if the wave front is convex as viewed from $z = \infty$. The lens transforms the phase fronts of laser beams in exactly the same way as those of spherical waves. As the diameter of a beam is the same immediately to the left and to the right of a thin lens, the q parameters of the incoming and outgoing beams are related by

$$1/q_2 = 1/q_1 - 1/f \tag{94}$$

where the q's are measured at the lens. If q_1 and q_2 are measured at distances d_1 and d_2 from the lens as indicated in Fig. 26, the relation between them becomes

$$q_2 = \frac{(1 - d_2/f)q_1 + (d_1 + d_2 - d_1 d_2/f)}{-(q_1/f) + (1 - d_1/f)} \tag{95}$$

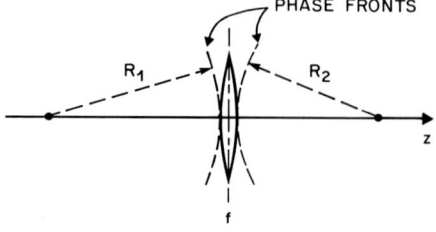

Fig. 25. Transformation of spherical wave fronts by a thin lens.

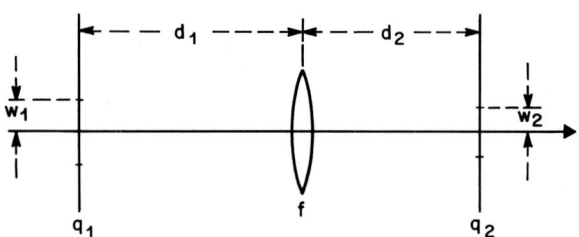

FIG. 26. Distances and parameters of a Gaussian laser beam transformed by a thin lens.

D. THE *ABCD* LAW

The complex beam parameters q_1 and q_2 of a Gaussian laser beam at the input and output of a general optical system described by the matrix elements *A*, *B*, *C*, and *D* of its ray transfer matrix (see Section I) are related in the simple form

$$q_2 = (Aq_1 + B)/(Cq_1 + D) \tag{96}$$

This is called the *ABCD* law. It allows the calculation of the parameters of laser beams propagating through any of the structures discussed in Section I from the matrix elements listed there. For more complicated systems the matrix elements are obtained by matrix multiplication.

E. MODE MATCHING

The modes of laser resonators can be characterized by Gaussian beams with certain properties and parameters which are defined by the resonator geometry. These beams are often injected into other optical structures with different sets of beam parameters. These optical structures can assume various physical forms, such as resonators used in scanning Fabry–Perot interferometers or regenerative amplifiers, or crystals of nonlinear dielectric material employed in parameteric optics experiments. To match the modes of one structure to those of another one must transform a given Gaussian beam (or higher order mode) into another beam with prescribed properties. This transformation is usually accomplished with a thin lens, but other more complex optical systems can be used. Although the present discussion is devoted to the simple case of the thin lens, it is also applicable to more complex systems, provided one measures the distances from the principal planes and uses the combined focal length *f* of the more complex system.

The location of the waists of the two beams to be transformed into each other and the beam diameters at the waists are usually known or can be computed. To match the beams one has to choose a lens of a focal length *f* that is larger than a characteristic length f_0 defined by the two beams and one has to adjust the distances between the lens and the two beam waists according to rules derived below.

In Fig. 26 the two beam waists are assumed to be located at distances d_1 and d_2 from the lens. The complex beam parameters at the waists are purely imaginary; they are

$$q_1 = j\pi w_1^2/\lambda, \qquad q_2 = j\pi w_2^2/\lambda \tag{97}$$

where $2w_1$ and $2w_2$ are the diameters of the two beams at their waists. The following relations hold

$$(d_1 - f)/(d_2 - f) = w_1^2/w_2^2 \tag{98}$$

$$(d_1 - f)(d_2 - f) = f^2 - f_0^2 \tag{99}$$

where

$$f_0 = \pi w_1 w_2/\lambda \tag{100}$$

Note that the characteristic length f_0 is defined by the waist diameters of the beams to be matched. Except for the term f_0^2, which goes to zero for infinitely small wavelengths, (99) resembles Newton's imaging formula of geometrical optics.

Any lens with a focal length $f > f_0$ can be used to perform the matching transformation. Once f is chosen, the distances d_1 and d_2 have to be adjusted to satisfy the matching formulas

$$d_1 = f \pm \frac{w_1}{w_2}(f^2 - f_0^2)^{1/2}, \qquad d_2 = f \pm \frac{w_2}{w_1}(f^2 - f_0^2)^{1/2} \tag{101}$$

Here one can choose either both plus signs or both minus signs for matching.

It is often useful to introduce the confocal parameters b_1 and b_2 into the matching formulas. They are defined by the waist diameters of the two systems to be matched

$$b_1 = 2\pi w_1^2/\lambda, \qquad b_2 = 2\pi w_2^2/\lambda \tag{102}$$

Using these parameters one gets for the characteristic length f_0

$$f_0^2 = \tfrac{1}{4}b_1 b_2, \tag{103}$$

and for the matching distances

$$\begin{aligned} d_1 &= f \pm \tfrac{1}{2}b_1[(f^2/f_0^2) - 1]^{1/2} \\ d_2 &= f \pm \tfrac{1}{2}b_2[(f^2/f_0^2) - 1]^{1/2} \end{aligned} \tag{104}$$

Note that in this form of the matching formulas, the wavelength does not appear explicitly.

Table I, reproduced from Kogelnik and Li (1966), lists for quick reference, formulas for the two important parameters of beams that emerge from various optical structures commonly encountered. They are the confocal parameter b and the distance t which gives the waist location of the emerging beam.

TABLE I
FORMULAS FOR THE CONFOCAL PARAMETER AND THE LOCATION
OF THE BEAM WAIST FOR A SELECTION OF OPTICAL STRUCTURES[a]

NO	OPTICAL SYSTEM	$\frac{1}{2}b = \pi w_0^2/\lambda$	t
1		$\sqrt{d(R-d)}$	—
2		$\frac{1}{2}\sqrt{d(2R-d)}$	$\frac{1}{2}d$
3		$\dfrac{\sqrt{d(R_1-d)(R_2-d)(R_1+R_2)-d}}{R_1+R_2-2d}$	$\dfrac{d(R_2-d)}{R_1+R_2-2d}$
4		$\dfrac{R\sqrt{d(2R-d)}}{2R+d(n^2-1)}$	$\dfrac{ndR}{2R+d(n^2-1)}$
5		$\frac{1}{2}\sqrt{d(4f-d)}$	$\frac{1}{2}d$
6		$\frac{1}{2}d$	$\frac{1}{2}d$
7		$\dfrac{d}{2n}$	$\dfrac{d}{2n}$
8		$\dfrac{nR\sqrt{d(2R-d)}}{2n^2R-d(n^2-1)}$	$\dfrac{dR}{2n^2R-d(n^2-1)}$

[a] From Kogelnik and Li (1966).

System No. 1 is a resonator formed by a flat mirror and a spherical mirror of radius R. System No. 2 is a resonator formed by two equal spherical mirrors. System No. 3 is a resonator formed by mirrors of unequal curvature. System No. 4 is, again, a resonator formed by two equal spherical mirrors, but with the reflecting surfaces deposited on planoconcave optical plates of index n. These plates act as negative lenses and change the characteristics of the emerging beam. This lens effect is assumed not present in systems Nos. 2 and 3. System No. 5 is a sequence of thin lenses of equal focal lengths f. System No. 6 is a system of two irises with equal apertures spaced at a distance d. Shown are the parameters of a beam that will pass through both irises with the least possible beam diameter. This is a beam which is "confocal" over the distance d. This beam will also pass through a tube of length d with the optimum clearance. (The tube is also indicated in the figure.) A similar situation is shown in system No. 7, which corresponds to a beam that is confocal over the length d of optical material of index n. System No. 8 is a spherical mirror resonator filled with material of index n or an optical material with curved end surfaces where the beam passing through it is assumed to have phase fronts that coincide with these surfaces.

When one designs a matching system, it is useful to know the accuracy required of the distance adjustment. The discussion below indicates how the parameters b_2 and d_2 change when b_1 and f are fixed and the lens spacing d_1 to the waist of the input beam is varied. One has (Kogelnik and Li, 1966)

$$b_2/f = \frac{b_1/f}{(1 - d_1/f)^2 + (b_1/2f)^2} \tag{105}$$

This means that the parameter b_2 of the beam emerging from the lens changes with d_1 according to a Lorentzian functional form as shown in Fig. 27. The

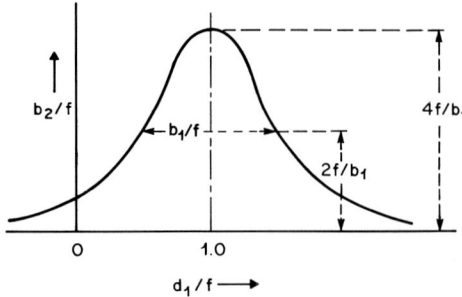

FIG. 27. The confocal parameter b_2 of the output laser beam as function of the lens-waist spacing d_1 at the input.

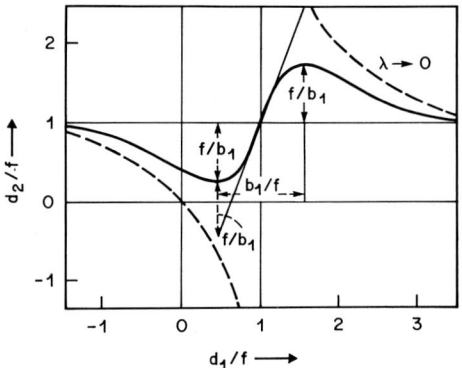

FIG. 28. The lens-waist spacing d_2 of the output laser beam as function of the lens-waist spacing d_1 at the input.

Lorentzian is centered at $d_1 = f$ and has a width of b_1. The maximum value of b_2 is $4f^2/b_1$. The relation

$$1 - d_2/f = \frac{1 - d_1/f}{(1 - d_1/f)^2 + (b_1/2f)^2} \tag{106}$$

shows the change of d_2 with d_1. The change is reminiscent of a dispersion curve associated with a Lorentizian as shown in Fig. 28. The extrema of this curve occurs at the half-power points of the Lorentzian. The slope of the curve at $d_1 = f$ is $(2f/b_1)^2$. The dashed curves in the figure correspond to the geometrical optics imaging relation between d_1, d_2, and f.

F. Circle Diagrams

The propagation of Gaussian laser beams can be represented graphically on a circle diagram. On such a diagram one can follow a beam as it propagates in free space or passes through lenses, thereby affording a graphic solution of the mode matching problem. The circle diagrams for beams are similar to the impedance charts, such as the Smith chart. In fact, there is a close analogy between transmission-line and laser-beam problems and there are analog electric networks for every optical system.

The first circle diagram for beams was proposed by Collins. A dual chart was discussed by Li. The basis for the derivation of these charts are the beam propagation laws discussed in Section II.B. The four quantities w, R, w_0, and z which are used to describe the propagation of Gaussian beams can be

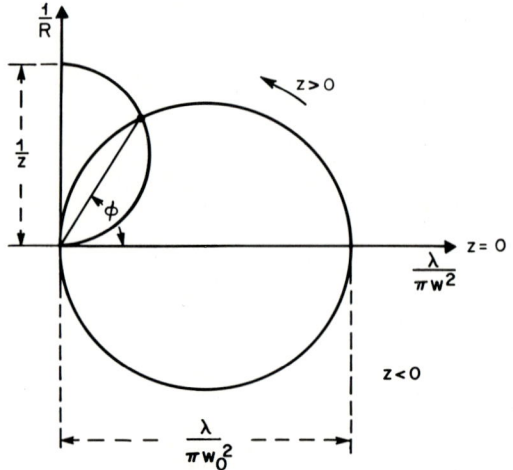

FIG. 29. Geometry of the circle diagram in the W plane.

expressed in complex variables W and Z:

$$W = \frac{\lambda}{\pi w^2} - j\frac{1}{R}, \qquad Z = \frac{\pi w_0^2}{\lambda} - jz = \frac{b}{2} - jz \qquad (107)$$

where b is the confocal parameter of the beam. For these variables one has the conformal transformation

$$W = 1/Z \qquad (108)$$

The two dual circle diagrams are plotted in the complex planes of W and Z, respectively. The W-plane diagram is shown in Fig. 29 where the variables $\lambda/\pi w^2$ and $1/R$ are plotted as axes. In this plane the lines of constant $b/2 = \pi w_0^2/\lambda$ and the lines of constant z of the Z plane appear as circles through the origin. A beam is represented by a circle of constant b and the beam parameters w and R at a distance z from the beam waist can be easily read from the diagram. When the beam passes through a lens the phase front is changed according to (93) and a new beam is formed, which implies that the incoming and outgoing beams are connected in the diagram by a vertical line of length $1/f$. The angle Φ shown in the figure is equal to the phase shift experienced by the beam as given by (84).

The dual diagram is plotted in the Z plane. The sets of circles in both diagrams have the same form and only the labeling of the axes and circles is different. In Fig. 30 both diagrams are unified in one chart. The labels in parentheses correspond to the z-plane diagram and b is a normalizing parameter which can be arbitrarily chosen for convenience.

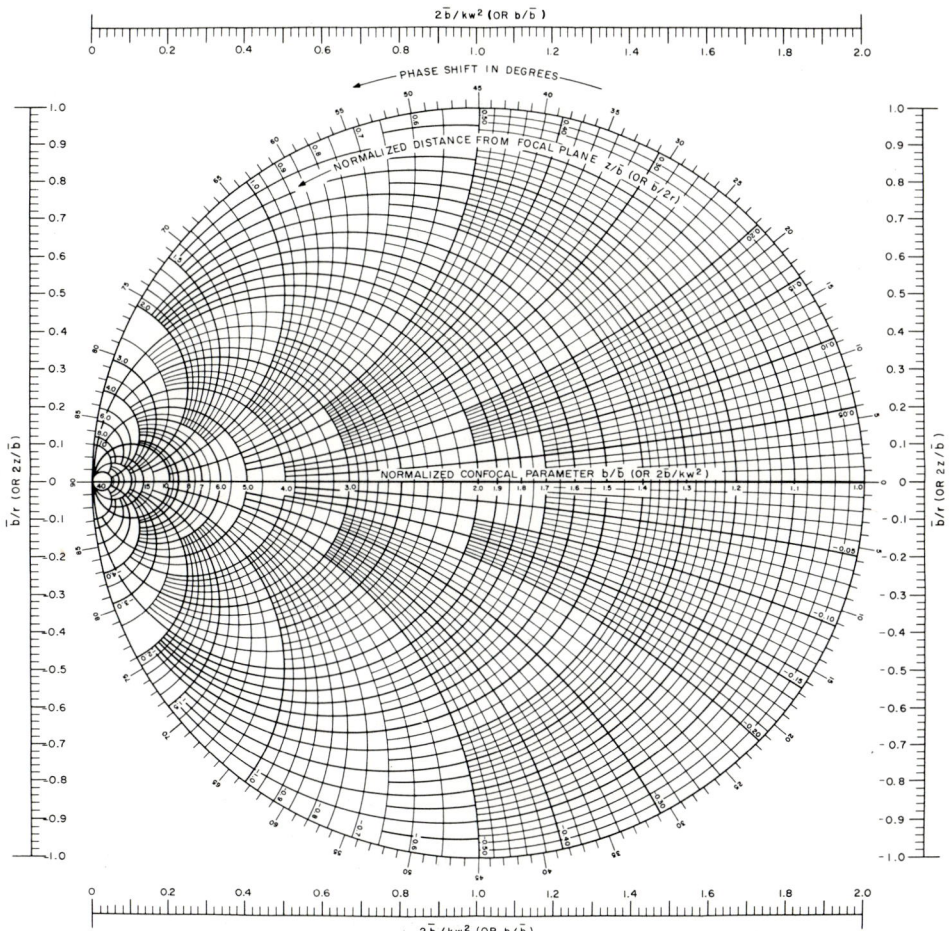

FIG. 30. Circle diagram for the graphical analysis of laser beam propagation. Both the W-plane and the Z-plane diagrams are combined into one. (From Kogelnik and Li, 1966.)

III. FRESNEL DIFFRACTION

The analysis of laser-beam propagation in terms of paraxial rays or Gaussian beams does not take account of diffraction effects due to the finite size of an aperture. These two methods neglect effects such as the formation of diffraction rings when a laser beam passes through a small aperture. The basis for an analysis of aperture diffraction effects is the Fresnel diffraction

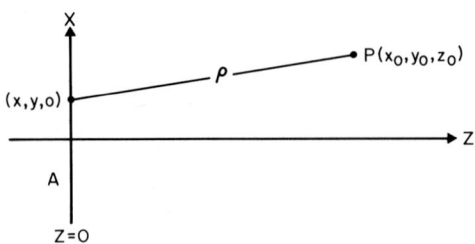

FIG. 31. Geometry for the analysis of Fresnel diffraction. The plane of the aperture is at $z = 0$, the observation point is P.

formula (Goodman, 1968; Kogelnik, 1966)

$$E(x_o, y_o, z_o) = \frac{jk \exp(-jkz_o)}{2\pi z_o} \int_A dA\, E(x, y, 0)$$
$$\cdot \exp\left\{-j\frac{k}{2z_o}[(x - x_o)^2 + (y - y_o)^2]\right\} \quad (109)$$

This formula relates the complex amplitude of the field $E(x_o, y_o, z_o)$ at an observation point P with the coordinates (x_o, y_o, z_o) to the complex amplitude of a given aperture field $E(x, y, z)$ in the plane $z = 0$, as sketched in Fig. 31. Here the region of integration is the whole aperture area A, x and y are coordinates in the aperture plane, and

$$k = 2\pi/\lambda \quad (110)$$

is the propagation constant of free space. The Fresnel diffraction formula is valid under the following conditions

(a) The aperture should be large compared to the wavelength λ and spatial variations in the aperture field should be smooth.

(b) The observation distance z_o should be large compared to the aperture opening.

(c) It is assumed that all electromagnetic fields of interest can be regarded as scalar fields.

An example of the use of the Fresnel diffraction formula to laser beams is its application to the analysis of laser resonators (Fox and Li, 1961; Kogelnik and Li, 1966).

A. RECTANGULAR SYMMETRY

For rectangular symmetry we have a square or rectangular aperture as well as aperture and observation fields that can be written as products of functions of one variable,

$$E(x, y, 0) = u(x) \cdot v(y), \qquad E(x_o, y_o, z_o) = u_o(x_o) \cdot v_o(y_o) \quad (111)$$

For this case the diffraction formula assumes the simplified form

$$u_0(x_o) = \left[\frac{jk\exp(-jkz_o)}{2\pi z_o}\right]^{1/2} \int_{-a}^{a} dx\, u(x)\exp\left[-j\frac{k}{2z_o}(x-x_o)^2\right] \quad (112)$$

$$v_0(y_o) = \left[\frac{jk\exp(-jkz_o)}{2\pi z_o}\right]^{1/2} \int_{-b}^{b} dy\, v(y)\exp\left[-j\frac{k}{2z_o}(y-y_o)^2\right] \quad (113)$$

where a and b indicate the aperture dimensions.

B. Circular Symmetry

When the aperture is circular and the aperture and observation fields are of circular symmetry, i.e.,

$$E(x,y,0) = w(r), \qquad E(x_o,y_o,z_o) = w_0(r_o) \quad (114)$$

where $r^2 = x^2 + y^2$, the diffraction formula takes the form

$$w_0(r_o) = \frac{jk\exp(-jkz_o)}{z_o}\int_0^{\rho} r\,dr\, w(r) J_0(krr_o)\exp\left[-j\frac{k}{2z_o}(r^2+r_o^2)\right] \quad (115)$$

Here J_0 is the zero order Bessel function, and ρ the aperture radius.

C. Fraunhofer Diffraction

Fraunhofer diffraction is a special case of Fresnel diffraction which occurs when the observation point is very far away from the aperture such that $x/z_o \ll 1$ and $y/z_o \ll 1$. The Fraunhofer diffraction formula (Goodman, 1968) is, thus, a special case of the Fresnel diffraction formula and has the form

$$E(x_0,y_0,z_o) = \frac{jk}{2\pi z_o}\exp(-jkz_o)\exp\left[\frac{-jk}{2z_o}(x_o^2+y_o^2)\right]$$
$$\cdot \int_A dA\, E(x,y,0)\exp\left[j\frac{k}{z_o}(xx_o+yy_o)\right] \quad (116)$$

It is used to determine the far field of a given aperture field distribution.

For rectangular symmetry the Fraunhofer diffraction formula assumes the form of a Fourier transform

$$u_0(x_o) = \left[\frac{jk\exp(-jkz_o)}{2\pi z_o}\right]^{1/2}\exp\left(-j\frac{kx_o^2}{2z_o}\right)\int_{-a}^{a} dx\, u(x)\exp\left(\frac{jkxx_o}{z_o}\right) \quad (117)$$

For circular symmetry the Fraunhofer diffraction formula assumes the form of a Fourier–Bessel or Hankel transform,

$$w_0(r_o) = \frac{jk}{z_o}\exp\left[-jkz_o\left(1+\frac{r_o^2}{2z_o^2}\right)\right]\int_0^{\rho} r\,dr\, w(r) J_0\left(\frac{krr_o}{z_o}\right) \quad (118)$$

D. Far Fields for Special Cases

For the special case of a rectangular aperture of width $2a$ with a uniform aperture field $u(x) = 1$ the Fraunhofer formula predicts a far field given by

$$u_0(x_o) = \left[\frac{jk\exp(-jkz_o)}{2\pi z_o}\right]^{1/2} \exp\left(\frac{-jkx_o^2}{2z_o}\right) \frac{2z_o}{kx_o} \sin\left(\frac{kax_o}{z_o}\right) \quad (119)$$

For the case of a circular aperture with uniform aperture field $w(r) = 1$ the formula predicts for the far field the Airy pattern given by

$$w_0(r_o) = j\exp[-jkz_o(1 + r_o^2/2z_o^2)](\rho/r_o)J_1(k\rho r_o/z_o) \quad (120)$$

Far field diffraction patterns of more complicated sector apertures are given in Asakura (1970).

For a Gaussian aperture distribution of effective width $2w$ and

$$E(x,y,0) = \exp[-(x^2 + y^2)/w^2] \quad (121)$$

which is not truncated by any other aperture, the far field is given by another Gaussian

$$E(x_o, y_o, z_o) = j\exp(-jkz_o)\exp\left[-j\frac{k}{2z_o}(x_o^2 + y_o^2)\right]$$
$$\cdot \frac{kw^2}{2z_o}\exp\left[-\left(\frac{kwr_0}{2z_o}\right)^2\right] \quad (122)$$

The far field patterns of a Gaussian aperture field truncated by circular apertures is given in Kauffman (1965), Buck (1967), and Bloom (1968) for several aperture diameters. Campbell and DeShazer (1969) and Schell and Tyras (1971) discuss, in addition, the Fresnel fields of truncated Gaussian aperture fields.

E. Effects of a Lens on Fresnel Fields

The effect of a lens on the Fresnel field of a laser beam passing through it is that of a simple transformation, as long as the lens is of sufficiently large diameter so that no significant truncation occurs.

A thin lens of focal length f adds a phase shift that increases quadratically with the distance from the optic axis (Kogelnik, 1965)

$$E_r(x,y) = E_1(x,y)\exp[jk(x^2 + y^2)/2f] \quad (123)$$

where it is assumed that the laser beam travels from the left to the right and E_1 is the complex amplitude of the field immediately to the left of the lens, and E_r the complex immediately to the right of the lens.

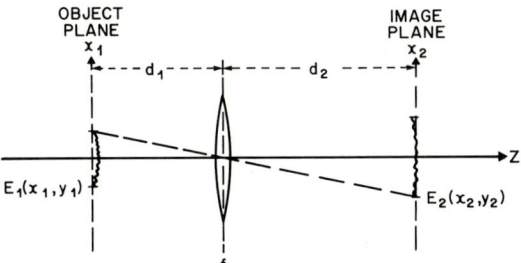

FIG. 32. Imaging of a given aperture field by a lens.

For distances d_1 and d_2 from the lens as sketched in Fig. 32 and related by the imaging formula

$$1/d_1 + 1/d_2 = 1/f \qquad (124)$$

the Fresnel formula predicts an image field E_2 related to the (complex) object field amplitude E_1 by (Kogelnik, 1965)

$$E_2(x_2,y_2) = -\frac{d_1}{d_2} E_1\left(-\frac{d_1}{d_2} x_2, -\frac{d_1}{d_2} y_2\right) \exp\left[-jk\left(d_1 + d_2 + \frac{r_2^2}{2f}\frac{d_1}{d_2}\right)\right] \qquad (125)$$

where $r_2^2 = x_2^2 + y_2^2$. The quadratic phase factor appearing in this formula indicates that a plane wave front in the object plane is transformed into a spherical wavefront in the image plane.

In the general case one has a laser beam with a spherical wave front in the object plane, say of radius R_1. This is sketched in Fig. 33. Equation (125) predicts for this case a beam in the image plane with radius of curvature R_2 given by (Kogelnik, 1965)

$$\frac{1}{d_1 + R_1} + \frac{1}{d_2 - R_2} = \frac{1}{f} \qquad (126)$$

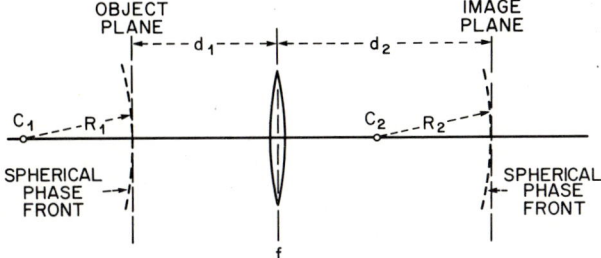

FIG. 33. The centers of curvature of the phase fronts of the object and image fields obey an object–image relation.

This means that even for the case of Fresnel diffraction, the centers of curvature of the image and object fields obey an image–object relationship as expected from geometric optics arguments. This relationship is sketched in Fig. 33.

REFERENCES

Asakura, T. (1970). *Jpn. J. Appl. Phys.* **9**, 652.
Bloom, A. L. (1968). "Gas Lasers." Wiley, New York.
Born, M., and Wolf, E. (1959). "Principles of Optics." Pergamon, New York.
Brower, W. (1964). "Matrix Methods in Optical Instrument Design." Benjamin, New York.
Buck, A. L. (1967). *Proc. IEEE* **55**, 450.
Campbell, J. P., and DeShazer, L. G. (1969). *J. Opt. Soc. Am.* **59**, 1427.
Fox, A. G., and Li, T. (1961). *Bell Syst. Tech. J.* **40**, 453.
Gerrard, A., and Burch, J. M. (1975). "Introduction to Matrix Methods in Optics." Wiley, New York.
Goodman, J. W. (1968). "Introduction to Fourier Optics." McGraw-Hill, New York.
Hanna, D. C. (1969). *IEEE J. Quantum Electron.* **QE-5**, 483.
Kauffman, J. F. (1965). *IEEE Trans. Antennas Propag.* **AP-13**, 473.
Kogelnik, H. (1965). *Bell Syst. Tech. J.* **44**, 455.
Kogelnik, H. (1966). *In* "Lasers" (A. K. Levine, ed.). Vol. 1. pp. 295–347. Dekker. New York.
Kogelnik, H., and Li, T. (1966). *Appl. Opt.* **5**, 1550.
Kogelnik, H., Ippen, E. P., Dienes, A., and Shank, C. V. (1972). *IEEE J. Quantum Electron.* **QE-8**, 373.
O'Neill, E. L. (1963). "Introduction to Statistical Optics." Addison-Wesley, Reading, Massachusetts.
Schell, R. G., and Tyras, G. (1971). *J. Opt. Soc. Am.* **61**, 31.

CHAPTER 7

Scattering from Optical Surfaces

J. M. ELSON, H. E. BENNETT, and J. M. BENNETT

*Michelson Laboratories, Naval Weapons Center
China Lake, California*

I. Introduction	191
II. Methods for Calculating Light Scattering	193
A. Geometrical Optics	194
B. Scalar Theory	194
C. Vector Theory	196
D. Particulate and Resonance Scattering	197
E. Numerical Methods	198
F. Comparison of A–E Methods	198
III. Total Integrated Scatter	199
A. Microirregularity Scattering	200
B. Surface Plasmon Excitation	206
C. Scattering in the Infrared	207
D. Scratches and Other Macroscopic Defects: The Scratch/Dig Specification	208
E. Particulate Scattering from Isolated Microirregularities	214
IV. Angular Dependence of Scattering	216
A. Angular Scattering from a Sinusoidal Grating	218
B. Angular Scattering from Nonsinusoidal Surfaces	221
C. Angular Scattering Theory for Polished Surfaces	222
D. Determination of the Spectral Density Function	226
E. Experimental Measurements of Angular Scattering and Surface Statistics	228
V. Scattering from Dielectric Multilayers	236
A. Polarization Effects in Low Efficiency Gratings	237
B. Angular Scattering from Surfaces Having Random Roughness	239
VI. Summary	241
References	242

I. INTRODUCTION

The presence of stray light is a continuing problem in the design and performance of optical systems. If the systems are well designed most of it comes from the optical components themselves. By proper baffling, it can be reduced in sensitive areas by orders of magnitude, but ultimately it is necessary to improve the components themselves. Most of the light is scattered by the component surfaces. Bulk scattering can also occur in windows or lenses, but it is typically one to two orders of magnitude less than surface scattering (Kozawa, 1962).

Surface scattering may arise from (1) irregularities such as scratches, digs, or particulates which are large relative to the wavelength of the incident light, (2) isolated irregularities which are comparable to or smaller than the wavelength of incident light, and (3) irregularities which are small in one or more dimensions relative to the wavelength but which are so closely spaced that they cannot be treated as independent scattering centers. The effect of each center is correlated with that of its neighbors; scattering from such ensembles is often called "microirregularity scattering."

The appropriate theoretical treatment of scattering depends on which of the above categories the scattering centers fall into. Historically, scattering from scratches and other macroscopic defects was probably recognized first. It is probably the easiest of the three to visualize and can be handled by using geometrical optics. Oddly enough, it is the most difficult one to handle quantitatively since the shapes of the macroirregularities must often be known in great detail. When the irregularities become comparable to or smaller than the wavelength of the incident light, the scattering problem becomes a diffraction problem, and geometrical optics is no longer adequate. A great simplification occurs, however, if the scattering centers can be considered to act independently. Mie scattering theory, of which Rayleigh scattering is a special case, can then be used. Historically, these theories were initially worked out near the turn of the century, although modifications and improvements are continually being made. Latest to be developed is the correlated scatterers theory. The first hint of such a theory of which we are aware was given by Chinmayanandam (1919). However, most of the work in this area has been done in the past 20 years and it is still continuing.

In the visible and ultraviolet regions of the spectrum, microirregularities are the principal source of scattered light for most optically polished surfaces. The average height of these irregularities is only a few nanometers and low scatter surfaces for these wavelength regions are often specified in terms of their rms roughness. Typical glass optical flats have rms roughnesses of 25 to 30 Å; good polished metal surfaces typically run from 30 to 50 or 60 Å; and, any surface under 15 Å rms is often called "superpolished." Roughnesses as low as 5 Å rms have been achieved; since the scatter goes as the square of the roughness, these surfaces scatter 25–35 times less than a well-figured conventional glass optical flat.

Microirregularity scattering drops exponentially with wavelength and in the near infrared scattering from surface blemishes, scratches, dust, etc., typically becomes the limiting factor. It is often nearly independent of wavelength. Pitting of the surface by sand or rain erosion increases the importance of defect scattering and it may become the dominant effect even in the visible region. Finally, scattering from isolated particulates such as dust must be considered. Usually dust is not nearly as much of a problem on optics as the damage which is frequently done to the surface in trying to remove it. It does affect the optical performance of low scatter components even when used in the visible and ultraviolet regions, however, and since dust particles have diameters of about 1 μm, it becomes particularly important in the near infrared.

In this chapter we will discuss the general methods which have been used to calculate light scattering. These include, principally, the scalar scattering theory and vector scattering theory, both of which treat scattering from correlated surface microirregularities. Mie theory deals with scattering from isolated surface defects such as dust or other particulates and can explain the experimentally observed increase in total integrated scattering (TIS) near a wavelength of 1 μm. A detailed discussion of TIS includes microirregularity scattering, surface plasmon excitation, scattering in the infrared, and scattering from scratches and other defects. The section on angular dependence of scattering deals primarily with scattering from correlated surface microirregularities. In the simplest case one can consider angular scattering from a sinusoidal grating, since a randomly rough surface can be considered to be composed of a two-dimensional superposition of sinusoidal gratings having different amplitudes, periods, and phases. The complete vector scattering theory will be discussed and it will be shown how the theory can be modified to use measured surface statistics to calculate the angular dependence of scattering. Scattering from dielectric multilayers with rough interfaces is a considerably more difficult problem. Polarization effects can cause large differences in the scattered light depending on whether or not the surfaces of the layers replicate the roughness of the substrate. Angular scattering curves for surfaces with replicating and nonreplicating roughness will be presented. The conclusion will summarize the present status of scattering from optical surfaces both from a theoretical and an experimental standpoint.

II. METHODS FOR CALCULATING LIGHT SCATTERING

Light scattering calculations have been handled using various approaches. These include geometrical optics, scalar theory, vector theory, particulate scattering from polarizable particles (Mie theory), and numerical calculational methods. In this section different approaches will be briefly discussed and then compared.

A. GEOMETRICAL OPTICS

Geometrical optics, or ray optics, can be applied to light scattering when the dimensions of the surface roughness are large compared to the wavelength of light. A typical geometrical optics approach assumes that the scattering surface consists of plane, flat facets having lateral dimensions that are large relative to the wavelength and which reflect like plane mirrors. Each facet has a surface normal and there will be a statistical distribution of the directions of the surface normals. Thus, scattering from a geometrical optics type of rough surface consists of adding up the contributions from each tilted facet of the surface. In order to actually treat a surface using this approach, one needs to have a characterization of the surface: either a model of the shapes and distribution of surface facets or an equivalent statistical characterization. A statistical model of a facet surface generated by a Markov chain has been given by Beckmann and Spizzichino (1963, see especially pp. 109–114). However, surface characterizations of geometrical optics-type surfaces are not easy to obtain. Furthermore, geometrical optics fails to describe the effects of interference, diffraction, and polarization. Fortunately, these phenomena are included in scalar and vector scattering theories, and scalar scattering theory can be extended to the geometrical optics limit to describe scattering from surfaces whose roughness is much larger than a wavelength of light. For this reason, geometrical optics calculations are rarely used in treating optical scattering problems.

B. SCALAR THEORY

The starting point of scalar theory treating scattering from rough surfaces is the Helmholtz–Kirchhoff diffraction integral.* This integral, and the resulting diffraction formula, is based on the fact that the solution to the wave equation at some point in space surrounded by an arbitrary closed surface (a mathematical construction) may be obtained if the solution is known at all points on the surface. When applied to scattering from rough surfaces, strictly speaking, the fields must be known at the rough surface itself. However, these fields are not generally known and to overcome this problem certain approximations, known as the Kirchhoff boundary conditions (Jackson, 1962, p. 282), are made. These approximations limit the validity of the scalar theory to scattering near the specular direction; polarization properties of the scattered light are not included. The major problem in scattering theory is to find the relationship between the statistics of the

---

* Several standard books on electromagnetic theory discuss this topic. See, for example, Jackson (1962).

scattering surface, those of the scattered wave front leaving the surface, and the far-field statistics of the scattered light. Surfaces whose roughness is much less than the wavelength are sometimes termed "weak scatterers." They produce phase variations in the near field of less than 2π. On the other hand, surfaces whose rms roughness is greater than a wavelength ("strong scatterers") produce phase variations in the near field much greater than 2π. Different types of statistical treatments are necessary for the two types of scatterers. An excellent review article on this subject has been written by Welford (1977) and the book by Beckmann and Spizzichino (1963) is also a primary reference.

Scalar theories have been applied to surfaces whose roughness is much less than the wavelength of light (Davies, 1954; Bennett and Porteus, 1961; Porteus, 1963). Davies (1954) has given a relation predicting the angular dependence of scattered light from rough surfaces at angles near the specular direction. He assumed that the surfaces were perfectly conducting, i.e., that the specular reflectance of a perfectly smooth surface of the material or the total reflectance of a rough surface of the same material, would be 100%. The surfaces were also assumed to have Gaussian height distribution functions and Gaussian autocovariance functions. Bennett and Porteus (1961) modified Davies' relation to include the actual reflectance of the material and then experimentally verified that the relations were valid for predicting the specular reflectance of rough surfaces when the roughness was much smaller than the wavelength. Porteus (1963) expanded the relation of Bennett and Porteus to include arbitrary height and autocovariance functions. The restriction that the roughness be much smaller than the wavelength was also removed.

Scalar theories based on the Helmholtz–Kirchhoff diffraction integral have also been applied to surfaces whose roughness is much greater than the wavelength of light (Davies, 1954; Chandley and Welford, 1975; Leader and Fung, 1977; Holzer and Sung, 1976, 1977; Chandley, 1976). Chandley (1976), Eastman and Baumeister (1974), and Leader (1971a) have measured surface statistics and angular scattering from rough surfaces, and have compared these measurements with predictions from theory. In the limit when the dimensions of the surface facets become very large relative to the wavelength, there is a transition between the scalar diffraction theory and geometrical optics. Hagfors (1966) has shown that when a rough surface is assumed to have a Gaussian autocovariance function, the limit of scalar diffraction theory yields the correct geometrical optics scattering result. On the other hand, Fung and Moore (1966) showed that when an exponential autocovariance function is assumed for the surface, the limit of the scalar diffraction theory is in disagreement with geometrical optics. This latter result is reasonable since an exponential autocovariance function has an unphysical discontinuous nonzero slope at the origin, which implies that the rms slope of such a surface would be infinite. Any real surface that approximated this condition would have a very jagged surface profile and would not appear facet- or mirror-like in the geometrical optics regime.

C. Vector Theory

To include the vector nature of the scattered field requires somewhat different techniques. Two dominant methods are considered here. The first is analogous to the Helmholtz–Kirchhoff scalar integral and is the vector equivalent or Stratton–Chu–Silver (SCS) integral (Silver, 1947). The other is a perturbation method, of which there are several variations. The SCS integral was originally applied to diffraction problems and is based on the calculation of radiation from a distribution of currents over a surface. The integral fully contains the vector nature of the currents (based on the surface fields) and yields the radiation from the distribution of surface currents. In principle, the integral can yield the scattering from any magnitude of surface roughness or shape, but in practice the results are often limited to roughnesses that are small compared to the wavelength. One major reason for this limitation is that the actual fields at the rough surface are not known (as is necessary to properly evaluate the integral) and, consequently, approximations are used. A number of authors have utilized the SCS integral as a starting point to calculate the scattering from rough surfaces, including Leader (1971b) and Fung and Chan (1969).

The perturbation technique, which was first used by Rayleigh for acoustical scattering (Rayleigh, 1945), depends on the surface roughness having a weak influence on the perfectly smooth situation. Under these assumptions it is justifiable to let the surface fields be approximated by the smooth surface fields. This assumption is generally made when the rms roughness δ is much less than the wavelength λ, so that only a small fraction of the total incident power is scattered out of the specular beam. Under these conditions the theory is termed first order because scattered field terms proportional to $(\delta/\lambda)^2$ or higher are dropped. Since the scattered fields are retained to an accuracy of δ/λ, then the boundary conditions need to be of similar accuracy, and the scattered power (Poynting vector) will be proportional to $(\delta/\lambda)^2$. In principle, perturbation theory may be used to calculate successively higher-order solutions, each of which are corrections to the zero-order or unperturbed solution. Normally, solutions beyond the first order are quite complicated.

There have been several variations of perturbation techniques. These include a plane wave expansion method by Peake (1959), a Dirac δ-function current model by Kröger and Kretschmann (1970), and Maradudin and

Mills (1975), and a coordinate transformation method by Elson and Ritchie (1974) and Elson (1975). Done properly, all these variations lead to the same results. Quantum mechanical perturbation methods are given and discussed by Elson and Ritchie (1971) and Celli *et al.* (1975).

D. Particulate and Resonance Scattering

Another possible contribution to scattering arises from surface contaminants such as dust particles or particulates. In the simplest case, the particulates may be approximated as dielectric spheres with no interaction between the particulates and the surface on which they rest. Mie (1908; see also Stratton, 1941; van de Hulst, 1957) has solved the problem of electromagnetic radiation interacting with spherical scatterers. Mie theory has also been used to calculate absorption by spheres at a metal interface by Beaglehole and Hunderi (1970). The calculation of absorption and scattering cross sections is a complicated boundary value problem soluble only for particle shapes with a high degree of symmetry. This problem, in addition to having been solved for spheres (Mie's solution), has also been solved for infinite cylinders and ellipsoids including as limiting cases thin disks and needles (van de Hulst, 1957, p. 70). When the particle dimensions are much smaller than the wavelength, the particle shape becomes less important. The particles then tend to behave like oscillating dipoles and produce dipole radiation characteristic of Rayleigh scattering (Jackson, 1962, pp. 573, 603).

In modeling a particulate-covered surface, one can consider that the particles are distributed in a random manner on a plane surface. Scattering may then be calculated by adding the phases of the light backscattered by the particles to the phases of forward scattered light which is subsequently reflected by the plane surface. In general, the contribution of the forward scattered light is much larger than the contribution of the backscattered light.

Spherical particles exhibit resonances in absorption (and also scattering) which may be distinguished as dielectric resonances and size resonances. Dielectric resonances may occur when the size of the particle is much less than the wavelength. In this Rayleigh region, the resonances are dipolelike, occurring when the real part of the dielectric constant of the sphere equals -2. The dipole resonance causes much higher intensity fields to be generated within the sphere and results in increased absorption and scattering. Size resonances may occur when the physical size of the scatterer and the wavelength relate in such a way as to produce internal standing waves similar to a resonant cavity. The standing waves constructively interfere to yield enhanced surface fields and hence scattered radiation.

Other efforts to explain scattering and related optical effects of particulates have considered ellipsoidal-shaped objects. The ellipsoids are assumed to model aggregates in metal films such as silver, where the aggregate sizes may vary from 100 to 4000 Å. There have been a limited number of theoretical and experimental investigations of scattering from such aggregated metal films (Truong and Scott, 1976, and references cited therein).

E. Numerical Methods

All of the surface scattering results discussed previously yield a solution in closed form. To obtain these solutions it is necessary to make various approximations and specify the ratio of the rms roughness to the wavelength. With the advent of high-speed computers, numerical solutions are often practical. The usual method involves a solution to the wave equation as an infinite series expansion. The coefficients are determined numerically by solving equations for the appropriate boundary conditions. The majority of the solutions that have been obtained by numerical methods are for scattering from grating surfaces. This is because a known surface profile is needed in order to specify the regions across which the electromagnetic theory boundary conditions must be satisfied. Numerical techniques are especially useful when applied to problems of high efficiency gratings where the grating amplitude is not small compared to the wavelength. It is usually possible to truncate the numerical algorithm after the desired accuracy has been attained. Some authors have calculated the diffraction expected from a colinear array of line currents (Zaki and Neureuther, 1971). Several different methods (Rayleigh, integral, differential) are outlined by Petit (1975). Berreman (1970) has calculated the expected change in coherent reflectance caused by a surface with hemispherical pits or bumps. The technique uses the computer to obtain a solution for general shapes; however, some simple examples are given in closed form. Because of the problem of specifying the surface profile, numerical methods have not been applied to scattering from surfaces having random roughness.

F. Comparison of A–E Methods

No one method can be singled out as best for calculating scattering from optical surfaces. When the surface roughness is much larger than the wavelength, geometrical optics may be applicable if the exact dimensions of the surface features can be specified. The Helmholtz–Kirchhoff diffraction integral provides the basis for a scalar theory which has been widely used to calculate scattering from surfaces which have correlated microirregularities. The theory does not provide polarization information, but it may be applied to situations where the roughness is much less than or much greater than the wavelength. It is most useful for predicting total integrated scatter (TIS), i.e., scattering in all directions, but has also been used in a limited way to predict the angular dependence of scattering. The accuracy of the angular dependence prediction is greatest for scattering near the specular direction and is most useful for highly reflecting metals where polarization effects are minimal. A widely used application of scalar theory has been the determination of the relation between the surface roughness and the decrease in specular reflectance of the rough surface (Bennett and Porteus, 1961; Porteus, 1963).

Vector theory is generally an improvement over scalar theory because polarization effects are included. The SCS integral is somewhat similar to the Helmholtz–Kirchhoff diffraction integral. The former is a surface integral which sums, at an observation point, the radiation from vector surface currents, while the latter sums the scalar phases from different points on the surface. When surface roughness is present, incoherent or scattered fields are produced. The SCS integral is formally exact but in practice cannot yield an exact solution. This is because the fields on the surface are not precisely known beyond reasonable approximations.

Perturbation methods also yield vector solutions for the incoherent field. These methods can, in principle, yield high-order solutions by iteration. Solutions beyond first order are generally not given because of complexity.

Single particle scattering, handled by the Mie theory, is especially useful for calculating scattering from a particulate-covered surface. Mie theory can handle absorption as well as scattering and treats both dielectric and metal particles. It is most easily applied to spherical objects and cannot treat scattering from correlated surface microirregularities.

While all of the previous approaches may offer closed form solutions, numerical approaches do not. In some cases the accuracy of solution may be quite good although the convergence of the solution may be poor. This could cause excessive computer run times. Usually the numerical method is especially chosen for the type of scattering surface to be evaluated. Random roughness is not considered in numerical techniques.

In this chapter, a perturbation method will be used to explore various aspects of angular scattering from a rough surface and from multilayers covering a rough surface. The results will be limited to first order, and are valid when the roughness is much less than the wavelength. The first topic to be discussed in detail is TIS from surfaces whose roughness is small compared to the wavelength. This will be handled principally by scalar scattering theory.

III. TOTAL INTEGRATED SCATTER

Total integrated scatter refers to the integrated sum of light scattered into all directions within the scattering hemisphere. It is not necessary to be aware of the angular distribution of the scattered light—only the fraction of the reflected light which is scattered out of the specular beam. This scattering may arise from (1) irregularities such as scratches, digs, or particulates which are large relative to the wavelength of light, (2) isolated irregularities which are comparable in size or smaller than the wavelength, and (3) correlated irregularities which have heights which are small relative to the wavelength but which cover the entire surface. Scattering from these correlated irregularities is often termed "microirregularity scattering," and is usually the dominant source of scattered light from optical components used in the near infrared, visible, and ultraviolet. At longer wavelengths scattering from scratches or particulates is often dominant.

A. Microirregularity Scattering

Figure 1 shows a micrograph of a well-polished metal surface with a measured value of the rms microroughness, $\delta = 35.5$ Å. The microirregu-

Fig. 1. Differential interference contrast micrograph of a polished copper surface with an rms roughness of 35.5 Å. (Bennett, 1978.)

larities are revealed by differential interference contrast (Nomarski) microscopy, an interferometric technique which is sensitive to height differences of fractions of a nanometer if the surfaces have jagged irregularities with steep slopes. These microirregularities cover the entire surface and are the dominant source of scattered light in the ultraviolet, visible, and near infrared regions of the spectrum.

Statistically, the surface microirregularities can be described by a height distribution function and an autocovariance function. It is only necessary to understand what the minimum effective lateral separation between surface features will be. For reasons to be described later, the minimum effective lateral distances are equal to the wavelength of the light incident on the surface. Irregularities having lateral separations smaller than this minimum value are averaged with their neighbors to produce an area having an average height with respect to the mean surface level. It is thus perfectly possible to have these small "building-block" areas differ in average height by a fraction of an angstrom, even though atomic separations are ~ 4 Å and the crystallites in the surface may have steps of several hundred angstroms when viewed under an electron microscope.

The heights of these building-block areas above and below the mean surface level can be described by a "height distribution function." Experimentally, it is found that in nearly all cases the height distribution functions are Gaussian to a very good approximation. (Examples of some measured height distribution functions will be shown later in this section.) We may then describe the surface heights by an "rms roughness" as defined for a Gaussian height distribution function. The fraction of the total reflected light (specular plus nonspecular) scattered away from the specular direction by microirregularities is described by a simple scalar scattering theory based on the Kirchhoff diffraction integral. The dependence of TIS on wavelength is given by (Bennett and Porteus, 1961)

$$\text{TIS} \equiv 1 - (R/R_0) = 1 - \exp[-(4\pi\delta \cos\theta_0/\lambda)^2] \cong (4\pi\delta \cos\theta_0/\lambda)^2 \quad (1)$$

where R_0 is the fraction of the incident light which is reflected into all angles including the specular direction, R is the fraction which is specularly reflected at an angle θ_0, the angle of incidence, and δ is the rms height of surface microirregularities. Since the TIS is a function of the height but not the lateral separations of the microirregularities (so long as they fall within an appropriate range) and since real optical surfaces typically have Gaussian height distribution functions with irregularities which are symmetric about the mean surface level, it is possible to determine the rms value of the microroughness uniquely from TIS measurements. The rms microroughness is thus a simple method for specifying the quality of an optical surface with regard to scattering by microirregularities.

Some materials can be polished more easily than others and yield surfaces with lower microroughness which scatter less light for an equivalent amount of polishing effort. Techniques can often be developed to reduce the surface microroughness for a given material, so there is no single roughness value which a given material polishes to. However, optical glasses typically can be more easily polished to low rms roughness values than can most metals and other crystalline materials. Neither glasses nor crystalline materials automatically polish to very low rms roughness values, say under 15 Å rms. Special techniques are required to produce such surfaces, which are frequently referred to as being "superpolished." Recently, techniques have been developed to polish some surfaces in such a way that the scattering from them is optically equivalent to microroughness of 5 Å rms or less. Such surfaces, which scatter only about $\frac{1}{25}$th as much as a typical polished glass surface, are sometimes called ultralow scatter surfaces.

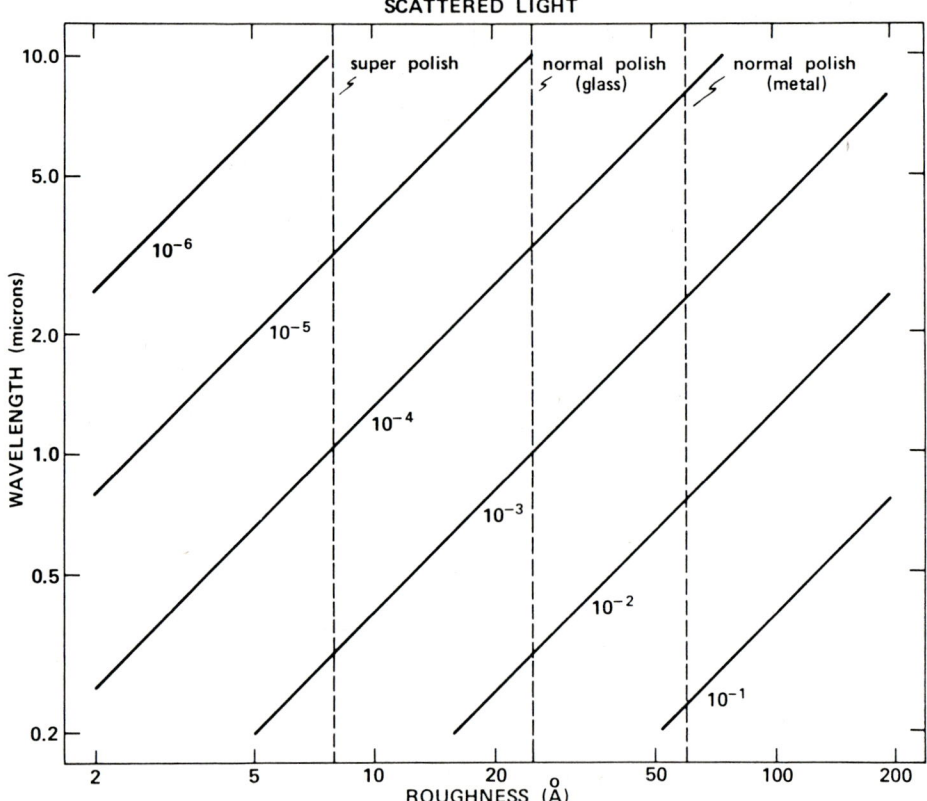

FIG. 2. Scattered light levels predicted theoretically (diagonal lines) for surfaces having rms roughnesses from 2 to 200 Å: The dashed lines indicate typical roughnesses of various kinds of polished surfaces. Wavelengths from the ultraviolet to the infrared are plotted logarithmically on the ordinate. (Bennett, 1978.)

If the rms roughness in Eq. (1) is plotted as a function of wavelength on log–log paper for a constant scattering level, a straight line results. A series of these scattering levels is shown in Fig. 2. By running a horizontal line across the graph at the wavelength of interest, it is possible to determine for a given roughness surface what the approximate scattering level will be. For example, at a wavelength of 5000 Å a superpolished surface will scatter less than 0.1%, whereas a conventionally polished glass surface will scatter several tenths of a percent and a typical polished metal surface will scatter over 1%. Recall that these percent values refer to the total reflected light which can vary markedly between dielectrics and metals (dielectrics can transmit much of the incident light).

In most cases, the scattered light obeys Eq. (1) quite well in the visible, ultraviolet, and near infrared regions of the spectrum. Figure 3 shows a typical plot of total scattered light as a function of wavelength in this wavelength region. In some cases, agreement is not so good. For example, if the sample is heavily scratched or coated with dust or particulates, the height distribution function is no longer Gaussian. Examples of measured height distribution functions are shown in Fig. 4. Many polished surfaces have

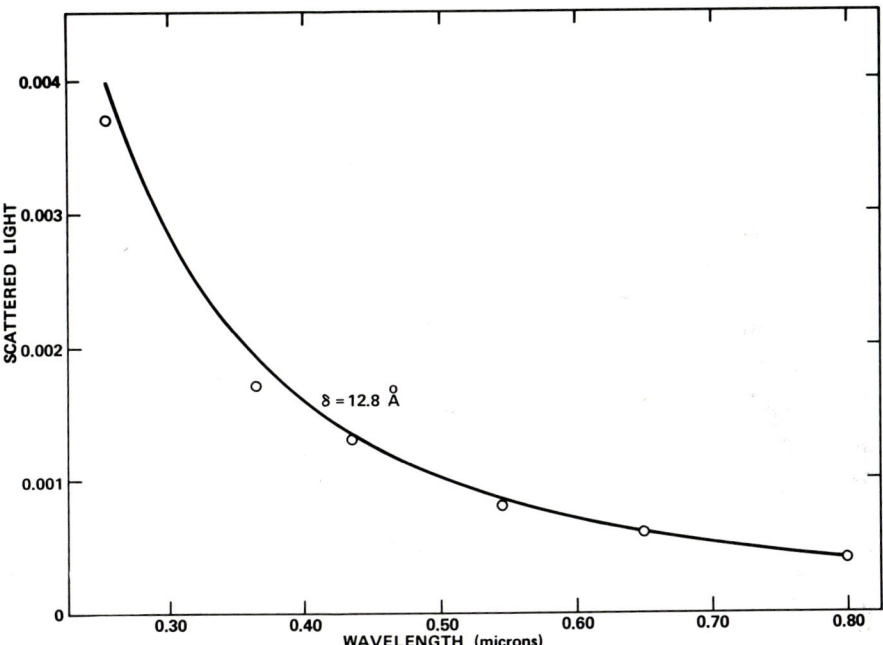

FIG. 3. Wavelength dependence of the total scattered light from an aluminum-coated superpolished fused quartz sample: (○) measured values; —— calculated from Eq. (1) assuming a value of 12.8 Å for δ. (Bennett, 1978.)

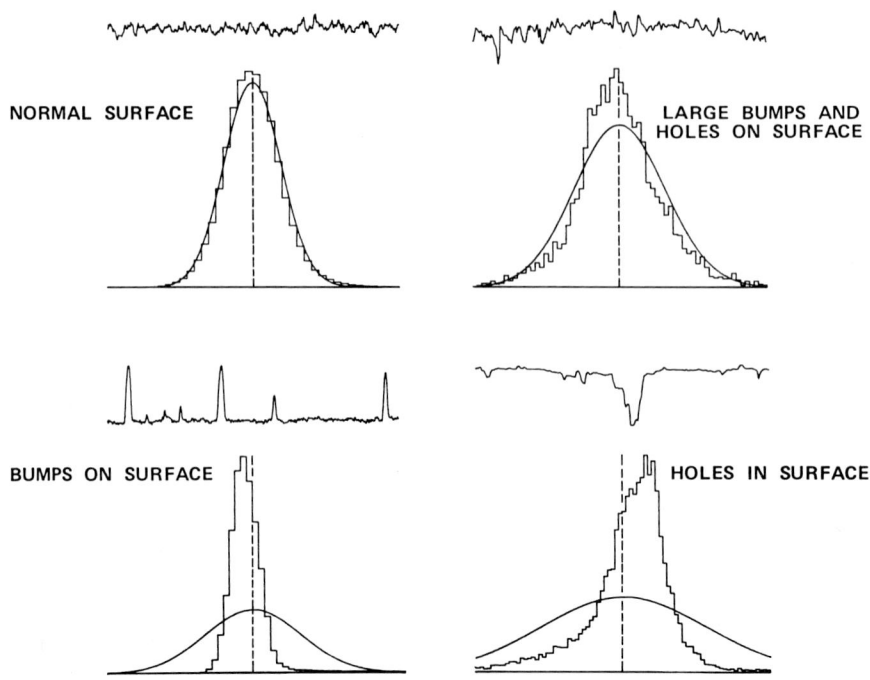

FIG. 4. Effect of the surface profile on the height distribution function for four types of surfaces. The surface profiles are shown above the corresponding histograms.

Gaussian height distribution functions similar to that shown in the upper left of the figure. (In this type of presentation the histogram is obtained from measured height data. The length of each bar represents the fraction of the total number of surface features which have heights equal to the value given on the abscissa, which is measured relative to the mean surface level (dashed vertical line). The smooth Gaussian curve encloses the same total area and has a half width derived from the measured rms roughness of the surface.) Some very soft materials such as polished KCl have proportionately too many large bumps or deep scratches. This distorts the measured histogram (upper right of Fig. 4), putting too much contribution in the tails. The influence of the extrema on the surface makes the rms roughness proportionately too large, so the half width of the Gaussian is also too large. TIS-derived values of the rms roughness tend to be higher than measured values for this type of surface. If a surface is extremely smooth but has large dust particles on it, a distorted histogram is obtained with the maximum shifted to negative values (because the particles have influenced the placement of the mean surface level). Similarly, an etched or pitted surface which is nominally smooth but has deep holes in it can have a distorted histogram with the peak shifted in the other direction. Scattering from particulate-covered or etched surfaces would not be expected to follow the relation in Eq. (1). Scattering from scratches also tends to be nearly independent of wavelength. If the autocovariance length of the surface is small enough so that it becomes shorter than some of the wavelengths tested, discrepancies will arise. It is thus useful when specifying quantitative surface microroughness values to specify the wavelength range in which the "equivalent surface microroughness" was measured. Convenient wavelengths are the visible mercury lines 4358 and 5461 Å, and the red HeNe laser line 6328 Å, or the red krypton laser line at 6471 Å.

Figure 5 shows typical roughness and the range of roughnesses, mostly derived from TIS measurements which were made at the Naval Weapons Center (NWC) on various well-polished substrate materials. The substrates were either polished in-house or were sent to NWC for evaluation by commercial vendors. In all cases, it is easy to obtain larger roughnesses, but reducing the lower limit is much more difficult. The average values shown in Fig. 5 are not representative of the quality of surface finish one would obtain by purchasing a polished mirror from any optical firm. For example, the copper mirror roughnesses are averages of samples polished by the Northrop Corporation, Battelle Northwest Laboratory, Perkin-Elmer Corporation, Spawr Optical, and NWC. They are not conventional copper mirrors but represent the best samples produced by these firms. The best bulk copper sample we have yet measured had an rms roughness of 15 Å rms, scattered only one quarter as much as the average good sample, and

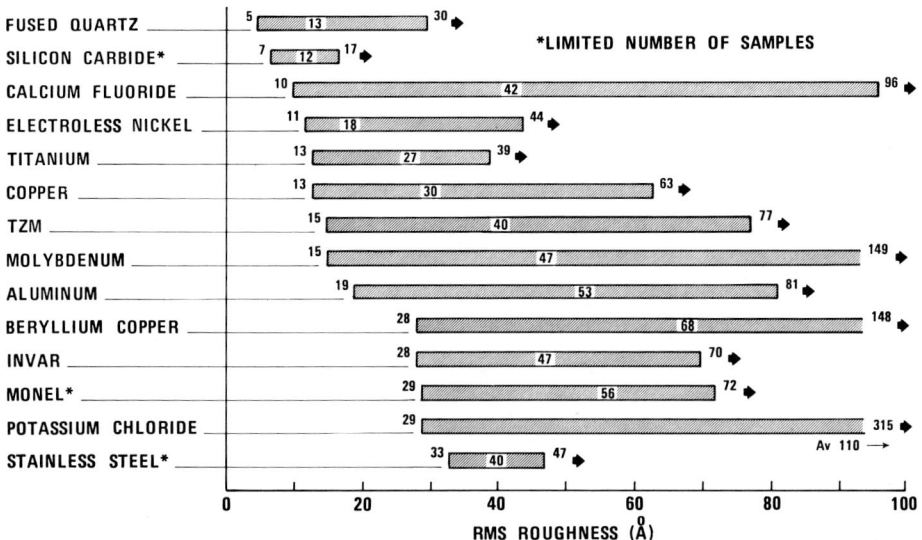

FIG. 5. Average rms roughness and range of roughnesses measured for polished optical surfaces.

over an order of magnitude less than conventionally polished mirrors. Because such a low microroughness has been achieved on copper is not a reason to specify bulk copper mirrors with an rms roughness of 15 Å unless the purchaser is ready to support a major development effort to push a technique from the laboratory one-of-a-kind stage to a production process. The same argument applies to the roughness values listed for the other materials.

B. Surface Plasmon Excitation

We have assumed in the above discussion that the optical constants of the coating do not affect the fraction of the reflected light which is scattered. This assumption is usually valid but is not justified in wavelength regions where surface plasmon excitations are important (Bennett and Stanford, 1976). Figure 6 shows the fraction of the incident light which is scattered from a polished glass surface coated with CaF_2 (to increase its microroughness) and then with silver. Dielectric overcoats of MgF_2 were then deposited on the silver. The long-dashed line shows the scattering level which is predicted from the scalar scattering theory, Eq. (1). The solid line shows the difference between the scattering level predicted from the simple theory and

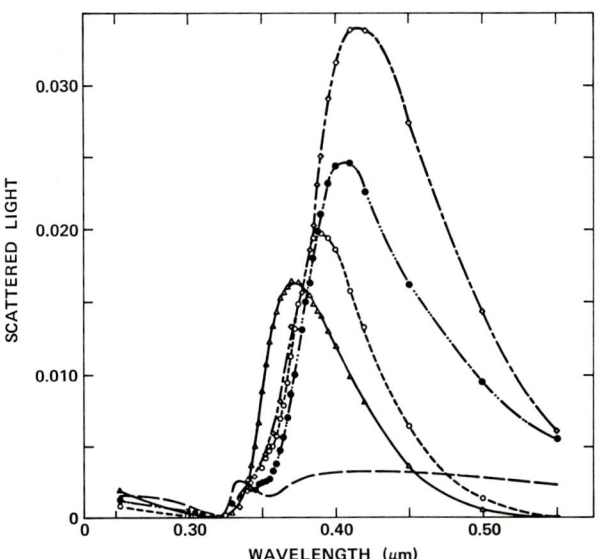

FIG. 6. Light scattered near the band edge in silver deposited on CaF_2-coated glass substrates: (———) scattering predicted from scalar theory; other curves show the increase in scattering caused by surface plasmon effects in (———) bare silver and silver coated with films of MgF_2 [(---) 150 Å MgF_2, (— — —) 700 Å MgF_2, (—··—) 120 Å MgF_2]. (Bennett and Stanford, 1976.)

that actually observed for the silver surface. The other dashed lines show the increase in scattering observed when the silver was overcoated with films of MgF_2 of various thicknesses. The maximum scattering level observed is an order of magnitude higher than would be predicted from the simple theory. If the metal coating had been aluminum, for example, instead of silver, scattering would have more nearly followed the simple theory. The character of the microroughness in this case made the effect unusually large; most silver-coated mirrors show much less dramatic effects in this wavelength region. The point, however, is that the optical constants of the surface can affect the scattered light levels observed in some cases and some of these are important technologically.

C. Scattering in the Infrared

In the infrared region scattering is usually higher than would be predicted from Eq. (1) for surfaces whose equivalent microroughness values are determined from visible measurements. A typical example of this effect is seen in Fig. 7. The scattering from an aluminized polished dense flint glass fits Eq. (1), which plots as a straight line on log–log paper, very well in the ultraviolet, visible, and near infrared. At a wavelength of 10 μm, however, the scattered light is an order of magnitude higher than predicted by the

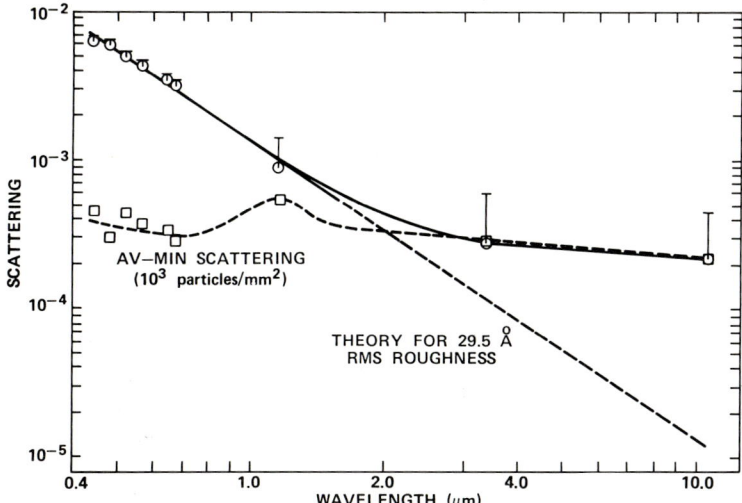

FIG. 7. Scattering from aluminized, polished dense flint glass. The diagonal line gives the contribution predicted for microirregularity scattering by a 29.5 Å rms roughness surface: (○) minimum scattering observed; bars and squares indicate the difference between average and minimum scattering observed at several points on the surface (10^3 particles/mm^2). (Bennett, 1978.)

theory. This discrepancy, which is commonly observed for surfaces in the infrared, may arise from several sources. The scattering angle as measured from the specular direction increases with increasing wavelength for correlated surface features having a given lateral size and separation (as can be seen from the grating equation). Long-range features with small average slopes, which at shorter wavelengths would produce light reflected in nearly the specular direction, will at longer wavelengths deflect light into larger angles where it will be detected as scattered light. A second and more important source of additional scattered light in the infrared is the presence of particulates, scratches, and various imperfections on the surface. The amount of light scattered from many of these imperfections is nearly independent of wavelength and since microirregularity scattering decreases exponentially with the square of the wavelength, at some point scattering from the larger surface imperfections will become the dominant source of scattered light. This effect is shown in Fig. 7. The squares are the differences between the average scattered light values at a given wavelength for a series of small areas on the surface and the minimum scattered light value observed from any of those small areas. These differences are nearly independent of wavelength. When the average scattering level reaches the difference value it stops decreasing and also becomes nearly independent of wavelength. This difference in scattering from point to point on the surface is believed to be caused by particulates and surface defects which are large compared to the wavelength and thus should be nearly independent of wavelength. A resonance should occur near a wavelength of 1 μm as a result of scattering by dust particles, which have a diameter of about 1 μm on the average. The surface tested was slightly dusty; dark field illumination revealed about 10^3 scattering sites/mm^2 of various sizes. It is thus not surprising that an increase in the difference values was observed in the 1-μm wavelength region.

D. Scratches and Other Macroscopic Defects: The Scratch/Dig Specification

Scratches and other macroscopic defects are widely recognized sources of scattered light from optical surfaces. The only official military specification on optical surface quality, MIL-13830A, refers to the allowed magnitude and density of these defects. As it is now formulated, both scratches and digs are defined by their width in micrometers. A dig having a diameter of 40 μm is a No. 40 dig, but a No. 40 scratch has a width of only 4 μm. Scratch and dig values are conveniently measured using a traveling microscope with Nomarski or dark field optics. Although digs (i.e., pits in the optical surface) have always been specified in terms of their diameter in

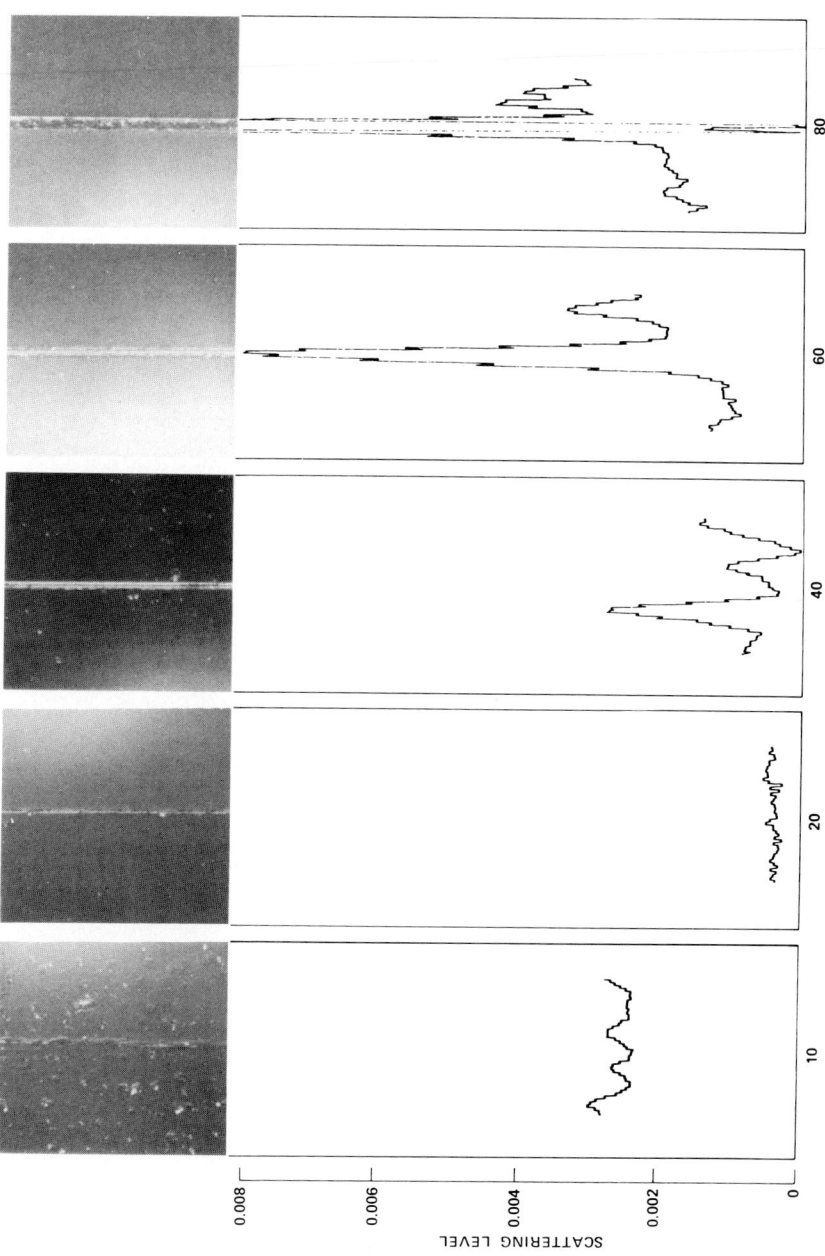

FIG. 8. Differential interference contrast micrographs and scattered light scans at 6328 Å for Frankford Arsenal Scratch Standard Serial No.F-66-1268. The apparent reversal at the center of the No.80 scratch is caused by overloading the electronics. The anomalously high scattering level for the No.10 scratch sample is caused by surface microroughness, which obscures the scattering from the scratch. (Bennett, 1978.)

micrometers, it is only recently that the width of scratches has been used to define them. Previously, sets of standard scratches were furnished. By visual comparison the inspector matched these standard scratches to scratches on the component to be tested. Although this test is quite subjective, it has been more or less successfully used for many years in the optical industry. Scattering at 0.6328 μm from a set of standard scratches furnished by Frankford Arsenal is shown in Fig. 8. The anomalously high scattering level observed for the No. 10 scratch sample is caused by microirregularity scattering, which completely masks the scattering from the fine scratch on this particular scratch standard.

Defining the scratch in terms of the scratch width makes the scratch specification significantly more quantitative than has been true before. There is not a unique relation between a scratch's width and its scattering behavior, however, as is illustrated by Fig. 9. The top set of scratches is an early standard set from Frankford Arsenal, which had cognizance of this particular military specification. These scratches have widths which are

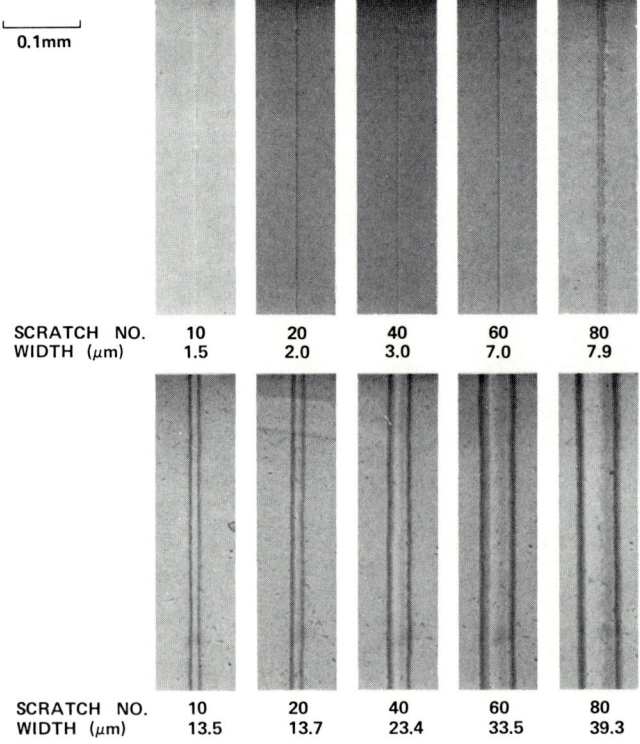

FIG. 9. Differential interference contrast micrographs of two sets of standard scratches. Although their widths differ by as much as an order of magnitude, the scattering levels from each pair are nearly equal. (Bennett, 1978.)

approximately in agreement with the new scratch identification scheme. The lower set of scratches are those on an experimental set of scratch standards produced under Frankford Arsenal control but having widths nearly an order of magnitude larger than those of the earlier standard. Scattering levels are similar to those found for the upper set of scratches, mainly because most of the scattered light comes from the sides of the scratches; the scratch bottoms are quite smooth. In practice the bottoms of scratches are not normally smooth and wide scratches tend to scatter more than narrow scratches. Since the width and scattering level of scratches can be adjusted separately, it would be possible to develop a set of scratch standards which not only had the specified widths but which scatter according to a predetermined scattering ratio. The narrow standard scratches would then look like weaker scatterers than the wider scratches. This scheme, in effect, was used in the original scratch standards; inspectors visually matched the appearance of the standard scratches to those of the sample tested. Unfortunately, visual observation is not a good way to detect differences in intensity. By using a more accurate detector it would be possible to quantify this procedure and arrive at a useful secondary test for surface quality which would be particularly useful for infrared optics. It has merit in the visible region also. Laser damage often occurs at scratches (Bloembergen, 1973), which must thus be eliminated for laser optics. In addition, in the process of polishing out the scratches a surface finish on glass surfaces is usually generated which is adequate for most noncritical applications. The main advantage of the scratch/dig specification is that it lends itself to volume production of parts and when properly set up is quick to use. Its main disadvantages are (1) it is not at present satisfyingly quantitative as far as scattering level goes and (2) it is not directly related to the source of most of the scattered light from optical surfaces in the visible, ultraviolet, and near infrared regions of the spectrum.

The reason scratches are so apparent visually is that scattering from them is localized in a small area of the surface where the scratch is. The total amount of light scattered per unit area of the surface by scratches is often negligible even though the scratches themselves are quite apparent, which gives rise to the opticians' admonition "optical components are made to be looked through, not at." For scratch dimensions large relative to the wavelength of light, scattering will be determined by geometrical optics and will be nearly independent of wavelength. The same is true of digs. Figure 10 shows the scattering observed for the standard and experimental No. 40 scratches as a function of wavelength. The wider experimental scratch shows almost no wavelength dependence. The narrower No. 40 standard scratch shows some decrease in scattering with increasing wavelength. This decrease results in part from the decrease in microirregularity scattering of the surface on which the scratch is made.

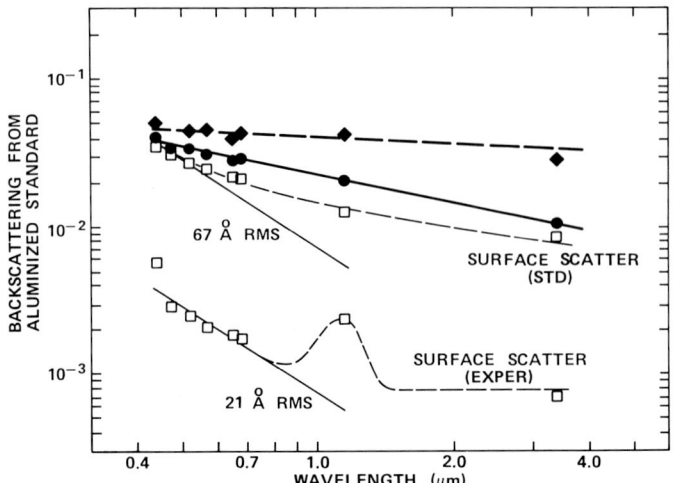

FIG. 10. Comparison of the scattering levels observed from: (◆) experimental 23-μm-wide No.40 scratch; (●) standard 3-μm wide No.40 scratch; (□) surface scattering adjacent to each scratch; the solid diagonal lines are predicted microirregularity scattering for roughnesses of 67 and 21 Å. The measuring beam diameter was 2 mm. (Bennett, 1978.)

Sand or rain erosion increases the number of pits on the surface and can significantly increase scattered light. Figure 11 shows the scattered light levels at 3.39 μm observed in the forward direction for two missile domes, one of which had been flown on an aircraft. Differences between the two were not apparent to a casual observer, but the scattering levels of the dome which

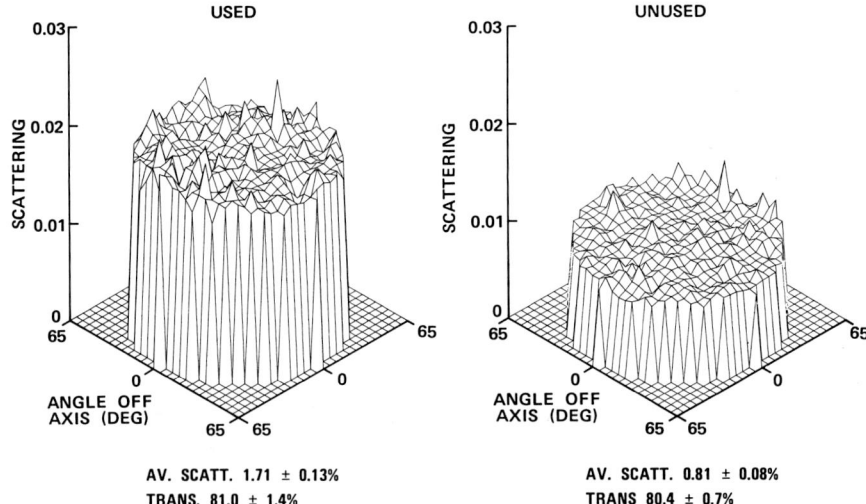

FIG. 11. Light scattered at a wavelength of 3.39 μm as a function of position for a missile dome which had been flown and one which had not. (Bennett, 1978.)

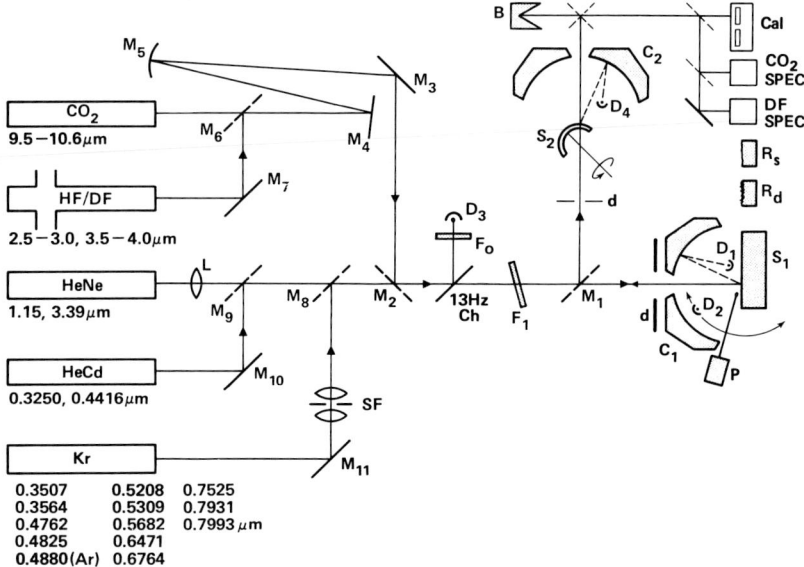

FIG. 12. Optical evaluation facility: mirrors M, lens L, spatial filter SF, chopper Ch, filters F, detectors D, calorimeter Cal, beam dump B, samples S, reference surfaces R, spectrometer monitors SPEC, Coblentz spheres C, and diaphragms d. Forward or back TIS, angular backscatter in plane of incidence, reflectance at normal incidence or 45°, transmission, and calorimetric measurements can be made with this instrument. (Bennett, 1978.)

had been flown were higher than those of the unused dome by over a factor of 2. If a rain field had been encountered, the differences would have been much more pronounced. Dust or sand erosion has a similar effect on exposed optics. These scattering measurements were made on the NWC Optical Evaluation Facility (Bennett and Stanford, 1976; Archibald and Bennett, 1978), a schematic of which is shown in Fig. 12. Either forward or backscattering can be measured at over 20 laser wavelengths extending from the ultraviolet to the infrared. Although the instrument is designed primarily to measure TIS, the angular dependence of scattering can also be investigated (Elson and Bennett, 1979). Reflectance and transmittance can be determined using this instrument, which is computer controlled so that measurements at several hundred points uniformly distributed over the sample surface can be taken automatically. The absorption of small samples which is sometimes related to surface effects can be determined in an adiabatic calorimeter, which has a sensitivity of about $3 \times 10^{-5}/W$ of input power, and which is also part of this instrument.

A comparison of forward and backscattering from an etched potassium chloride sample as a function of wavelength is shown in Fig. 13. The forward scattering level drops about 15 times and the backscattering level nearly 40 times in going from the visible to the infrared for this sample, and the

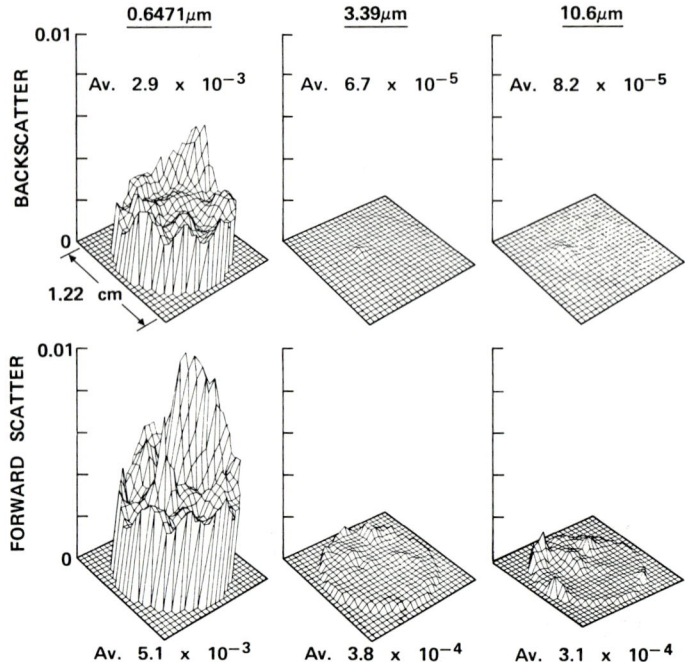

FIG. 13. Light scattered from etched potassium chloride as a function of wavelength and position on the surface. (Bennett, 1978.)

positions where maximum scattering occurs also shift somewhat between the visible and the infrared. The forward scattering level is always higher than the backscattering level, as would be expected for a dielectric material if the scattering were caused by particulates (van de Hulst, 1957, p. 145).

E. Particulate Scattering from Isolated Microirregularities

When the surface irregularities become comparable in size to the wavelength of light scattered from the surface, the scattering process can no longer be described by geometrical optics. It is convenient in this case to separate the scattering into two categories: scattering from isolated particulates and the previously described scattering from correlated microirregularities. In particulate scattering each scattering center behaves independently of all others, so that the total scattered light arising from particulate scattering is the sum of the contributions from each particle. We can then use Mie scattering theory to describe the resultant scattered light. These scattering functions become very complicated, but if we consider only dipole scattering from transparent spherical particles and disregard the angular dependence

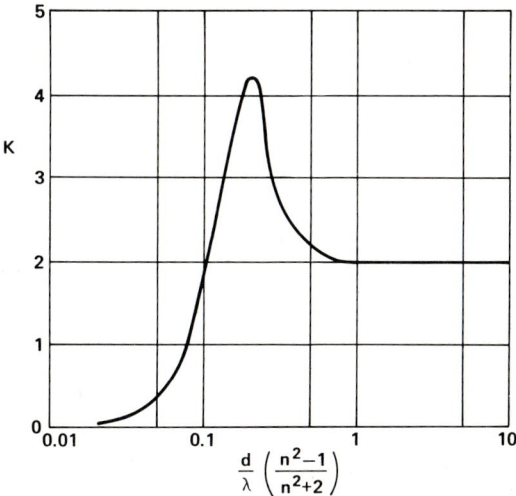

FIG. 14. "Universal" scattering curve for nonabsorbing spherical particles of refractive index n and diameter d at wavelength λ. The scattering coefficient is K. (Bennett, 1978.)

of the scattered light, the TIS from isolated particulates is described approximately (Orr and Dallavalle, 1959) by the "universal scattering curve" shown in Fig. 14. It can also be applied to slightly absorbing particles. The essential features of this particulate scattering are seen to be:

(a) When the particle size is large compared to the wavelength λ, the scattering cross section is independent of wavelength and is twice the geometrical cross section of the particle (the extinction paradox) (van de Hulst, 1957, p. 107);

(b) As the wavelength approaches the particle diameter, the scattering cross section k increases abruptly and a resonance occurs near the point at which the two are equal; and

(c) As the wavelength continues to increase the scattering cross section falls rapidly and for diameters much smaller than a wavelength we have Rayleigh scattering, which decreases as the inverse fourth power of the wavelength.

Another characteristic of particulate scattering is that forward scattering is always larger than backscattering for transparent particles (van de Hulst, 1957, p. 145). The two approach each other in the Rayleigh limit. If the particles were metallic, the reverse would be true and forward scattering would be less than backscattering.

The increased scattering predicted by Fig. 14 in the neighborhood of $d/\lambda = 1$ is probably the explanation for the increased scattering observed near a wavelength of 1 μm in Fig. 7. An increase of this kind is frequently seen for smooth surfaces and, as mentioned earlier, is thought to be caused by dust which has an average diameter of about 1 μm.

IV. ANGULAR DEPENDENCE OF SCATTERING

Although TIS measurements are relatively simple to make and are generally satisfactory for assessing optical surfaces, some applications require a knowledge of scattering into a particular angle or range of angles. For example, in a laser gyro system where light is incident on mirrors at 30 or 45°, the retroscattered light, i.e., light that is backscattered along the incident beam direction, is of crucial importance because it causes the gyro to malfunction. In trying to distinguish weak objects near a very bright object, such as in a reflecting solar coronograph, scattered light within a few degrees or even a fraction of a degree of the specular direction can completely mask the desired image. In high power laser systems where the light from the laser goes through a series of amplifiers, any light backscattered into the laser can depump it and cause catastrophic damage. Finally, in many types of optical systems it is not possible to baffle the system adequately to prevent all angles of scattered light from reaching the detector and thus degrading the image quality.

Scattering, whether from the viewpoint of angular scattering or TIS, can be caused by (1) scratches, digs, and other surface features whose dimensions are large compared to the wavelength of light, (2) isolated particulates on a surface, or (3) microroughness whose heights are much smaller than the wavelength. Angular scattering from the first type of surface features can, in principle, be calculated using the laws of geometrical optics. In practice, however, this is quite difficult because of the lack of a good model for the shapes and sizes of the facets making up the surface features. Scratches, for example, are often cut on glass surfaces with a diamond scribe and their edges consist of a series of conchoidal fractures. The steeply sloped facets on many scratches produce a considerable amount of large angle scattering, causing the scratches to appear distinctly against the smaller background scattering from the microirregularities. Ground or matte surfaces, whose facets are also large compared to the wavelength, produce a considerable amount of large angle scattering. Ditchburn (1964) has shown that if a ground glass surface has a Gaussian autocorrelation function, its scattering is independent of angle, i.e., it is a perfect diffuser. Such a surface is called a Lambertian surface since it follows Lambert's law (Ditchburn, 1964).

Angular dependence of scattering from dust or other isolated particulates on a surface can, in principle, be handled by Mie theory (Mie, 1908; see also Stratton, 1941; Kerker, 1969). In practice, this is also difficult because one must know (1) the distribution of sizes and shapes of the particulates, (2) their (complex) refractive indices, and (3) the appropriate interaction terms for the effect of the surface on which the particles are sitting. In a simplified experiment where the angular scattering was measured from a low scatter mirror contaminated with spherical silver particles whose diameters were in the 9–18 μm range, Young (1976) found good agreement between the predictions of Mie scattering theory and the measured angular scattering at wavelengths of 0.6328 and 10.6 μm. Most of the observed scattering could be explained by assuming that the radiation scattered in the forward direction was reflected from the mirror surface unaffected by the presence of the particle. The scattering calculations assumed that the particles were in free space and were illuminated by polarized radiation. While the results of this study are very encouraging, they do not duplicate actual laboratory conditions where the dust particles are more likely in the size range around 1 μm, probably not spherical, and have optical constants more like slightly absorbing dielectrics, such as impure SiO_2. A general rule of thumb for angular scattering from particulates is that they can produce much more large angle scattering than the surface on which they are resting. For this reason one can see glints from very small dust particles on a surface when that surface is illuminated at oblique incidence by a microscope illuminator or laser beam.

Angular scattering that is most tractable to handle theoretically is the third type mentioned above, i.e., scattering from correlated surface microirregularities whose heights are much smaller than a wavelength of light. Even in this case, however, there is no simple relation giving the angular distribution of scattered light as a function of surface roughness, similar to the TIS relation in Eq. (1). The reason is that the angular distribution of the scattering depends not only on the rms height of the surface irregularities but also on their lateral separations. A convenient measure of the separation of surface features is the autocorrelation or autocovariance function which will be defined rigorously later in this section. For the moment we will assume an autocovariance length a (which is the standard deviation of the Gaussian) at which value the autocovariance has dropped to $1/e$ of the maximum value.

The scalar scattering theory can give information about the angular scattering at angles close to the specular direction. From this theory it can be shown that the scattering into an angle α measured from the specular direction is (Porteus, 1963) $16\pi^4\delta^2 a^2\alpha^2/\lambda^4$, where δ is the rms roughness and λ the wavelength. This expression gives the scattering (ratio of scattered light to total reflected light from the surface) into a cone of half angle α. (The scattering expressions $dP/d\Omega$ given later in this section are for scattering into a unit solid angle $d\Omega$ when the incident intensity is unity and include the total reflectance of the surface.) The simple expression above holds only for small scattering angles since it is the first term of an expansion (Porteus, 1963).

The small angle approximation makes the expression independent of the functional form assumed for the autocovariance function, so it should work for real surfaces which do not have Gaussian autocovariance functions.

In order to understand angular scattering from an isotropic rough surface having random roughness, we can consider that the surface is made up of a superposition of sinusoidal gratings having different amplitudes, phases, and periods (Church *et al.*, 1977; Stover, 1975). In this approach one loses the phase information about how the various sinusoidal gratings are superimposed, but if the rough surface can be adequately characterized in this manner, one can obtain a correct expression for the angular scattering. Another approach is to actually measure the autocovariance function for the rough surface and use the vector scattering theory given later in this section to determine the angular scattering. Although in principle it is possible to calculate the two-dimensional angular scattering from the surface autocovariance function and to obtain the autocovariance function from the measured angular scattering, one cannot unambiguously obtain a representation of the actual surface profile because of the missing phase information mentioned above.

In this section we will first discuss the angular scattering from a sinusoidal grating (one component of surface roughness). Then we will discuss the complete vector theory which can be used to calculate the angular scattering, given the surface autocovariance function. This vector includes polarization effects and can be used for all angles of incidence. In the experimental part of this section we will describe three methods for obtaining surface statistics and will give the advantages and disadvantages of each. We will then show some autocovariance functions measured for actual optical surfaces, including a diamond-turned surface. Finally, we will compare angular scattering calculated from a measured autocovariance function with the actual angular scattering measured for the same surface.

A. Angular Scattering from a Sinusoidal Grating

Since a surface having random roughness can be considered to be composed of a two-dimensional superposition of sinusoidal gratings having different amplitudes, periods, and phases, an insight into angular scattering can be obtained by considering the angular scattering from a single sinusoidal grating having an amplitude A, measured from the mean surface level, and grating spacing d. For normally incident light the angle of diffraction (scattering) is given by the grating equation:

$$\sin \theta = m(\lambda/d) \tag{2}$$

where m is the order of interference and λ the wavelength. The intensity I of light diffracted into the mth order is proportional to (Church and Zavada, 1975)

$$I \sim (A/\lambda)^{2|m|} \qquad (3)$$

For microrough surfaces, $A/\lambda \ll 1$, so that only the first diffracted order need be considered. It is clear from Eq. (2) that the *angle* into which light is diffracted or scattered depends *only* on the lateral spacing between the grooves and the scattered *intensity* depends primarily on the groove depth, which can be related to the rms surface roughness δ. A plot of the geometry for Eqs. (2) and (3) and the relation between d and θ is given in Fig. 15. Irregularities which have separations which are large compared to a wavelength will scatter light near the specular direction; those with smaller separations will scatter at larger angles and those whose separations approach the wavelength will scatter at angles approaching 90°. Figure 16 is another way of showing the relation between the wavelength, scattering angle, and separation of surface irregularities for normally incident light on a rough surface. The dashed lines are an example of how to use the nomogram. For example, for a wavelength of 0.5 μm, scattering at an angle of 5° from the specular direction is produced by surface features having a lateral separation of about

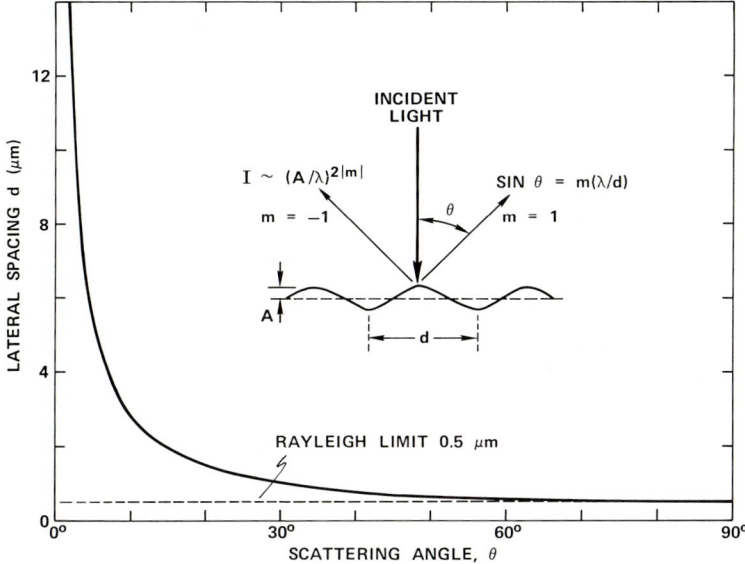

FIG. 15. Scattering angle from a coherent scatterer at normal incidence as a function of grating groove separation for light having a wavelength of 5000 Å. (Bennett, 1978.)

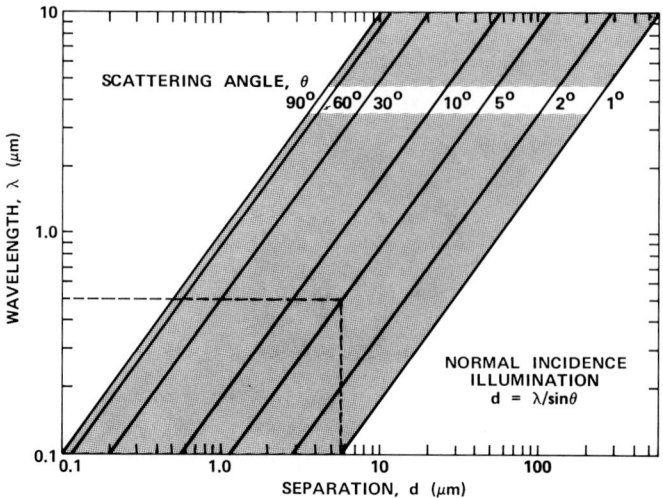

FIG. 16. Nomogram giving separation of surface features that produce scattering at a particular angle and wavelength. The dashed lines are an illustrative example. (Elson and Bennett, 1979.)

5.7 μm. Again, note that scattering into the maximum angle, 90°, is accomplished by microirregularities having a minimum separation, equal to the wavelength, and that irregularities having smaller separations cannot produce visible scattering when light is normally incident on a surface. Therefore, evaporated or sputtered films that consist of particles having diameters in the 400–1000 Å range should not produce any scattering in the visible spectral region. Films tend to contour the underlying substrate so that the scattering is produced by features on the substrate having larger lateral separations. This situation for *correlated* microirregularities is in marked contrast to that for particulate scattering, where particles whose diameters are much smaller than a wavelength can produce scattered light.

If surfaces are anisotropic, such as diamond-turned surfaces, the light scattering depends on the polarization of the incident beam and the orientation of the surface features. Returning to the one-dimensional case of angular scattering from a sinusoidal grating, the complete expression for the diffracted (or scattered) intensity at an angle θ assuming a normally incident beam polarized parallel to the plane of diffraction (perpendicular to the grating grooves) is (Church and Zavada, 1975)

$$I_p/R_0 = (2\pi A/\lambda)^2 \sec\theta \tag{4}$$

and for a beam polarized perpendicular to the plane of diffraction (parallel to the grating grooves) is

$$I_s/R_0 = (2\pi A/\lambda)^2 \cos\theta \tag{5}$$

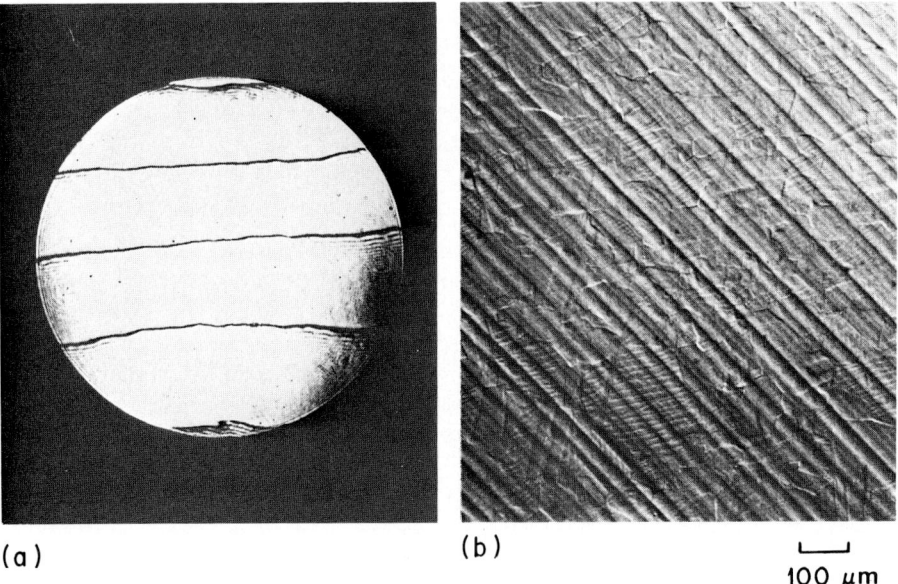

FIG. 17. (a) Optical figure (1/20th wave) and (b) surface finish (21.0 Å rms) of a 4-in. diameter, diamond-turned copper mirror produced by the Y-12 Plant at Oak Ridge, Tenn. (Bennett, 1978.)

Note that Eqs. (4) and (5) give the *intensity* of light diffracted into the first order at *angles* specified by the grating equation, Eq. (2), so that the θ's apply only to *specific angles*. When the groove spacing d approaches the wavelength λ (θ approaches 90°), the intensities of the p and s diffracted components are very different because of the $\sec\theta$ and $\cos\theta$ terms, and depend strongly on d as well as A. However, if the incident light is unpolarized, this effect is negligible except for grating spacings which are almost identical to the wavelength. The diffraction effects are particularly important for diamond-turned surfaces which exhibit shallow parallel grooves cut by the diamond point. A Nomarski micrograph of the surface of one such copper mirror is shown on the right side of Fig. 17. This mirror, which was produced by the Oak Ridge Y-12 Plant, has an excellent figure in spite of the grooves, as is shown on the left side of the figure. The light scattered from this mirror is comparable to that from a good, conventionally polished, glass optical flat.

B. Angular Scattering from Nonsinusoidal Surfaces

In the preceding section scattering from a surface having a simple sinusoidal profile was discussed. Although very few optical surfaces are of this type, surfaces having nonsinusoidal profiles may be reconstructed as a Fourier superposition of sinusoidal profiles. Hence, the concept of diffraction from a single sinusoidal profile may be used to give a qualitative picture of scattering from a more complex surface.

Surfaces having nonsinusoidal profiles may be divided into two types: those having periodic and random surface irregularities. Any general surface profile $z(\boldsymbol{\rho})$ can be considered as a Fourier superposition of sinusoidal profiles as follows:

$$z(\boldsymbol{\rho}) = \int d^2k\, Z(\mathbf{k}) e^{i\mathbf{k}\cdot\boldsymbol{\rho}} \tag{6}$$

The quantity $\boldsymbol{\rho} = (x, y)$ is a position vector of a point in the (x, y) plane from which the height $z(\boldsymbol{\rho})$ is measured. Since the surface is Fourier-analyzed in two dimensions, $\mathbf{k} = (k_x, k_y)$ is the wavevector of the Fourier component whose amplitude is $Z(\mathbf{k})$. Equation (6) may represent random or periodic roughness. If $z(\boldsymbol{\rho})$ refers to random (periodic) roughness, then $Z(\mathbf{k})$ is generally a continuous (discrete) function of $\mathbf{k}$. Light incident on a surface having a nonsinusoidal profile will scatter according to the sum of the scattering from the sinusoidal components making up its profile. Thus, scattering from periodic roughness components will be in discrete directions, while scattering from random roughness will be continuously distributed in angle because the distribution of sinusoidal surface components is continuous. To see how the sinusoidal distribution enters into the scattering formulas, the equations obtained from first-order perturbation theory will now be considered.

C. Angular Scattering Theory for Polished Surfaces

Polished surfaces whose rms roughnesses δ are much less than the wavelength λ and gratings whose amplitudes are small compared to λ are ideally suited for scattering calculations using first-order perturbation theory. A number of authors (Silver, 1947; Leader, 1971b; Fung and Chan, 1969; Rayleigh, 1945) have treated the problem of scattering from rough surfaces with no overcoats in the case where $\delta \ll \lambda$ and vector properties of the scattered light are included. The differential fraction dP of the incident energy scattered per unit solid angle $d\Omega = \sin\theta\, d\theta\, d\phi$ may be written as (Elson, 1975)

$$\frac{dP}{d\Omega} = \frac{(\omega/c)^4}{\pi^2} \cos\theta_0 \cos^2\theta |1 - \varepsilon|^2 \frac{|Z(\mathbf{k} - \mathbf{k}_0)|^2}{L^2} \left[\frac{|\chi_\theta|^2}{|v - iq\varepsilon|^2} + \frac{|\chi_\phi|^2}{|v - iq|^2} \right] \tag{7}$$

where

$$\chi_\theta = \frac{(vv_0 \cos\phi + kk_0\varepsilon)\cos\phi'}{v_0 - iq_0\varepsilon} - \frac{i(\omega/c)v \sin\phi \sin\phi'}{v_0 - iq_0} \tag{8a}$$

and

$$\chi_\phi = \frac{\omega}{c}\left[\frac{(\omega/c)\cos\phi\sin\phi'}{v_0 - iq_0} - \frac{iv_0\sin\phi\cos\phi'}{v_0 - iq_0\varepsilon}\right] \quad (8b)$$

The light is scattered into a direction specified by (θ, ϕ). This expression can be applied to scattering from randomly rough surfaces or diffraction from low efficiency gratings. The component of the wave vector parallel (perpendicular) to the mean surface is $k_0 = (\omega/c)\sin\theta_0$ [$q_0 = (\omega/c)\cos\theta_0$], and $k = (\omega/c)\sin\theta$ [$q = (\omega/c)\cos\theta$] for the incident and scattered light, respectively. Also, $v_0 = [k_0^2 - \varepsilon(\omega/c)^2]^{1/2}$ and $v = [k^2 - \varepsilon(\omega/c)^2]^{1/2}$, where ε is the complex dielectric constant of the scattering medium for angular frequency ω. The quantity $Z(\mathbf{k} - \mathbf{k}_0)$ is the Fourier transform of the surface profile. The electric field polarization angle ϕ', measured relative to the plane of incidence, may be set equal to 0 (or $\pi/2$) to specialize to the case of p-polarized (s-polarized) incident light. Averaging over ϕ' from 0 to 2π yields the result for unpolarized incident light. The first and second terms in the bracket of Eq. (7) represent scattered light which is p- and s-polarized, respectively (measured relative to the plane of scattering). Since $\omega/c = 2\pi/\lambda$, it is seen that Eq. (7) varies as λ^{-4}. The distribution of Fourier components enters as the magnitude squared of $Z(\mathbf{k} - \mathbf{k}_0)$, as given in Eq. (6). Recall that $Z(\mathbf{k} - \mathbf{k}_0)$ may represent either periodic or random surface irregularities. In the case of periodic profiles, the surface shape is known *a priori*, and consequently $Z(\mathbf{k} - \mathbf{k}_0)$ may be readily calculated. In the case of random roughness, $z(\mathbf{\rho})$ is not known and statistical properties for the surface based on ensemble averages must be used.

There are several observations that can be made from Eq. (7). These include the relation of Eq. (7) to the expression for TIS, Eq. (1), the bidirectional reflectance distribution function (BRDF), diffuse scattering from surfaces having random roughness, and diffraction from low efficiency gratings. First we will show how Eq. (7) can be used to obtain the expression for TIS as given in Eq. (1). The following assumptions are made: (1) the rough surface is assumed to be perfectly reflecting, i.e., the limit of $|\varepsilon|$ approaches ∞; (2) the angle of incidence θ_0 is 0; (3) the scattering angle θ is limited to angles near the specular direction; and (4) the autocovariance function is assumed to have zero slope at the origin. A convenient form is a Gaussian, which yields, for normal incidence ($\mathbf{k}_0 = 0$),

$$\langle |Z(\mathbf{k})|^2\rangle/L^2 = \pi a^2 \delta^2 \exp(-k^2 a^2/4) \quad (9)$$

where a is the autocovariance length. Equation (9) is obtained by assuming that an ensemble average, denoted by $\langle\ \rangle$, yields a Gaussian autocovariance function of the form $G(\tau) = \delta^2 \exp(-\tau^2/a^2)$. The Fourier transform of $G(\tau)$ yields Eq. (9). We have assumed that the autocovariance length a is much greater than the wavelength, so the surface is gently undulating and has a region about the origin comparable to λ, where the autocovariance function would be quite flat, as is a Gaussian. Physically, this means that the surface height does not vary much over a distance of a wavelength or less. This assumption is also consistent with the Kirchhoff boundary conditions where the surface is assumed to be locally flat to make the Fresnel reflection coefficients applicable.

To calculate the TIS from Eqs. (7) and (9), an integration over the total scattering hemisphere must be performed. Letting $\theta_0 = 0$, $\theta \ll 1$, and $|\varepsilon| \to \infty$, Eq. (7) becomes

$$\frac{dP}{d\Omega} = \frac{(\omega/c)^4}{\pi^2} a^2 \delta^2 \exp\left(\frac{-k^2 a^2}{4}\right) \tag{10}$$

which may be integrated to yield

$$P = (4\pi\delta/\lambda)^2 \tag{11}$$

Equation (11) is the same as Eq. (1) for a perfectly conducting surface ($R_0 = 1$, $|\varepsilon| \to \infty$). Note that because of the condition $a/\lambda \gg 1$ the range of integration on θ has been extended to include the range $(0, \infty)$ because the exponential term becomes negligible several degrees away from the specular direction.

The BRDF is sometimes used in connection with scattering measurements or calculations. It is simply related to Eq. (7), being obtained by dividing Eq. (7) by $\cos \theta$, i.e., BRDF $= (dP/d\Omega)/\cos \theta$. The BRDF is completely symmetrical with respect to interchanging θ_0 and θ. This means that a surface which scatters light incident at θ_0 into an angle θ also scatters reciprocally, so that light incident at an angle θ is scattered into θ_0. For example, s-polarized light incident at θ_0 is scattered into p-polarized light at θ, on the one hand, and p-polarized light incident at θ is scattered into s-polarized light at θ_0. The BRDF is defined by Nicodemus (1970) and is proposed to be a function from which various scattering relationships can be obtained.

Diffuse scattering occurs because the polished surface is not perfectly smooth. The residual microroughness is random and the profile shape is not known in detail *a priori*. To handle random roughness, the surface profile must be measured directly or treated statistically. One statistical method of treating random processes, in principle, is to employ ensemble averages. The angular scattering from random roughness is ensemble averaged and applying this to Eq. (7) leads to the abbreviation

$$\langle|Z(\mathbf{k} - \mathbf{k}_0)|^2\rangle/L^2 = \delta^2 g(\mathbf{k} - \mathbf{k}_0) \tag{12}$$

where $\langle \ \rangle$ denote an ensemble average and $g(\mathbf{k} - \mathbf{k}_0)$, the spectral density function, is a continuous function of $\mathbf{k}$. Equation (12) is the Fourier transform of the autocovariance function, δ^2 is the mean value of the square of the roughness, and L^2 is the area of illumination. If the spectral density function is not evaluated by experimental measurement, one can assume a form for the autocovariance function (such as a Gaussian or exponential) and take the Fourier transform, or assume a form for the spectral density function itself. Experimentally, the spectral density function may be determined from a measurement of the angular distribution of scattered light using Eq. (7), or by measuring surface height data directly (see Section D, following).

Diffraction from low efficiency gratings is much easier to handle analytically because one can calculate $Z(\mathbf{k} - \mathbf{k}_0)$ directly from the periodic profile $z(\mathbf{\rho})$, and an ensemble average is not required. The height profile may be expanded in a Fourier series. Rather than keeping the plane of incidence perpendicular to the direction of the grating grooves as is usually done, more generality is obtained if the angle between the plane of incidence and groove direction is allowed to vary, as shown in Fig. 18. When the area of illumination on the surface is much larger than the incident wavelength,

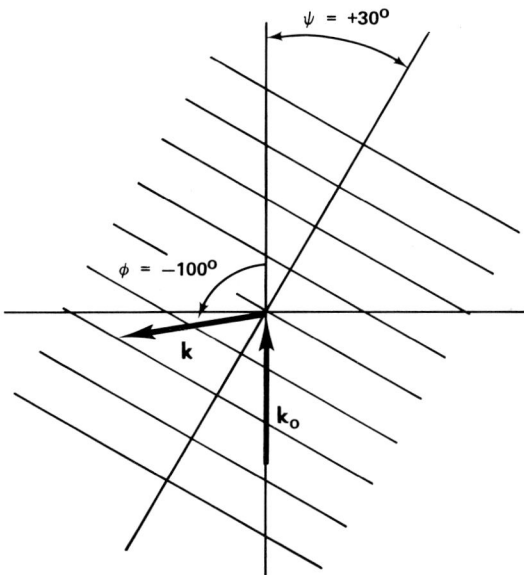

FIG. 18. Schematic representation of diffraction from a grating when there is an angle Ψ between the direction perpendicular to the grooves (shown as parallel lines) and the incident wave-vector component $\mathbf{k}_0$ parallel to the surface. The plane of the paper is the plane of the scattering surface. The wave-vector component $\mathbf{k}$, parallel to the surface, of the -1 diffracted order is at an angle Φ relative to the plane of incidence. The angles Φ and Ψ are positive in the clockwise direction. In this example the grating spacing is 1.1λ, θ_0 (angle of incidence) is $45°$ and θ (angle of diffraction) is $27.5°$. (Elson, 1977.)

the distribution of surface wave vector components is given by

$$\frac{|Z(\mathbf{k} - \mathbf{k}_0)|^2}{L^2} = \pi^2 \sum_{n=-\infty}^{\infty} |C_n|^2 \delta(k_x - k_0 - k_n \cos \psi) \delta(k_y - k_n \sin \psi) \quad (13)$$

where $\delta(x)$ are Dirac δ functions, $k_0 = (\omega/c) \sin \theta_0$ (the plane of incidence is the x–z plane), $k_n = 2\pi n/d$, and d is the grating spacing. The coefficient of expansion in the Fourier series of the periodic profile is C_n and L^2 is the area of illumination. The angle ψ is the angle between the plane of incidence and the direction perpendicular to the grating grooves, as shown in Fig. 18. In Eq. (13) the wave vector component parallel to the surface of the diffracted light is $k = (k_x, k_y)$, where $k_x = (\omega/c) \sin \theta \cos \phi$ and $k_y = (\omega/c) \sin \theta \sin \phi$. The angle θ is the polar diffraction angle (measured relative to the mean surface normal) and ϕ is the azimuthal diffraction angle (measured from the incident plane, the clockwise direction being positive). It is seen that

$$\cos \phi = (k_0 + k_n \cos \psi)/k \quad (14a)$$

$$\sin \phi = k_n \sin \psi/k \quad (14b)$$

where $k = [(k_0 + k_n \cos \psi)^2 + k_n^2 \sin^2 \psi]^{1/2}$ and $\theta = \sin^{-1}(kc/\omega)$. Using Eq. (13), Eq. (7) may be integrated over the scattering hemisphere to yield the result for the fractional amount of incident energy diffracted into the nth grating order:

$$P^{(n)} = \frac{16\pi^2}{\lambda^2} |C_n|^2 \cos \theta \cos \theta_0 |1 - \varepsilon|^2 \left[\frac{|\chi_\theta|^2}{|v - iq\varepsilon|^2} + \frac{|\chi_\theta|^2}{|v - iq|^2} \right] \quad (15)$$

This equation is applicable to low efficiency gratings where the grating amplitude is much less than the incident wavelength. The direction of the diffracted beam (θ, ϕ) is given by Eqs. (14), and the first and second terms in the bracket of Eq. (15) denote diffracted light polarized parallel (p) and perpendicular (s) to the plane of diffraction.

D. Determination of the Spectral Density Function

As discussed in the next section, interferometric or profilometer techniques may be used to measure surface height data directly. These height data yield autocovariance functions $G(\tau)$ and spectral density functions $g(\mathbf{k} - \mathbf{k}_0)$. To see how measured height data are used, we first assume that a continuous one-dimensional record of surface height data $z(\rho)$ is available and that the mean surface level is zero, i.e., $\langle z(\rho) \rangle = 0$, where $\langle \ \rangle$ denote an ensemble average (Lee, 1960). The autocovariance function $G(\tau)$ may be defined as (Lee, 1960)

$$G(\tau) = \langle z(\rho) z(\rho + \tau) \rangle \quad (16)$$

or, under the assumption that the data are stationary and ergodic (Lee, 1960), equivalently as

$$G(\tau) = \lim_{L \to \infty} \frac{1}{L} \int_{-L/2}^{L/2} z(\rho) z(\rho + \tau) \, d\rho \tag{17}$$

From Eq. (7) it is seen that the basic quantity relating to the surface factor is $Z(k)$, which is the Fourier transform of the surface profile, defined as

$$Z(k) = \int d\rho \, z(\rho) e^{-ik\rho} \tag{18}$$

Only one dimension is considered in Eq. (18). Forming the expression $\langle |Z(k)|^2 \rangle$ yields, after some algebraic manipulation, the surface factor $g(k)$ as

$$\delta^2 g(k) = \frac{\langle |Z(k)|^2 \rangle}{L} = \int G(\tau) e^{ik\tau} \, d\tau \tag{19}$$

where L results from integration over ρ. We now need to relate the continuous height data, assumed above, to experimental height data which are discrete and are measured over a finite length interval. Discrete data imply that there are discrete units of lag length τ such that $\tau = m\tau_0 \leq M\tau_0$. Here m is an integer ranging from 0 to M, and τ_0 is the digitization interval. Also, the height data are given as $z(n)$ [corresponding to $z(\rho)$], where $\rho = n\tau_0$. The total number of data points is $N > M$, where M indicates the maximum lag length chosen. The discrete counterpart to Eq. (17) is (Bendat and Piersol, 1971)

$$G_M(m) = \frac{1}{N - m} \sum_{n=1}^{N-m} z(n) z(n + m) \tag{20}$$

and the discrete counterpart to Eq. (19) becomes

$$g(k) = \tau_0 \left(G_M(0) + 2 \sum_{m=1}^{M-1} D(m) G_M(m) \cos km\tau_0 \right) \tag{21}$$

where $D(m)$ is a lag window such that $D(0) = 1$ and $D(M) = 0$. Also, the wave vector k, because of the digitization of the data, has a maximum allowable magnitude $k_{\max} = \pi/\tau_0$. It is convenient to choose $k = K k_{\max}/M$, where $K = 0, \pm 1, \pm 2, \ldots, \pm M$. The product $D(m) G_M(m)$ is often called a modified autocovariance function. The advantage of modifying the autocovariance function by introducing the window function $D(m)$ is so that the autocovariance function can be defined over the infinite range of its argument in order to permit its Fourier transform to be taken. Complete details of analyzing discrete data records of finite length are given elsewhere (Elson and Bennett, 1979). Examples of measured autocovariance functions are given in the next section.

E. Experimental Measurements of Angular Scattering and Surface Statistics

Angular scattering from surfaces whose roughness is small compared to the wavelength is frequently measured by illuminating the surface at normal incidence or near normal incidence and moving a detector in a plane at a constant distance from the surface. If the source is polarized, s- or p-polarized scattered light can be measured, depending on whether the source is polarized perpendicular or parallel to the plane of incidence. However, to avoid surface plasmon coupling at the surface, s-polarized light is more straightforward to measure and interpret. Elson and Bennett (1979), Young (1976), and Stover (1975) have described instruments of the type mentioned above to measure angular scattering. Church et al. (1977) have an arrangement in which the angle between the source and detector is fixed at 90° and the sample rotates, causing the angles of incidence and scattering to vary between 0 and 90°. When the angle of incidence is 0°, the scattering angle is 90°, and vice versa. Other scattering experiments are referenced in the paper by Elson and Bennett (1979). In all scattering experiments it is important to have a well-defined incident beam with a small divergence angle, either focused on the sample or collimated. If the beam profile is Gaussian, it may have to be measured with the sample removed, and then subtracted from the intensity profile of the scattered light. The collection angle should also be well defined and the detector response needs to be linear over several decades. When measuring very low intensities of scattered light, it is imperative to track down and eliminate spurious scattered light from other sources. Since dust in the air is a very effective scatterer, it is desirable to make scattering measurements in a clean room or similar environment.

Statistical properties of optical surfaces may be measured using stereo electron microscopy (Dancy and Bennett, 1976), multiple-beam interferometry (Bennett, 1976), or a surface profiling instrument (Elson and Bennett, 1979). Stereo electron microscopy involves making a replica of the surface for observation in a transmission electron microscope, taking stereo pairs of electron micrographs, and then measuring heights of surface features using a stereo viewer and parallax bar. Figure 19 shows examples of one of a pair of electron micrographs used for stereo measurements. The surfaces have been shadowed obliquely with a platinum–carbon mixture to make surface detail such as the fine scratch visible. Since the surfaces have an overall texture, it is not possible to determine heights of surface features by

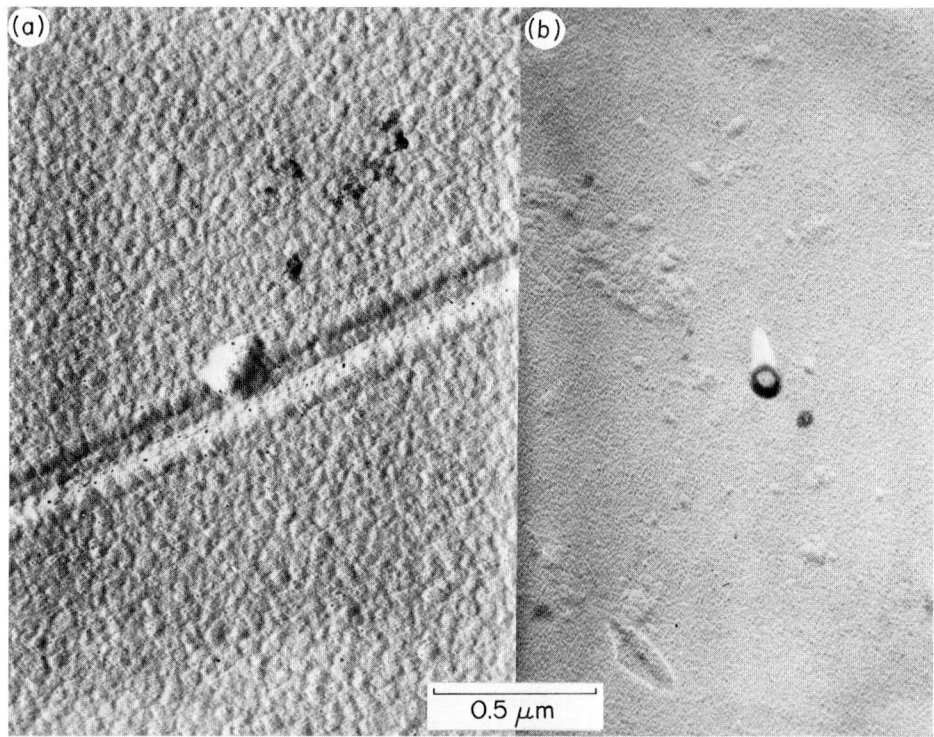

Fig. 19. One print of a stereo pair showing single-stage replicas of (a) silver rapidly sputtered onto a room temperature fused quartz substrate and (b) a bare, bowl feed polished fused quartz substrate. Original total magnification is 60,000X. The scratch in (a) is approximately 0.18-μm wide and 215-Å deep, and could not be seen in an optical microscope.

measuring shadow lengths. Further, the heights of the features are too small to be observable in the scanning electron microscope. Heights of features are measured relative to a reference level, which must also be chosen on the micrograph, and under favorable conditions can be determined with an uncertainty of ±20 Å. In order to obtain surface slope information from stereo electron micrographs, heights of surface features need to be measured, preferably at equally spaced intervals, height differences calculated, and divided by the separations between points. This type of data is tedious to obtain since all measurements must be made visually. Fortunately, surface features which scatter light in the visible and infrared spectral regions have lateral separations which can be detected with a profilometer or interferometer. Thus, stereo electron microscopy is only needed for determining surface statistics for vacuum ultraviolet or x-ray scattering.

Multiple-beam interferometry is a method for obtaining statistics of surface features which have lateral separations of the order of several micrometers. Although the nominal resolution of one such system, a scanning interferometer employing multiple-beam fringes of equal chromatic order (FECO) (Bennett, 1976), was about 2 μm, the measured lateral resolution was somewhat larger (Elson and Bennett, 1979). FECO interferometers have the advantage that all portions of the surface can be observed interferometrically, so a representative place can be selected for the statistical measurements. The height sensitivity obtainable with the FECO scanning interferometer is about 3 Å rms. Figure 20 shows an autocovariance function of a polished fused quartz surface obtained with the scanning interferometer. The curve is an average of 51 scans taken at different places on the surface. It can be compared with Fig. 21 taken on the same surface with a profilometer, as described in a following paragraph. A scanning Fizeau interferometer has also been used to measure surface statistics (Eastman and Baumeister, 1974). Its lateral resolution, which was determined by the effective detector size was 7 μm and the height sensitivity was ~20 Å rms. Fizeau interferometers can give information about the surface roughness along the length of the interference fringe in the field of view. In order to inspect other portions of the surface, the wedge angle of the interferometer must be adjusted, which is a tedious operation.

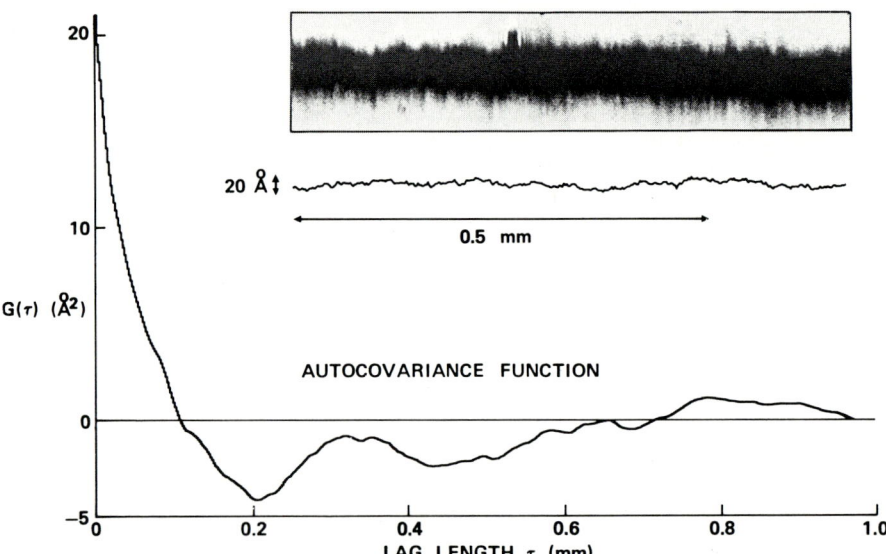

FIG. 20. Autocovariance function obtained from interferometric data for an 18.6 Å rms roughness polished fused quartz surface. Curve is an average of 51 scans. A photograph of a FECO interference fringe and TV scanning camera trace are also shown. (Bennett, 1976.)

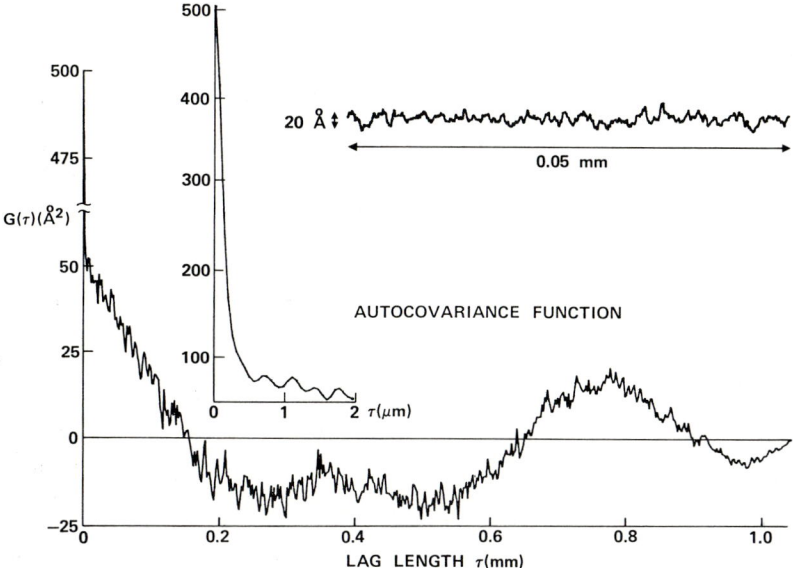

FIG. 21. Autocovariance function and surface scan obtained from profilometer data on the same surface as in Fig. 20. Data are for a single surface scan. (Elson and Bennett, 1979.)

A surface profiling instrument is probably the best method for obtaining surface statistics to be used in the vector scattering theory for calculating angular scattering because of its excellent height sensitivity and good lateral resolution. One such instrument (Elson and Bennett, 1979), whose output is digitized and fed into a minicomputer for statistical calculations, has a height sensitivity of about 1 Å rms and a lateral resolution of about 0.1 μm. The lateral resolution, which is a small fraction of the stylus radius, is possible because the slopes on the polished surfaces are small. The loading on the stylus is adjustable and can be set so that no permanent marks are made on the surface (Dancy and Bennett, 1977). The main drawback to all stylus-type instruments is that one cannot preselect a typical area on the sample except by visual observation using a low power microscope. Table I

TABLE I
COMPARISON OF METHODS FOR OBTAINING
STATISTICS OF OPTICAL SURFACES

	Height sensitivity	Lateral resolution	Maximum length
FECO interferometer	~3 Å rms	~2.0 μm	1 mm
Talystep profilometer	~1 Å rms	~0.1 μm	2 mm
Stereo microscopy	±20 Å	~5 Å	~1 μm

summarizes the characteristics of the three methods of determining surface statistics.

The digitized surface scan data obtained with the profilometer are used to calculate height distribution functions, such as those shown in Fig. 4, slope distribution functions, rms roughnesses, rms slopes, and autocovariance functions. An autocovariance function obtained from a single 1-mm long scan on a polished fused quartz surface is shown in Fig. 21. (This is the same surface whose autocovariance function obtained with the FECO scanning interferometer was shown in Fig. 20.) The profilometer shows fine structure in the autocovariance function, such as the initial sharp spike produced by closely spaced scratches on the surface that was not resolved by the scanning interferometer. However, the general shape of both curves is similar for larger values of the lag length τ. The initial spike in the data in Fig. 21 has an approximately exponential shape because the surface data points were not taken close enough together to show the continuous nature of the surface. Exponential autocovariance functions are suggestive of random surface data (Elson and Bennett, 1979). Figure 22 shows the autocovariance function of a chemically polished sapphire sample that was locally very smooth but had a long-range waviness. In this case, the initial portion of the autocovariance function is a very good Gaussian, as indicated by the dashed

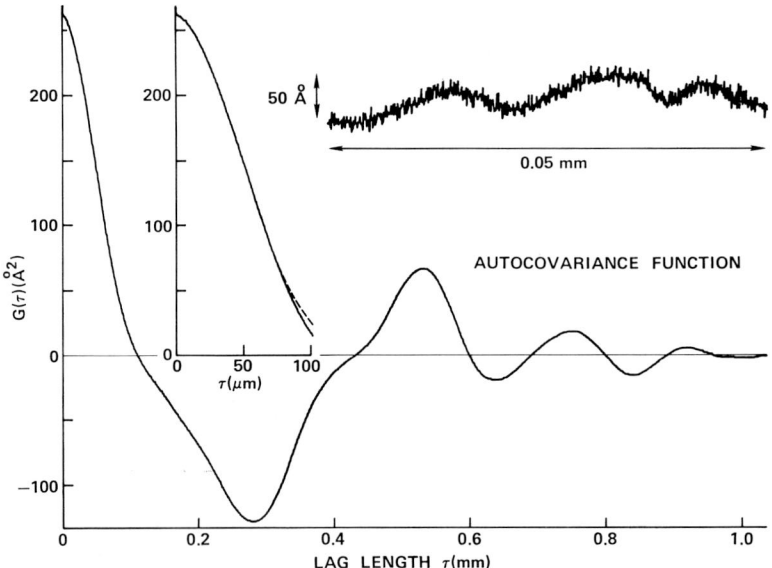

FIG. 22. Autocovariance function and surface scan obtained from profilometer data on a chemically polished sapphire surface: initial portion of the curve is an excellent fit to a Gaussian (---). Data are for a single surface scan. (Elson and Bennett, 1979.)

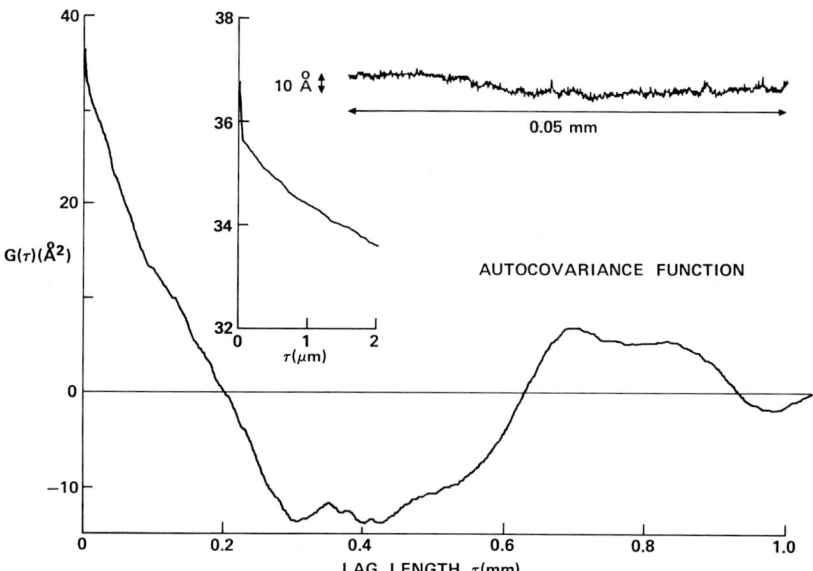

FIG. 23. Autocovariance function and surface scan obtained from profilometer data on a polished silicon carbide surface. Data are for a single surface scan. (Elson and Bennett, 1979.)

curve which almost completely duplicates the measured curve in the insert. This is one of a very few samples we have found with Gaussian autocovariance functions. The initial portions of most of the curves are closer to exponentials, but generally cannot be fit to any simple analytic function. Figure 23 shows a curve of this type for silicon carbide. This material can be polished to an extremely low scatter surface devoid of pits and scratches, similar to the surface on bowl feed polished fused quartz (Dietz and Bennett, 1966), and is a promising material for ultraviolet and x-ray mirrors (Rehn et al., 1977; Choyke et al., 1977). The shape of the autocovariance function is similar to that for other scratch-free but slightly wavy surfaces.

Surfaces with periodic components, such as diamond-turned surfaces or gratings have oscillatory autocovariance functions. Figure 24 shows the initial portion of a curve for one such surface, a diamond-turned copper sample made at the Lawrence Livermore Laboratory. The periodicity of about 14 μm appearing on this surface is not the basic groove separation, but is a longer-range effect probably caused by vibration between the tool and workpiece. An excellent article on scattering from diamond-turned surfaces has recently been published by Church et al. (1977), who show that the angular scattering profile consists of scattering peaks caused by periodic components of surface roughness superimposed on a continuous scattering background produced by the residual random roughness on the surface.

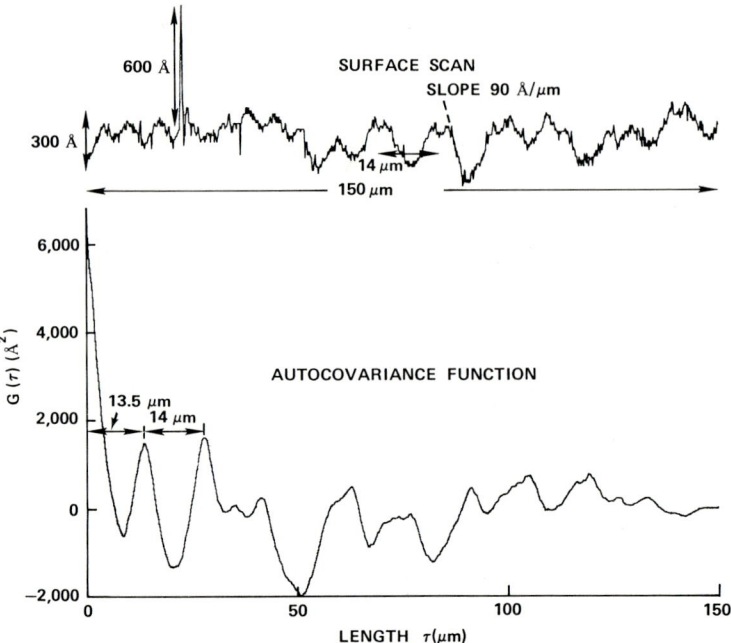

FIG. 24. Initial portion of autocovariance function and surface scan obtained from profilometer data on a diamond-turned copper sample made at the Lawrence Livermore Laboratory. Data are for a single surface scan. (Becker *et al.*, 1978.)

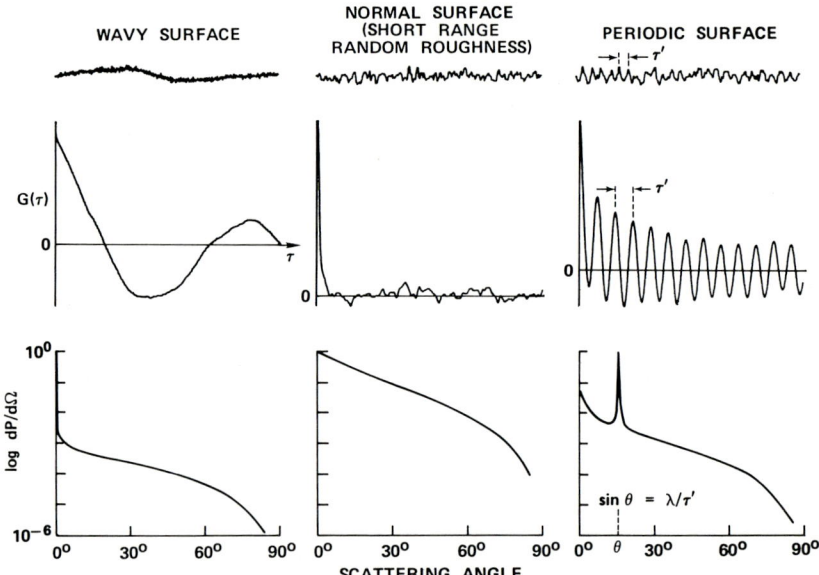

FIG. 25. Three basic types of surface profiles and their corresponding autocovariance functions and angular scattering curves. (Elson and Bennett, 1979.)

By examining the autocovariance functions in the preceding four figures, one can see common components of surface structure (Elson and Bennett, 1979). These are summarized in Fig. 25 and consist of (1) long-range waviness, (2) short-range random roughness, and (3) periodicity. The autocovariance functions and angular scattering curves for these components are also shown. Surfaces having long-range waviness scatter light predominantly very close to the specular direction, while those with short-range random roughness (most polished optical surfaces) scatter over a wider range of angles. Surfaces having only discrete periodic components will scatter light only at angles given by the grating equation, Eq. (2), as discussed in Section IV.A. To calculate angular scattering from surfaces that have more than one component of surface structure, each type of roughness can be treated separately, the angular scattering calculated, and then the scattering curves superposed (to first order). This type of approach is possible because any rough surface can be treated as a superposition of different types of surface structure.

Figure 26 shows two angular scattering curves calculated from measured surface statistics compared with angular scattering curves measured on the same samples (Elson and Bennett, 1979). The measured (dashed) curve for the rougher surface was matched to the calculated curve in the 20–40° range of scattering angles, but no other adjustments were made. The agreement between theory and experiment is encouraging, but not yet quantitative.

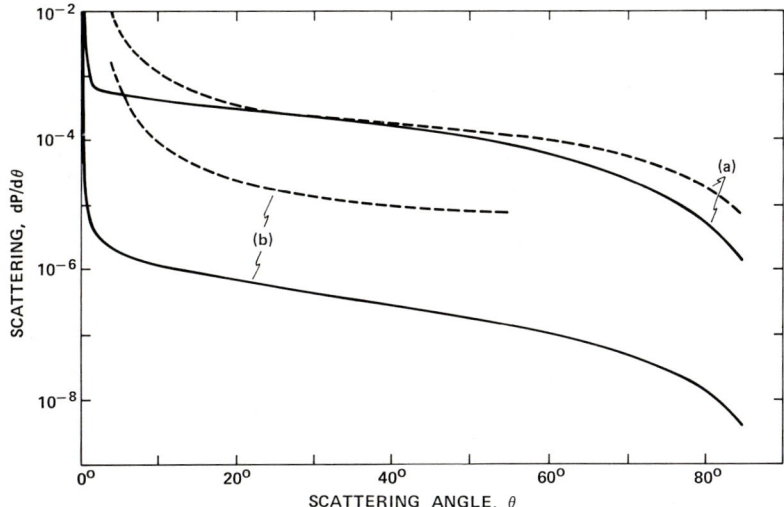

FIG. 26. (———) Calculated and (– – –) measured curves showing the angular scattering for polished fused quartz surfaces whose rms roughnesses are (a) 18.6 Å and (b) 5.2 Å. The measured curve for the rougher surface was matched to the calculated curve in the 20–40° range of scattering angles; no other adjustments were made. (Elson and Bennett, 1979.)

In particular, it appears that there may be another unaccounted for scattering mechanism operating in the case of the smoother sample. One possibility is dipole scattering from isolated particulates which may increase the entire scattering level without affecting the shape of the curve.

V. SCATTERING FROM DIELECTRIC MULTILAYERS

Optical surfaces having one or more overcoats are very common. These include tarnish or oxide layers, protective coatings, antireflection coatings, reflection increasing coatings, band pass or band elimination filters, polarizing beam splitters, etc. Along with the increased versatility to be gained by using multilayers comes the possibility of increased scattering, either from within the individual layers or at the interfaces between the layers. Although scattering within a film may be important for certain types of materials, it generally can be neglected relative to the scattering at the interfaces. For this reason we will consider only scattering at the interfaces of the multilayers. Possibilities of complicated interference effects in the scattered light from the multilayers arise because (1) the optical thicknesses of the multilayers can enhance or reduce the scattered light by interference effects just as they affect the specularly reflected or transmitted light and (2) correlation effects of the roughness across the surface of a layer or cross correlation effects of the roughness between layers can modify the angular distribution of the scattered light. Effects of correlated roughness on a single surface have already been discussed in Section IV, where we have shown how different autocovariance functions for rough surfaces produce different angular distributions of the scattered light.

Only a few people have considered scattering from surfaces having a single dielectric overlayer (Elson, 1976b, Mills and Maradudin, 1975) and even fewer have considered scattering from multilayers. Eastman (1974) used a scalar theory to investigate the TIS from surfaces covered with multilayers and Elson (1976a, 1977) used a vector perturbation method which yields the angular distribution of the scattered light. Details of the vector theory to be used here have already been published (Elson, 1977; Scott and Elson, 1978) and will not be repeated. The multilayer scattering formulas are rather complicated algebraically but are in closed form. The only assumption is that the roughness of any interface must be much less than the wavelength. In the theory the dielectric constants of the substrate and/or multilayers may be complex and the thicknesses and number of layers, the incident beam polarization, and the angles of incidence and scattering can all be specified as inputs.

We have applied the multilayer scattering formulas to two situations: (1) diffraction from a low efficiency grating covered with a reflectance enhancing dielectric stack and (2) diffuse scattering from a metallic mirror covered with a dielectric multilayer stack. In the first case we will consider the polarization properties of the light diffracted into the -1 grating order and in the second case we will compare the angular dependence of the light scattered by a single rough surface. Autocorrelation and cross correlation effects of the roughness, polarization of the incident beam, and the thicknesses of the layers will be shown to be important.

A. Polarization Effects in Low Efficiency Gratings

The periodic rough surfaces are assumed to have a rectangular profile with the groove width equal to half of the groove spacing (duty cycle 0.5), and the groove depth H is assumed to be much less than the wavelength. The grating surface is covered with silver, with dielectric constant $\varepsilon_s = -16.4 + i0.53$. The dielectric stack on top of the silver is assumed to be composed of alternating layers of MgF_2 with $\varepsilon_l = 1.9$ and ZnS with $\varepsilon_h = 5.29$, each of $\lambda/4$ optical thickness at the design wavelength of 6328 Å. The grating surface has a spacing d of 1.1λ, or 6961 Å, and the physical thicknesses of the MgF_2 and ZnS layers are 1148 and 673 Å, respectively. Thus, a 10 (20) layer dielectric stack would have a total physical thickness of 0.91 μm (1.82 μm). Since the physical thicknesses of the mutilayer coatings are so large, there is a question of whether the rectangular profile on the silver grating surface replicates through the multilayers. Fortunately, there is experimental evidence which indicates that the degree of replication in multilayers on grating surfaces may be quite good (Elson, 1976b, 1977). Thus, we will assume that replication does occur, so that there is perfect correlation between all the surfaces of the multilayers, as well as perfect correlation between the grooves on any one surface.

In the following example we will consider the polarization ratio of the p- and s-reflected components in the -1 diffracted order, assuming unpolarized incident light. This information is of importance if, for example, one is using a low efficiency grating as a beam sampler and wants the sampled light in the -1 diffracted order to be independent of the polarization of the incident beam. Although the following calculations are for a grating with a rectangular profile, calculated polarization ratios (to first order) also hold for other shape profiles such as triangular and sinusoidal. However, the *intensities* of the diffracted orders do depend on the profile shape.

Figure 27 shows the calculated polarization ratio in the -1 diffracted order as a function of angle of incidence for a value of $\psi = 20°$, where ψ is the angle between the plane of incidence and the groove direction, as shown in Fig. 18. We are assuming that the light is p polarized (s polarized) when the electric field is parallel (perpendicular) to the plane of incidence or diffraction. The plane of incidence (diffraction) is defined by the incoming (diffracted)

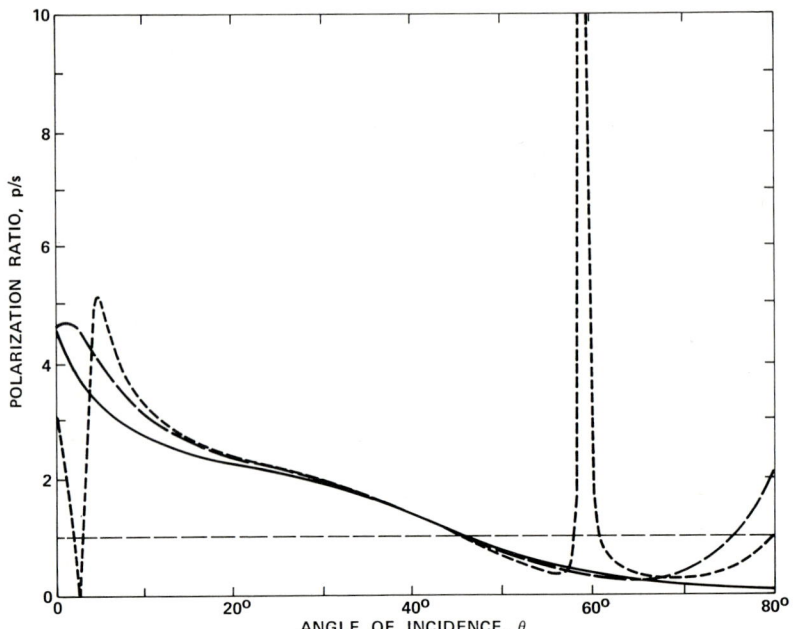

FIG. 27. Calculated polarization ratio in the -1 diffracted order of a low efficiency grating with a rectangular profile as a function of angle of incidence, for $\Psi = 20°$: (———) bare silver grating surface; the dashed curves are for a silver grating surface overcoated with (— —) 10 or (- - -) 20 alternating $\lambda/4$ layers of ZnS and MgF_2. Both sets of multilayers perfectly replicate the grating groove profile. The horizontal dashed line indicates a polarization ratio of unity. Other conditions are given in the text.

wavevector and the normal to the grating surface. The polar and azimuthal angles θ and ϕ of the diffracted beams in the -1 order can be calculated from Eqs. (14). In Fig. 27 are shown curves for the polarization ratio of (1) a bare silver grating surface, (2) a silver grating surface overcoated with 10 alternating layers of ZnS and MgF_2, and (3) a silver surface overcoated with 20 alternating layers of ZnS and MgF_2. All dielectric layers are assumed to have $\lambda/4$ optical thickness at normal incidence. It is clear from Fig. 9 that the polarization ratios in the 20-layer example behave quite differently from the other two cases, exhibiting anomalies at angles of incidence θ_0 of 2.4 and 59.5°. For an explanation of this anomalous behavior, consider for the moment the polarizing angle for a simple dielectric stack on a transparent substrate without any grating grooves present (planar interfaces). For p-polarized light incident at the polarizing angle, 59.5°, the specular reflectance of the 10 (20) layer stack is about 54% (0.025%), while both stacks have nearly 100% reflectance for s-polarized light. Hence, the 20-layer stack could serve as an excellent polarizer with the p-polarized light being transmitted and the s-polarized light reflected. At the polarizing angle the 20-layer stack does not allow p-polarized light to be specularly reflected. If the transparent substrate were replaced by a metallic substrate with grating grooves present, then light incident in the 2.0–3.0° range yields a -1 diffracted order which coincides with the polarizing angle. Thus, the dip in the p/s ratio at $\theta_0 = 2.4°$ in Fig. 27 is caused by the polarizing angle effect. The anomaly appearing at $\theta_0 = 59.5°$ is also a direct result of the polarizing angle effect. The reason for this anomalous increase in p-polarized light stems from the deep penetration of the p-polarized field into the stack when incident at the polarizing angle. The large p-polarized field strength yields large scattering currents which consequently generate an anomalous increase in the -1 order. A more detailed discussion of these effects is given elsewhere (Elson, 1979).

Two other important points are illustrated in Fig. 27. First, there are situations where it is possible to obtain polarization ratios of unity and hence the -1 diffracted order of the grating can be made polarization independent for an unpolarized incident beam. This result is also true when the optical thicknesses of the layers in the stack are $\lambda/4$ at the angle of incidence at which the grating is being used instead of at normal incidence, as is assumed in Fig. 27. Second, the choice of the incident wavelength does not affect the appearance of anomalies or the ψ dependence if the layer thicknesses are adjusted appropriately. In other words, the anomalies illustrated in Fig. 27 will also be seen at 10.6 μm at angles related to the correct polarizing angle for 10.6 μm.

B. ANGULAR SCATTERING FROM SURFACES HAVING RANDOM ROUGHNESS

The anomalies noted in the preceding section for gratings coated with multilayer dielectric films have analogues in the case of scattering from rough silver surfaces coated with multilayer dielectric films, as will be discussed in this section. Here we will consider a rough polished surface coated with silver and then overcoated with 20 alternating layers of ZnS and MgF_2, each of $\lambda/4$ optical thickness at the design wavelength, 6328 Å. Two cases will be considered: (1) uncorrelated roughness where the surfaces of the multilayers are all rough and have the same autocovariance function, but the roughness does not replicate through the stack, and (2) correlated roughness where it does. An exponential autocovariance function is assumed for each surface in both cases, with an autocovariance length of 0.35 μm. Autocovariance functions of similar form have been measured for polished quartz surfaces. The situation for real multilayer films is probably somewhere between the two extremes of correlated and uncorrelated roughness. There is evidence that evaporated silver films can replicate steps on surfaces (Koehler and Eberstein, 1953) and multilayer films can replicate grating profiles (Elson, 1976b, 1977), so presumably the substrate roughness will be replicated by multilayer dielectric films evaporated on it. However, for very smooth substrates, i.e., those having roughnesses ~5 Å rms or less, the crystallites of the evaporated or sputtered films which generally have lateral dimensions of 1000 Å or less, may modify the roughness of the substrate and produce a partially correlated situation.

Figure 28 shows $dP/d\Omega$, the scattered light per unit solid angle (incident intensity assumed to be unity), plotted as a function of polar scattering angle θ for (1) a rough silver surface, (2) a rough silver surface coated with 20 dielectric layers having correlated roughness, and (3) a rough silver surface coated with 20 dielectric layers having no roughness correlation. The light is assumed to be normally incident and the scattered light is p polarized, i.e., polarized in the plane defined by the incident wave vector and the polarization of the incident beam. In this case the azimuthal angle ϕ is 0° for positive values of θ, denoting scattering in the forward quadrant, and 180° for negative values of θ, denoting scattering into the backward quadrant. It is seen that there is very little difference in the general shapes of the three curves; all are peaked in the specular direction. However, the magnitude of the scattering for

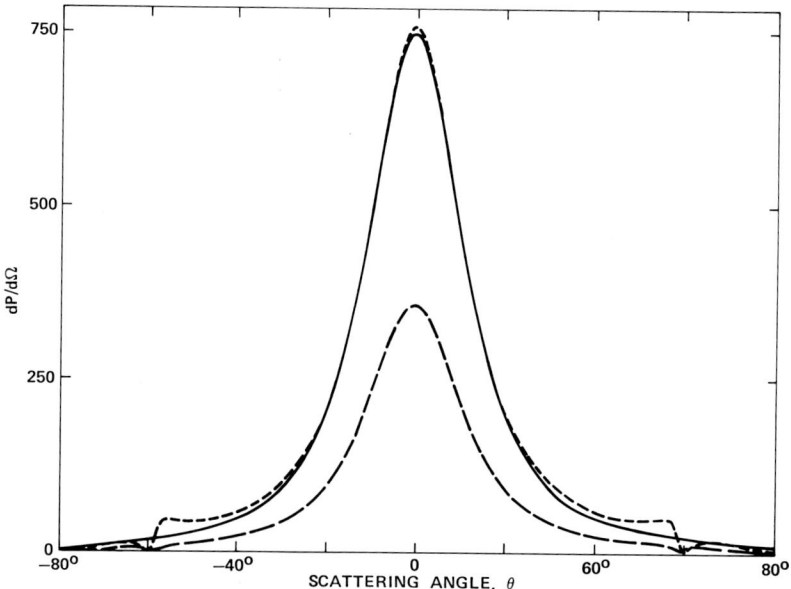

FIG. 28. Calculated scattered light per unit solid angle $dP/d\Omega$ plotted as a function of polar scattering angle θ assuming normally incident light and p-polarized scattered light: (——) rough bare silver surface; the dashed curves are for a rough silver surface overcoated with 20 alternating $\lambda/4$ layers of ZnS and MgF$_2$ whose roughnesses at the film interfaces are (– – –) correlated or (— —) uncorrelated. Other conditions are given in the text.

the uncorrelated roughness is much lower than for the correlated roughness. Evidently the $\lambda/4$ optical thicknesses of the dielectric films suppress the scattered light when the roughnesses on the surfaces are uncorrelated since the scattered light from the bare silver surface is almost identical to the scattered light from the correlated case. The anomalies associated with the polarizing angle that were discussed in the preceding section are also present here, but to a much lesser extent. It is seen that there is an abrupt discontinuity in the scattered light for the correlated case at scattering angles near $\pm 60°$. The effect is shown to a lesser extent for the uncorrelated case. It appears that it is possible that a $\lambda/4$ dielectric stack may not only enhance the specular reflectance of a metal surface, but at the same time decrease the amount of scattering if the stack can be made so that the roughnesses at the interfaces of the layers are as small and as uncorrelated as possible. The subject of scattering from multilayers having uncorrelated and correlated roughnesses is discussed further by Elson (1979) and Scott and Elson (1978). They show that the anomalous effects at scattering angles of $\pm 60°$ do not appear for s-polarized scattered light and that the corresponding s-polarization curves are nearly identical to the curves in Fig. 28 except for the anomalies.

VI. SUMMARY

In this chapter we have described the types of defects that generally occur on optical surfaces and the kinds of scattering they produce. Theoretical treatments of scattering are grouped into categories, depending on the nature of the scatterers (isolated particulates, scratches, correlated random microroughness, or periodic surface profiles), and their sizes (large or small compared to the wavelength of light).

(1) The scalar theory relating TIS to rms microroughness for Gaussian distributed surface irregularities whose heights are small compared to the wavelength is in good agreement with experiment for most polished optical surfaces. As would be expected, predictions based on the scalar theory disagree with experiment in cases where the surface is particulate covered, heavily scratched, etched, or pitted. These surfaces do not have Gaussian height distributions. There is also disagreement between theory and experiment for many diamond-turned surfaces whose rms slopes are quite different from those on polished glass surfaces. Surfaces on which the predominant lateral separation of microirregularities is smaller than a wavelength also scatter less than is predicted by a simple application of the scalar theory.

(2) Vector scattering theory can, in principle, predict the angular distribution of scattered light from a surface having correlated microroughness provided that the two-dimensional spectral density function for the surface

is known. Experimentally, only one-dimensional surface statistics have been measured and there are mathematical difficulties in using these data to accurately calculate two-dimensional scattering from areas on surfaces. Qualitative agreement between theory and experiment has, however, been obtained.

(3) Mie theory can, in principle, be used to calculate scattering from isolated, uncorrelated particulates. In practice, this theory is only tractable when the particles are assumed to have simple shapes, such as spheres, ellipsoids, platelets, etc., sizes in a limited range, and known dielectric constants. There are difficulties therefore in using it to calculate accurate scattering levels from real surfaces covered by dust particles whose shapes, size distribution, and dielectric constants are largely unknown. Also, interaction effects between the particulates and the surface on which they rest are difficult to handle theoretically. Qualitative agreement between theory and experiment for dust-covered surfaces is, however, probable.

(4) Scattering from scratches, digs, and other surface defects whose dimensions are large compared to the wavelength can, in principle, be handled by geometrical optics. However, in practice this is difficult because the exact shapes, sizes and orientations of facets making up the surface defect must be known in detail. Typically, there are not enough of these defects on the surface to be able to apply statistics with any degree of confidence. The TIS from scratches and digs is in many cases related approximately to their widths, making some quantification of the scratch–dig standards that are used for inspection and quality control of optical surfaces possible.

(5) The scalar theory predicting the scattering from densely packed, correlated surface microirregularities whose dimensions and separations are comparable to or larger than a wavelength is in qualitative agreement with experiment, but the height distribution function and autocovariance function for the rough surface must both be known.

(6) Finally, both TIS and the angular dependence of scattering from microrough surfaces covered with dielectric multilayers can now be handled theoretically for any angle of incidence provided that the roughness is small compared to the wavelength, scattering within the layers is neglected, the autocovariance function for the surface roughness at the interfaces is known, and the correlation between the roughnesses of the different multilayers is either total or absent. Predictions from this multilayer vector scattering theory have not yet been verified experimentally.

REFERENCES

Archibald, P. C., and Bennett, H. E. (1978). *Opt. Eng.* **17**, 480.
Beaglehole, D., and Hunderi, O. (1970). *Phys. Rev. B* **2**, 309.

Becker, D. L., Bennett, J. M., Foileau, M. J., Porteus, J. O., and Bennett, H. E. (1978). Surface and optical studies of diamond turned and other metal mirrors, *Opt. Eng.* **17**, 160.
Beckmann, P., and Spizzichino, A. (1963). "The Scattering of Electromagnetic Waves From Rough Surfaces." Macmillan, New York.
Bendat, J. S., and Piersol, A. G. (1971). "Random Data: Analysis and Measurement Procedures," p. 311. Wiley (Interscience), New York.
Bennett, H. E. (1978). Scattering characteristics of optical materials, *Opt. Eng.* **17**, 480.
Bennett, H. E., and Porteus, J. O. (1961). *J. Opt. Soc. Am.* **51**, 123.
Bennett, H. E., and Stanford, J. L. (1976). *J. Res. Natl. Bur. Stand., Sect. A* **80**, 643.
Bennett, J. M. (1976). *Appl. Opt.* **15**, 2705.
Berreman, D. W. (1970). *Phys. Rev. B* **1**, 381.
Bloembergen, N. (1973). *Appl. Opt.* **12**, 661.
Celli, V., Marvin, A., and Toigo, F. (1975). *Phys. Rev. B* **11**, 1779.
Chandley, P. J. (1976). *Opt. Quantum Electron.* **8**, 323, 329.
Chandley, P. J., and Welford, W. T. (1975). *Opt. Quantum Electron.* **7**, 393.
Chinmayanandam, T. K. (1919). *Phys. Rev.* **13**, 96.
Choyke, W. J., Partlow, W. D., Supertzi, E. P., Venskytis, F. J., and Brandt, G. B. (1977). *Appl. Opt.* **16**, 2013.
Church, E. L., and Zavada, J. M. (1975). *Appl. Opt.* **14**, 1788.
Church, E. L., Jenkinson, H. A., and Zavada, J. M. (1977). *Opt. Eng.* **16**, 360.
Dancy, J. H., and Bennett, J. M. (1976). *In* "High Energy Laser Mirrors and Windows," Semiannu. Rep. Nos. 7 and 8, pp. 166–176. NWC TP 5845 (July). Naval Weapons Center, China Lake, California.
Dancy, J. H., and Bennett, J. M. (1977). *In* "High Energy Laser Mirrors and Windows," Annu. Rep. No. 9, pp. 133–144. NWC TP 5988 (Nov.). Naval Weapons Center, China Lake, California.
Davies, H. (1954). *Proc. Inst. Electr. Eng.* **101**, 209.
Dietz, R. W., and Bennett, J. M. (1966). *Appl. Opt.* **5**, 881.
Ditchburn, R. W. (1964). "Light," 2nd Ed., pp. 209–210, 394. Wiley (Interscience), New York.
Eastman, J. M. (1974). Ph D Thesis, Univ. of Rochester, Rochester, New York.
Eastman, J. M., and Baumeister, P. W. (1974). *Opt. Commun.* **12**, 418.
Elson, J. M. (1975). *Phys. Rev. B* **12**, 2541.
Elson, J. M. (1976a). "Low Efficiency Diffraction Grating Theory," AFWL-TR-75-210 (Mar). Kirtland Air Force Base, Albuquerque, New Mexico.
Elson, J. M. (1976b). *J. Opt. Soc. Am.* **66**, 682.
Elson, J. M. (1977). *Appl. Opt.* **16**, 2872.
Elson, J. M. (1979). *J. Opt. Soc. Am.* **69**, 48.
Elson, J. M., and Bennett, J. M. (1979). *J. Opt. Soc. Am.* **69**, 31.
Elson, J. M., and Ritchie, R. H. (1971). *Phys. Rev. B* **4**, 4129.
Elson, J. M., and Ritchie, R. H. (1974). *Phys. Stat. Solidi B* **62**, 461.
Fung, A. K., and Chan, H. (1969). *IEEE Trans. Antennas Propag.* **AP-17**, 590.
Fung, A. K., and Moore, R. K. (1966). *J. Geophys. Res.* **71**, 2939.
Hagfors, T. (1966). *J. Geophys. Res.* **71**, 379.
Holzer, J. A., and Sung, C. C. (1976). *J. Appl. Phys.* **47**, 3363.
Holzer, J. A., and Sung, C. C. (1977). *J. Appl. Phys.* **48**, 1739.
Jackson, J. D. (1962). "Classical Electrodynamics." Wiley, New York.
Kerker, M. (1969). "The Scattering of Light and Other Electromagnetic Radiation." Academic Press, New York.
Koehler, W. F., and Eberstein, A. (1953). *J. Opt. Soc. Am.* **43**, 747.
Kozawa, S. (1962). *In* "Proceedings of the Conference on Optical Instruments and Techniques" (K. J. Habell, ed.), pp. 410–428. Chapman and Hall, Ltd., London.

Kröger, E., and Kretschmann, E. (1970). *Z. Phys.* **237**, 1.
Leader, J. C. (1971a). "Spectral and Angular Dependence of Specular Scattering from Rough Surfaces," MCAIR 71-013 (Apr). McDonnell Aircraft Company, Saint Louis, Missouri.
Leader, J. C. (1971b). *J. Appl. Phys.* **42**, 4808.
Leader, J. C., and Fung, A. K. (1977). *J. Appl. Phys.* **48**, 1736.
Lee, Y. W. (1960). "Statistical Theory of Communication," pp. 200–218. Wiley, New York.
Maradudin, A. A., and Mills, D. L. (1975). *Phys. Rev. B* **11**, 1392.
Mie, G. (1908). *Ann. Phys. (Leipzig)* **25**, 377.
Mills, D. L., and Maradudin, A. A. (1975). *Phys. Rev. B* **12**, 2943.
Nicodemus, F. E. (1970). *Appl. Opt.* **9**, 1474.
Orr, C., Jr., and Dallavalle, J. M. (1959). "Fine Particle Measurement," p. 119. Macmillan, New York.
Peake, W. H. (1959). *IRE Trans. Antennas Propag.* **AP-7**, Spec. Suppl., S324.
Petit, R. (1975). *Nouv. Rev. Opt.* **6**, 134.
Porteus, J. O. (1963). *J. Opt. Soc. Am.* **53**, 1394.
Rayleigh, Lord (1945). "The Theory of Sound," Vol. 2. Dover, New York.
Rehn, V., Stanford, J. L., Baer, A. D., Jones, V. O., and Choyke, W. J. (1977). *Appl. Opt.* **16**, 1111.
Scott, M. L., and Elson, J. M. (1978). *Appl. Phys. Lett.* **32**(3), 158.
Silver, S. (1947). "Microwave Antenna Theory and Design," p. 161. McGraw-Hill, New York.
Stover, J. C. (1975). *Appl. Opt.* **14**, 1796.
Stratton, J. A. (1941). "Electromagnetic Theory," pp. 415, 566. McGraw-Hill, New York.
Truong, V. V., and Scott, G. D. (1976). *J. Opt. Soc. Am.* **66**, 124.
van de Hulst, H. C. (1957). "Light Scattering by Small Particles." Wiley, New York.
Welford, W. T. (1977). *Opt. Quantum Electron.* **9**, 269.
Young, R. P. (1976). *Opt. Eng.* **15**, 516.
Zaki, K. A., and Neureuther, A. R. (1971). *IEEE Trans. Antennas Propag.* **AP-19**, 208.

CHAPTER 8

Adaptive Optical Techniques for Wave-Front Correction

JAMES E. PEARSON, R. H. FREEMAN,
and
HAROLD C. REYNOLDS, JR.

United Technologies Research Center
Optics and Applied Technology Laboratory
West Palm Beach, Florida

I. Introduction	246
A. Problem Definition	246
B. Sources of Phase Degradation	248
C. Correction Concepts	254
D. Elements of an Active Optical System	259
II. Adaptive System Types	262
A. Active versus Adaptive Optics	262
B. Modal versus Zonal Correction	264
C. Error Sensing Algorithms	268
III. Phase Shifting Elements	281
A. Introduction	281
B. Acousto-Optical/Electro-Optical Elements	284
C. Segmented Mirrors	284
D. Continuous Surface Mirrors	285
IV. System Design Considerations	299
A. Control Bandwidth Requirements	299
B. Phase Correction Requirements	302
C. Signal-to-Noise Considerations	307
V. Compensation Performance	312
A. Turbulence Compensation	312
B. Thermal Blooming Compensation	322
C. Other Compensation Techniques	327
VI. Target Considerations	329
A. Introduction	329
B. Potential Solutions to Target Problems	330
C. Isoplanatism and Extended Target Referencing	331
D. Speckle Modulation	332
E. Range Considerations	336
References	337

I. INTRODUCTION

A. Problem Definition

From an engineering point of view, the entire history of optics has been an effort to increase the quality of components and systems, that is, to minimize or eliminate distortions that prevent achievement of diffraction-limited performance. Great advances have been made in developing optical forms, design techniques, and manufacturing methods that minimize fixed distortions. Achievement of diffraction-limited performance is usually limited by high manufacturing cost, however, and standard techniques can do nothing for time-varying distortions. The potential solution to these limitations is the development of "adaptive" optical techniques: techniques for measuring and removing fixed and time-varying distortions in real time. Successful development of such techniques is expected to have a major impact on the design and performance of both noncoherent (imaging) systems and coherent (laser) systems. Some of the potential application areas are listed in Table I.

In any optical system, there are two quantities that can be controlled: intensity and phase. As is well known, however, changes in the phase distribution within an optical system have a substantially greater impact on performance than do variations in the intensity. The primary role of adaptive optics is thus measurement and removal of deleterious phase perturbations. Although systems that operate on the intensity have been conceived and may be important in some applications, in this chapter we will discuss only techniques for measuring and controlling phase.

At this point, it is worthwhile to define what we mean by the "performance" of an optical system. For imaging systems, the performance will be measured by the width or height of the point-spread function in the image plane. For laser systems where the goal is to focus a beam onto a point, the measure will be the width or height of the focal-plane irradiance distribution.

TABLE I

Application Areas for Adaptive Optical Techniques

High energy lasers
Astronomical imaging
Large aperture imaging systems
Multiple beam pointing, phasing, and alignment (e.g., laser fusion)
Large optical systems (to relieve manufacturing and alignment tolerances)
Power transmission via lasers
Laser communications
Photolithography

---

8. ADAPTIVE TECHNIQUES FOR WAVE-FRONT CORRECTION

In each case, the most commonly used quantity is the Strehl ratio:

$$S = \text{Strehl ratio} = \frac{\text{observed peak focal plane irradiance}}{\text{diffraction-limited peak irradiance}} \quad (1)$$

Adaptive optical techniques can be applied to both noncoherent (passive imaging) and coherent (active laser) systems. The basic concepts for both types of systems are similar, although the detailed implementations can be quite different. In this chapter, we will concentrate on the use of coherent optical adaptive techniques (COAT) in laser systems because the operation of transmitted-beam systems is somewhat easier to understand and because most of the research accomplished to data has been with such COAT systems. Ample references are given, however, so that further reading can be done on adaptive techniques for imaging applications.

Cast in terms of a coherent laser transmitter, the problem that a COAT system addresses is shown in Fig. 1. Figure 1a indicates that in the absence of any distortions, a diffraction-limited beam can be focused by a laser array. Any real system, however, must cope with phase distortions; the usual result is indicated in Fig. 1b: a distorted beam. If we could somehow measure the phase distortions in the propagation path and then apply the inverse (conjugate) phase as in Fig. 1c, a nearly diffraction-limited beam can again be achieved. This is what a COAT system tries to do: determine the phase distortions automatically in real time and apply a time-dependent "predistortion" to the transmitted beam wave front so that the net result after propagation through the atmosphere is a diffraction-limited wave front, even for time-varying distortions. It is in this sense that we employ the term "adaptive" for the systems considered here.

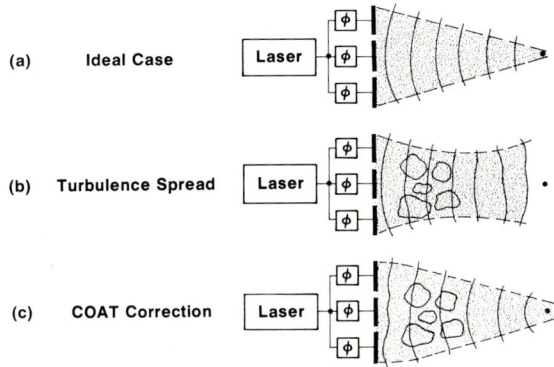

FIG. 1. Schematic representation of: (a) a phased laser array focused through an ideal propagation medium; (b) array defocusing caused by atmospheric or other phase errors; (c) adaptive refocusing by proper "predistortion" of the radiated wave front.

B. Sources of Phase Degradation

1. General Formulation

In designing an adaptive system to correct phase errors, both the spatial and temporal properties of the wave-front errors must be accounted for. A useful way to express the optical phase error ϕ_e at a given location (the wave-front error) is

$$\phi_e(\mathbf{r}, t) = \sum_{n=1}^{\infty} a_n(t) f_n(\mathbf{r}) \qquad (2)$$

where $f_n(\mathbf{r})$ is a function set orthogonal over the aperture of interest and $a_n(t)$ are time-dependent amplitude coefficients. It will occasionally be convenient for us to deal with wave-front or path length variations rather than actual optical phase. The two quantities are related, of course, by

$$\text{phase variations} = \Delta\phi = k \cdot (\text{path length variations}) = k\,\Delta L \qquad (3)$$

where k is the optical wave vector, $k = 2\pi n/\lambda$, with n the index of refraction, and λ the optical wavelength.

Two quantities that determine the requirements placed on the adaptive optical system are the spatial power spectrum $P_s(\kappa)$ and the temporal power spectrum $P_t(\nu)$, of the phase error function ϕ (the spatial frequency is κ and the temporal frequency is ν). These two quantities can be found from the appropriate transform integral of Eq. (2) and are given by

$$P_s(\kappa) = \langle \Phi_e^{\,2}(\kappa) \rangle = \sum_{n=1}^{\infty} \langle a_n^{\,2} \rangle F_n^{\,2}(\kappa) \qquad (4)$$

$$P_t(\nu) = \langle \Phi_e^{\,2}(\nu) \rangle = \sum_{n=1}^{\infty} A_n^{\,2}(\nu) \langle f_n^{\,2} \rangle = \sum_{n=1}^{\infty} A_n^{\,2}(\nu) \qquad (5)$$

where the brackets $\langle \ \rangle$ represent a time average in Eq. (4) and a spatial average over the aperture in Eq. (5). An assumption is made that $\langle f_n^{\,2} \rangle = 1$ and the A_n (Fourier transform of a_n) are statistically independent with zero mean. The capital letter functions in Eqs. (4) and (5) are Fourier transforms of the corresponding quantities in Eq. (2). The width of the spatial frequency spectrum P_s will determine the required number of phase control elements across the optical aperture and the height will set the phase excursion requirement. The width of temporal spectrum P_t will determine the required control system bandwidth and the height will set the error rejection (servoloop gain) requirement. There are thus four quantities of interest: (i) N_a, the number of correction degrees of freedom; (ii) $(\Delta\phi_c)_{\max}$, the maximum phase correction excursion; (iii) f_s, the servosystem control bandwidth; (iv) G_s, the servosystem gain. Section IV discusses the requirements for each of these quantities as set by the various distortions discussed in the next section.

TABLE II
PROPAGATION DISTORTIONS THAT CAN LIMIT THE
PERFORMANCE OF OPTICAL SYSTEMS

Coherent laser transmitting systems
    Atmospheric turbulence
    Nonlinear atmospheric lensing (thermal blooming)
    Laser medium phase distortions
    Optical train errors and distortions (fixed and flux-induced)
    Pointing and tracking jitter

Passive imaging systems
    Atmospheric turbulence
    Optical train errors and distortions (increasingly important for large optics)
    Mechanical mount vibration and jitter

---

2. Phase Distortion Sources

The phase distortions that limit the performance of an optical system can originate in the device (for laser systems), in the optical train, or in the propagation medium (usually the atmosphere). Table II lists the primary types of distortion that adaptive systems must deal with. Most of the distortions are linear, in that they do not depend on the optical system state or on the laser power. The primary exception is thermal blooming (for a thorough review of thermal blooming, see Smith, 1977; see also Whinnery et al., 1967; Gebhardt and Smith, 1969; Smith, 1969), which is an intensity-dependent phase error that is caused by absorption in the atmosphere and that produces a negative-lens effect to spread the beam. Under certain conditions, another nonlinear phenomenon known as kinetic cooling can occur (Wallace and Camac, 1970; Wood et al., 1971). In this case, the atmospheric absorption acts like a *positive* lens rather than a negative lens. In some cases, this effect can produce a "channeling" of the beam that can *increase* the focal-plane irradiance over its diffraction-limited value. It can also offset the negative effects of thermal blooming. Kinetic cooling has been demonstrated in the laboratory (Gebhardt and Smith, 1972), but its beneficial (or deleterious) effects have not been observed under actual atmospheric propagation conditions.

Each of the distortions listed in Table II places different requirements on an adaptive optical system designed to remove them. These requirements are discussed in Section IV. A fundamental requirement in all cases, however, is that the distortion must be equivalent to a phase-only error at the adaptive correction plane. In other words, if the distortion is propagated* or transformed to the adaptive correction plane and the Fresnel diffraction effects are removed, then the amplitude changes in the resultant wave must be negligible

---

* By propagating a diffraction-limited wave through the distortion to the correction plane.

for full correction to be possible. This requirement is related to the earlier statement that a COAT system operates only on the *phase* of the beam. This restriction places a limit on the amplitude of distortions at a given location that can be corrected. Qualitatively, large amplitude phase errors near the transmitter can be fully corrected, while those near the focal plane cannot. On the other hand, the closer a fixed amplitude phase error is to the focal plane, the less effect it has on the focal-plane irradiance or Strehl ratio.

The qualitative discussion presented above can be quantified by following an analysis developed by Herrmann (1977). The propagation phase errors are approximated by a single thin lens of strength K placed some distance Z from the transmitter as shown in Fig. 2. For an undisturbed beam ($K = 0$), a wave-front curvature at the transmitter of $C = R$ gives the maximum target plane irradiance (minimum beam radius at the target). The correction limitation with a lenslike distortion is illustrated by assuming a Gaussian beam that has a $1/e$ intensity radius of a_t. The beam radius a_f in the target plane is given by

$$a_f^2 = a_t^2[1 + CR + KR(1 - z) + KR_c^2 z(1 - z)]^2 \\ + a_t^2[1 + KRz(1 - z)]^2/N_f^2 \tag{6}$$

where $N_f = ka_t^2/R$ is the transmitted beam Fresnel number,* $z = Z/R$, and all other quantities are defined in Fig. 2.

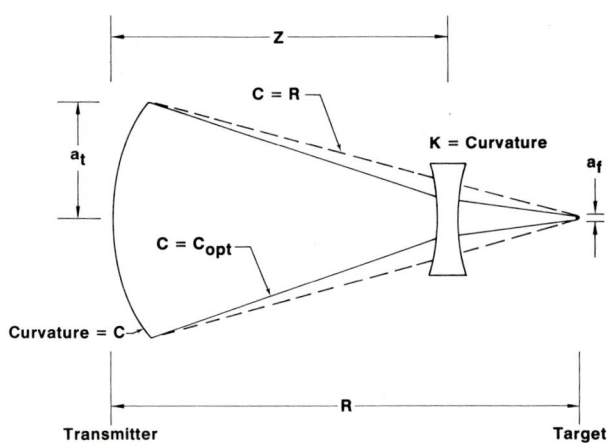

FIG. 2. Geometry of transmitted beam whose curvature C is adjusted to correct for a thin lens placed a distance Z from the transmitter that changes the beam curvature by K. The target range is R, the transmitter radius is a_t, and the focal plane beam radius is a_f.

---

* Sometimes the Fresnel number is written as $N_f = a_t^2/R\lambda$.

8. ADAPTIVE TECHNIQUES FOR WAVE-FRONT CORRECTION

A COAT system will attempt to minimize a_f by adjusting the curvature C of the transmitted beam. For linear distortions (K = constant) such as turbulence, the minimum value of a_f and the value of C at which it occurs are

$$(a_f)_{\min} = a_t[1 + KRz(1 - z)]/N_f \tag{7}$$

$$C_{\text{opt}} = -[1 + KR(1 - z)]/R[1 + KRz(1 - z)] \tag{8}$$

These quantities are plotted in Fig. 3 as functions of z. Note that the lens is least correctable when it is located at $z = \frac{1}{2}$, has no effect at $z = 1$, and can be perfectly corrected if it is at $z = 0$. For nonlinear distortions such as thermal blooming, the situation is qualitatively the same in the sense that errors near the transmitter can be corrected better than those near the target. Quantitatively, however, the situation is very different. Unlike a fixed lens, the nonlinear blooming lens gets stronger closer to the target diameter, because the curvature K is dependent on the beam and is given approximately as (Herrmann, 1977)

$$K(z) = S[a_t/a(z)]^3 \tag{9}$$

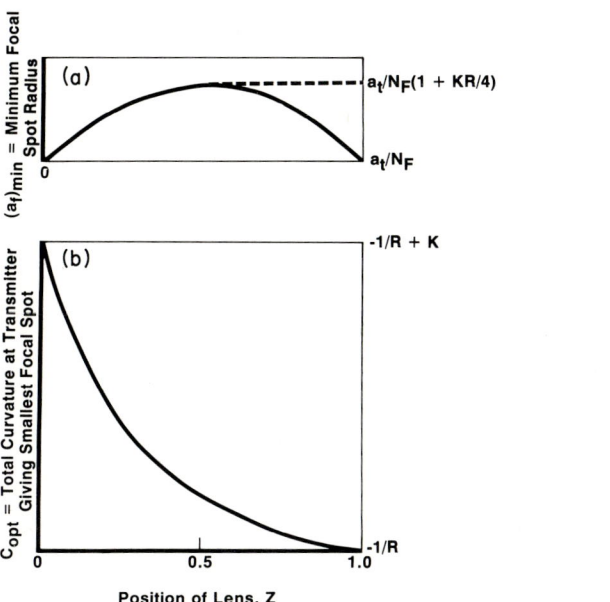

FIG. 3. Phase correction for a Gaussian beam when a single fixed lens of strength K located at a distance $z = Z/R$ between the transmitter and the target. (a) minimum focal spot radius, Eq. (7), after optimally choosing the transmitted beam curvature; (b) optimum transmitter curvature, Eq. (8).

where $a(z)$ is the beam diameter at the lens. The shorter optical lever arm is thus offset by a stronger lens, with the net result that the reduction in focal-plane irradiance is relatively independent of the lens location. This observation is also borne out in calculations by Hogge (1974). The ability of a COAT system to compensate the phase distortion is reduced, however, the farther the distortion is from the transmitter. This point is discussed further in Section II.D and the reader is referred to the article by Herrmann (1977) for further elaboration on this point.

In most situations, the propagation distortions do not occur in a single plane, but instead are distributed continuously from the transmitter to the focal plane. Because of the effects described above, a COAT system will be able to compensate only for those phase errors that occur at some fraction of the focal-plane distance. The fraction depends on the strength and character of the errors. For turbulence, if the phase is perfectly compensated, the amplitude effects, known as scintillation, will limit the Strehl ratio to (Tatarski, 1961)

$$S = \exp(-4\sigma_X^2) \qquad (10)$$

where σ_X^2 is the log–amplitude variance given for spherical waves by

$$\sigma_X^2 = 0.124 k^{7/6} C_N^2 Z^{11/6} \qquad (11)$$

An important parameter for turbulence is the atmospheric correlation length that has been defined by Fried (1965) for plane wave propagation as

$$r_0 = 1.68(C_N^2 Z k^2)^{-3/5} \qquad (12)$$

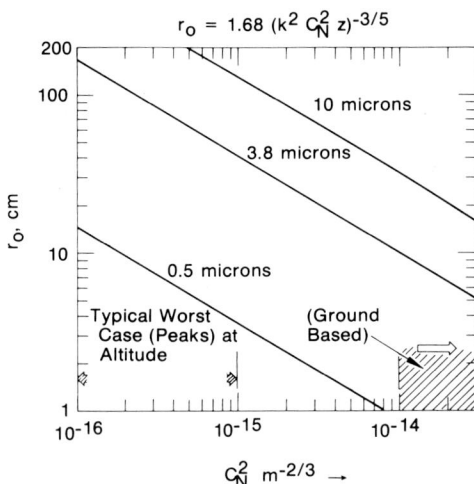

FIG. 4. Atomspheric coherence length r_0 as a function of the turbulence structure constant for $z = 4$ km. Plane wave propagation is assumed. For focused beams, (r_0) spherical wave $= 1.8 \, (r_0)_{\text{plane wave}}$. From Bufton (1973) (worst case) and Dowling and Livingston (1973) (ground based).

where C_N is the refractive index structure constant, Z is the range, and k is the optical wave vector. Equation (12) holds for plane waves; for spherical waves (focused beams), r_0 is increased (Fried, 1966) by a factor of 1.8. Figure 4 plots r_0 as a function of C_N^2 for several laser wavelengths of interest and for $Z = 4$ km.

When thermal blooming is present throughout the propagation path, the situation is more complicated because of the dependence of the distortion on the beam and the atmospheric parameters. The characteristics of thermal blooming are conveniently described for a gas using a distortion parameter, N_D, defined by Bradley and Herrmann (1974) as

$$N_D = \frac{\alpha P}{\rho C_p V_0 \varepsilon_0} \frac{d\varepsilon}{dT} \frac{kZ}{a} \tag{13}$$

where α is the absorption coefficient, P the total power, C_p the specific heat of medium, ρ the medium density, a the initial radius of beam ($1/e$ intensity for Gaussian, $D_t/2\sqrt{2}$ for aperture of diameter D_t), k the wave number, V_0 the wind velocity, ε_0 the free space dielectric constant, $d\varepsilon/dT$ the change of dielectric constant with temperature, and Z the range. The quantity N_ω is a beam slewing parameter given by

$$N_\omega = \omega Z / V_0 \tag{14}$$

where ω is the angular slew rate.

The characteristics of thermal blooming are illustrated in Fig. 5, taken from Bradley and Hermann (1974). For fixed conditions, as the total transmitted power increases, the peak target irradiance increases to a maximum

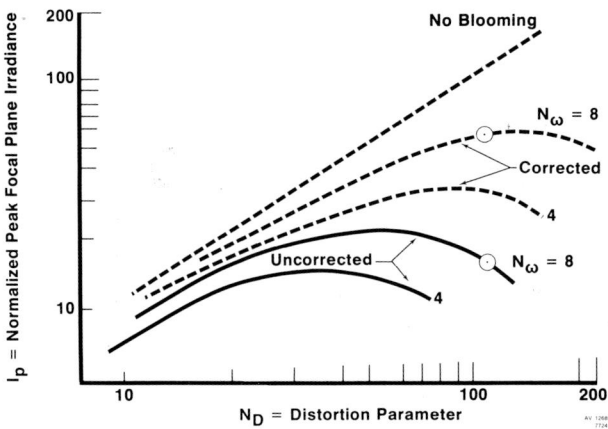

FIG. 5. Normalized peak focal plane irradiance $I_p = N_D A_o/A$, where A_o is the diffraction-limited focal area and A the focal area with thermal blooming present, as function of the distortion number N_D.

at $P = P_c$, the "critical power," and then decreases for further increases in transmitted power. Curve fitting to computer simulation studies have shown (Bradley and Herrmann, 1974; Hayes and Ulrich, 1975) that the Strehl ratio in the presence of blooming is given by

$$S_B = (1 + R_B^2)^{-1} = [1 + 0.72(P/P_c) + (P/P_c)^2]^{-1} \tag{15}$$

where the critical power in watts can be found from (Takken and Cordray, 1974; Brown 1975b)

$$P_c \approx \frac{(4 \times 10^8) V (D_t)^{1/2} (\lambda N_f)^{5/4} \exp(-2\alpha Z/3)(1 + 0.6\omega Z/V)}{\alpha Z^{3/4}} \tag{16}$$

The form of Eq. (15) suggests one way to evaluate the effect of combined sources of phase error by first defining the Strehl ratio when each distortion is considered separately as

$$S_i = (1 + R_i^2)^{-1} \tag{17}$$

where R_i is the ratio of an effective increase in beam area, caused by the ith distortion, to the diffraction-limited beam area. The net Strehl ratio from many distortions is then found from

$$S = (1 + \sum R_i^2)^{-1} \tag{18}$$

For turbulence and thermal blooming, computer simulations (Brown, 1975b) have shown this to be a valid procedure. The Strehl ratio after correction with a COAT system is then evaluated by calculating new values of R_i for each distortion and then applying Eq. (18). This procedure will be used in Section IV to evaluate system performance.

C. Correction Concepts

There are numerous ways to classify COAT systems. One way is to distinguish whether the phase errors are sensed by the beam going to the target or by the energy returned from the target. With this distinction, there are only two types of adaptive optical systems—outgoing-wave and return-wave—as illustrated in Fig. 6. In the outgoing-wave system of Fig. 6a, the phase errors are sensed by the outgoing beam as it propagates to the target. The peak focal-plane energy is sensed directly by a detector in the target plane or indirectly by maximizing the power reflected back to a receiver. In any case, the effect of the phase errors on the *return* energy is of no consequence in the control loop. The multidither control algorithm discussed in Section II.C is the only known outgoing-wave system that includes the target in the control loop.

(a)

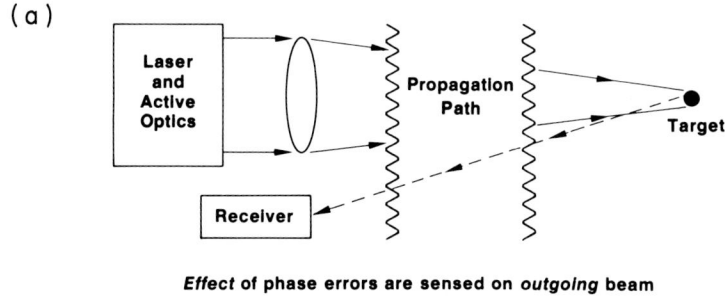

Effect of phase errors are sensed on *outgoing* beam
Coherent only (multidither only known technique)

(b)

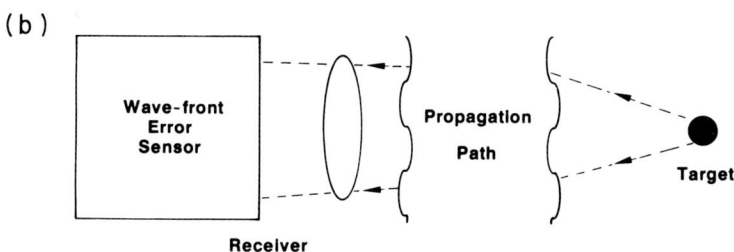

Phase errors sensed on wave fronts *returning* from the target
Coherent or nonconerent operation

FIG. 6. Two basic types of adaptive optical system: (a) outgoing wave (b) return wave.

In return-wave systems (Fig. 6b), the phase errors are sensed by the energy returned from the target. A point source on the target will return an undistorted spherical wave to the receiver. The error sensing system makes a measure of the deviations between the actual return wave and the assumed ideal wave. If the conjugate of these phase errors is impressed on the outgoing beam, then by reciprocity the beam should focus to a diffraction-limited spot (subject to the restrictions discussed in Section I.B). Return-wave systems are often called "phase conjugate" systems, but we prefer to use this term to describe the heterodyne detection system discussed in Section II.B.

The energy used by return-wave systems can be reflected laser light, reflected sunlight, or self-emissions from the target. Without a laser, these systems become "compensated imaging" systems (Muller and Buffington, 1974; Buffington *et al.*, 1977a,b; Hardy *et al.*, 1977). For laser systems, there are three basic types of target-return COAT systems as illustrated in Fig. 7. The fundamental difference between these systems is the manner of determining the wave-front propagation phase errors; for linear (not power-dependent) distortions, each system ultimately attempts to impress the

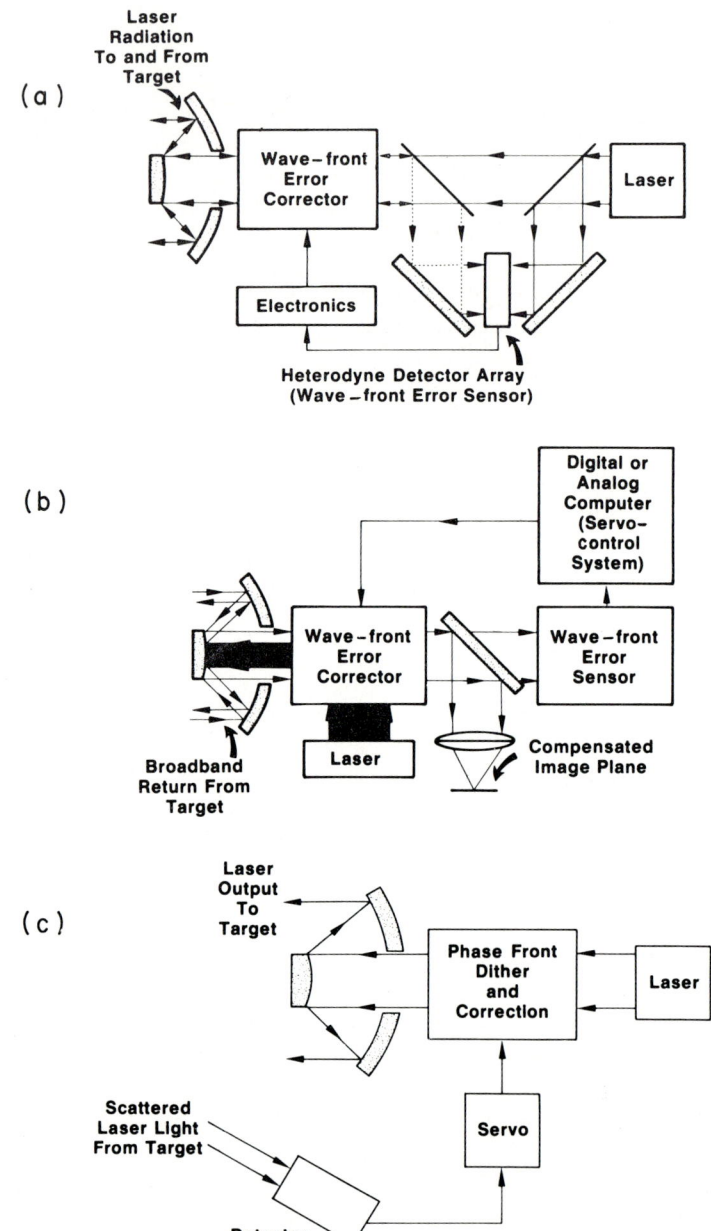

FIG. 7. Three types of transmitting-COAT systems. Each one is basically a closed-loop servomechanism capable of sensing and removing wave-front phase errors: (a) phase conjugate; (b) image compensation or imaging-COAT system used to correct a transmitted beam (TRIM-COAT); (c) multidither, outgoing wave.

conjugate of the phase distortion onto the transmitted wave front. The phase conjugate system in Fig. 7a uses heterodyne detection to determine the actual phase error, interfering a portion of the transmitted beam with the laser radiation reflected from a bright spot (a "glint") on the target. The image compensation system (sometimes called imaging-COAT or "I-COAT") in Fig. 7b produces optical path length compensation using scattered laser light, broadband scattered light, or self-emission from the target. Such systems can make an actual phase measurement (Hardy *et al.*, 1977) or can use intensity-only techniques (Muller and Buffington, 1974; Buffington *et al.*, 1977a,b). Neglecting any dispersion, a laser beam can then be sent along the same compensated optical path and will experience no net distortion. The systems in Figs. 7a and 7b are both return-wave systems. The one in Fig. 7b is sometimes referred to as "TRIM-COAT" since it combines a *tr*ansmitted beam with *im*aging COAT. The third system is an outgoing-wave multidither system. Table III compares these three types of transmitting-COAT systems.

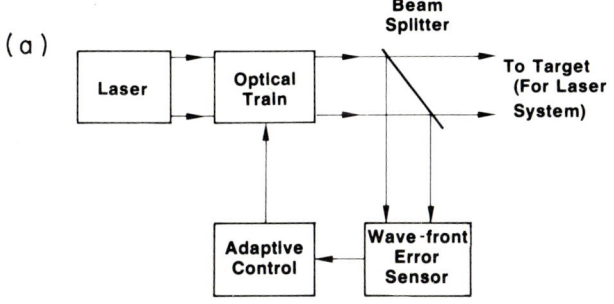

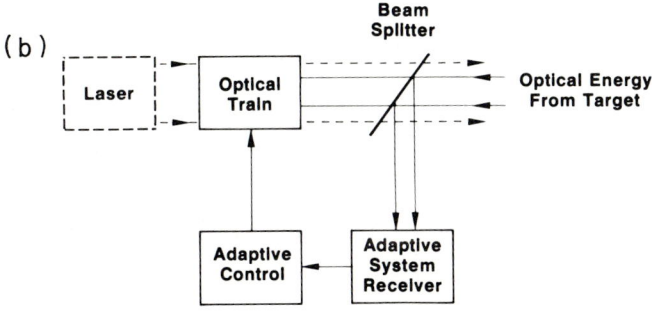

FIG. 8. Adaptive optical systems classified by the location of the sensed phase errors: (a) local-loop control (b) target-loop control.

TABLE III
A COMPARISON OF THREE BASIC CLASSES OF COAT SYSTEMS

System	Common path optical backscatter interference	Required number of receivers	Internal path length matching required	Type of detection	Doppler tracking required	Compensation for laser internal distortions	Operation with broad-band laser sources	Referencing options
Multidither outgoing-wave	None to minimal	1	No	Direct	No	Yes	Yes	Glint, edge, black hole; extended targets
Phase conjugate	Yes	N	Yes	Heterodyne	Yes	No[c]	Severe problems	Glint only
Imaging COAT or TRIM[a]-COAT	None for incoherent referencing	$1-N^b$	No	Direct	No	No[c]	Yes	Glints, features, on extended targets incoherent or coherent illumination or self-emission

[a] Hybrid system consisting of transmitting–imaging COAT systems.
[b] Many types of such systems are possible.
[c] Not possible on a continuous basis with one servosystem. It is possible using two servosystems or by time-sharing a single system.

There is another general way to classify COAT systems: by the location of the phase errors they sense and correct. Figure 8 schematically shows that a COAT system can operate either in a "local-loop" mode or in a "target-loop" or target-return mode. A local-loop system samples the outgoing-beam and removes distortions that occur up to the sampling plane. This system is also referred to as a "beam clean-up" system. A target-loop system corrects for propagation-path errors. If the target-loop system is also an outgoing-wave system, it can sense and correct for *all* phase errors, including any produced in the laser. If it is a return-wave system the laser device is not corrected and the optical train can be corrected only from the aperture-sharing element outward. A return-wave system thus needs to be used with a local-loop system if the entire path from inside the device to the target is to be corrected. The primary advantage of a return-wave system is that it can work with more distant targets than can the outgoing-wave system. This point is discussed further in Section VI. Its primary disadvantage is a more complex receiver and its inability to correct *all* distortions including those in the laser device. Further comparisons of outgoing and return-wave systems are presented in Section II.C.

D. Elements of an Active Optical System

The details of the hardware in a COAT system depends on which of the types illustrated in Fig. 7 are being considered and on the application. There are four elements, however, that are common to every system. These elements shown schematically in Fig. 9 are: (1) optical train and receiver; (2) phase error sensor; (3) servocontrol system; (4) optical phase shifting element. In this section, we will discuss the optical train and receiver; Section II discusses phase error sensors; Section III deals with the phase corrector element; and Section IV treats issues relating to the servocontrol system.

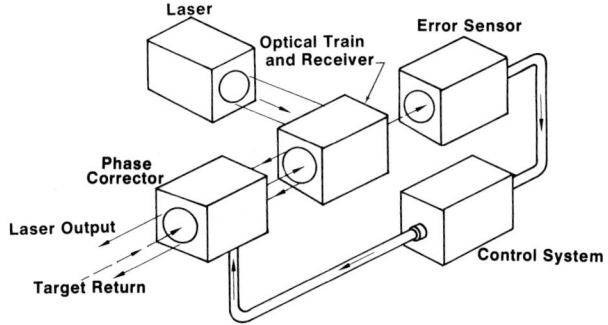

FIG. 9. Major elements of an adaptive optical system.

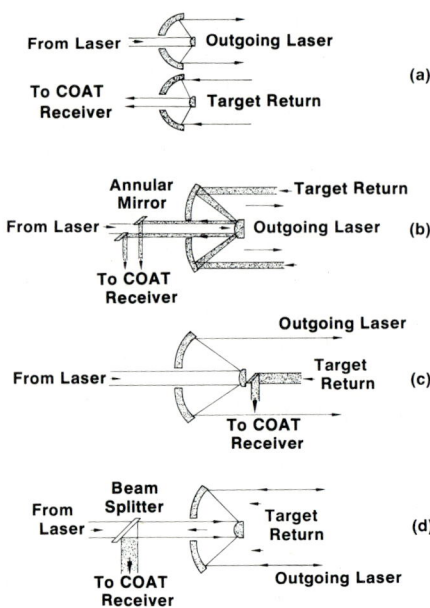

FIG. 10. Alternate receiver aperture configurations: (a) separate transmitting and receiving apertures; (b) annular receiver around the transmitting aperture; (c) circular receiver in the obscuration of an annular transmitted beam; (d) fully shared common receiving and transmitting apertures.

The critical element in the optical train for an adaptive system is the aperture-sharing element—the element that separates the target return energy from the outgoing beam or that obtains a low-power sample of the outgoing beam for a local-loop control system. The four basic aperture-sharing configurations that can be used are illustrated in Fig. 10. In selecting a receiver, the distinction between outgoing-wave and return-wave systems comes into play. One of the characteristics of an outgoing-wave system is that its ability to correct for phase errors is independent of the location of the phase errors, the optical receiver, and the phase correction element. Any of the configurations in Fig. 10 could in principle be used with equal performance since the receiver is just a photon "bucket." The best practical choice will depend on factors such as collecting aperture (signal to noise), backscatter suppression, and target effects (see Section VI).

Return-wave systems cannot use the side-by-side arrangement of Fig. 10a, since the energy returning from the target will traverse a completely different path from the outgoing beam. Return-wave systems rely on reciprocity and therefore work most effectively with the fully shared aperture in Fig. 10d. For some applications, however, an annular receiver or an on-axis receiver

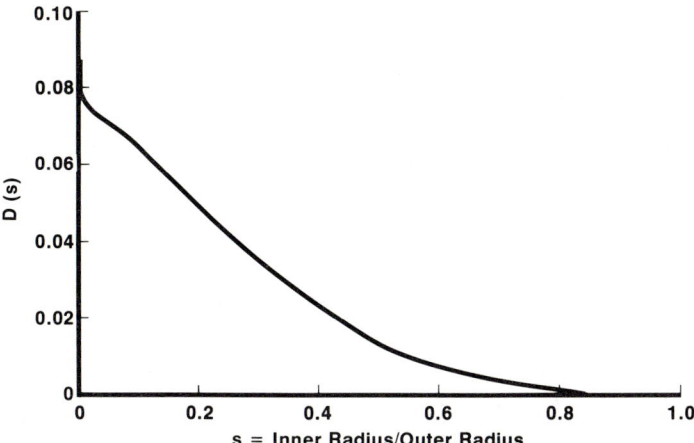

FIG. 11. Residual error coefficient $D(s)$ versus $s = a/b$ for phase compensation of an annular aperture of outer radius b and inner radius b using a phase measurement along only the outer perimeter of the aperture [see Eq. (12)].

in the transmitter obscuration can be used. Butts and Hogge (1977) have investigated the use of a thin annular receiver (Fig. 10b) together with an interpolation formula that calculates the interior phase based on a perimeter measurement. They calculate the residual phase error for the specific case of atmospheric turbulence compensation using a thin annular receiver of mean radius b and an annular transmitter aperture having an inner radius of a and an outer radius of b. Their result is

$$\langle \Delta\phi_e^2 \rangle = D(s)(2b/r_0)^{5/3} \qquad (19)$$

where r_0 is given by Eq. (12), $s = a/b$, and $D(s)$ is the function shown in Fig. 11. The value of $D(s)$ for an uncompensated aperture varies from 1.032 for a circular aperture to 1.839 for an infinitesimally thin annulus. The residual phase error for the compensated beam is thus reduced to 8.5% (0.088/1.032) or less of the uncompensated phase error for a *full* aperture. For a full-aperture receiver, the compensation would be perfect within the approximations of this analysis. As Fig. 11 shows, the perimeter measurement plus interpolation becomes more effective as the transmitter aperture becomes a narrower annulus.

A comparison of the various aperture geometries shown in Fig. 10 has been made by Greenwood (1977a) for the case of compensating turbulence-induced tilt errors only. As expected, the off-axis separate receiver is the least effective while the performance of all the "shared" aperture configurations is nearly identical.

II. ADAPTIVE SYSTEM TYPES

A. Active versus Adaptive Optics

The terms "active optics" and "adaptive optics" are frequently used interchangeably, but the distinction is an important one. An *active* optical system or the active components thereof, is one capable of reshaping a wave front by adding a set of controllable path differences. No closed loop servo-control that operates on the actual wave-front error information is present, so an alternative terminology could be predictive active optics or open-loop active optics. An *adaptive* optical system includes a closed-loop phase error sensor that acts to drive a set of path differences to zero. With this definition, compensation systems such as holography (Kogelnik and Pennington, 1968) and speckle interferometry (Knox, 1976), are not adaptive optics.

Having made this distinction, we now consider the system shown in Fig. 12, where a single active optic (a deformable mirror) is shown controlled by any one of three possible sensors. The predictive-loop system (loop A) can be used to compensate for a predicted change in the aberration of the laser media, to predictively compensate thermal blooming (Primmerman and Fouche, 1976) or to correct for mirror deformations caused by beam-induced heating or by degradation of the optical coatings. The closed-loop control system in this case acts to maintain a desired figure on the deformable mirror. The local-loop (loop B) and target-loop (loop C) systems operate as discussed in Section I.C. By our definition, only loops B and C would constitute *adaptive* optical systems.

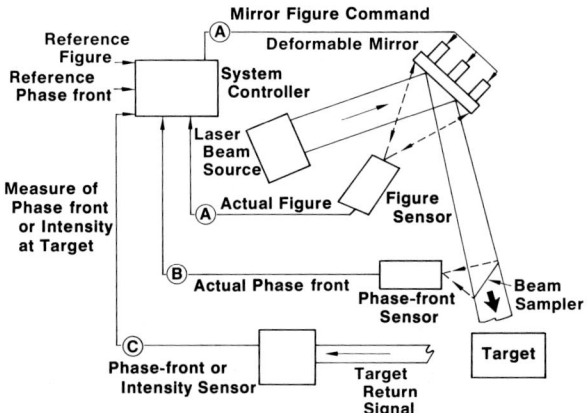

FIG. 12. Wave-front control system that can be used as either a predictive system (loop Ⓐ with figure sensor) or as a local-loop adaptive optical system (loop Ⓑ with phase-front sensor) or as a target-loop system (loop Ⓒ).

If a deformable mirror figure sensor is used, the corrections are no better than the *a priori* knowledge of the desired figure and the precision with which the deformable surface can be controlled. In this case, the figure sensor measures the figure at described points on the deformable surface corresponding to the actuator locations. This information when combined with the desired surface contour provides the command signals for each actuator. Since this scheme requires calibration or knowledge of all the systematic error sources and only senses the mirror figure, it is the least accurate of all the schemes.

The performance requirements for an open-loop figure sensor system are defined in terms of seven variables: (1) beam size, (2) frame rate, (3) maximum phase deviation (or dynamic range), (4) number and arrangement of actuators, (5) spatial resolution, (6) measurement accuracy, and (7) data error rate.

The size of the beam, or the diameter of the mirror to be controlled, affects mostly the size of the input aperture to the sensor system. For very large beam sizes this requires a similarly large sensor system, a fact frequently overlooked in early system considerations.

The frame rate requirement is set by the frequency content of the phenomenon that is to be controlled. Although Shannon's theorem states that the field must be sampled at only twice the rate of the highest frequency content of the signal, in practice, sampling 3 to 5 times per cycle is required for effective control.

The maximum dynamic range of the quantity to be measured affects both the dynamic range required in the detector and electronics and, in the case of phase measurement, the bandwidth of the system.

The number and arrangement of actuators and the spatial resolution are, of course, related. For phase measurement, however, resolution sufficient to resolve the spatial frequency associated with the actuators may not be enough. Phase is measured interferometrically and the system must resolve the fringes in the interferogram.

Accuracy and data error rate are related to signal-to-noise ratio. It is tempting, when dealing with high-power laser systems, to assume that the fraction of power sampled can be increased indefinitely so that signal-to-noise ratio can be improved as required. As the sampled power is increased, however, distortion of optical elements such as beam splitters can occur, which will reduce accuracy.

Measurements of optical figure employ interferometric phase measurement techniques used to experimentally analyze laser beams. The optical figure of the mirror to be measured and controlled is probed with a visible laser beam and the unit of measurement is in wavelengths of the probe beam. By using a visible wavelength probe beam, high resolution is attainable and the high speed and high sensitivity of photomultiplier technology can be applied.

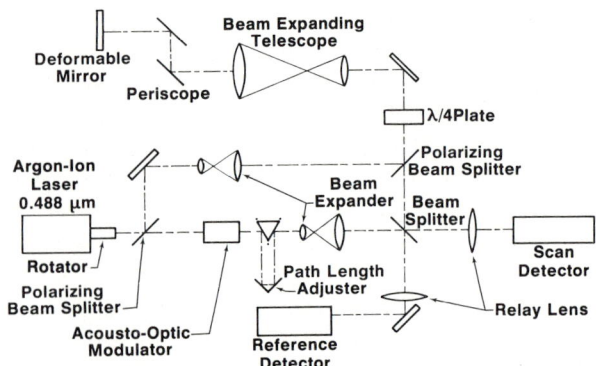

FIG. 13. Figure sensor optical schematic. (For further details, see Angelbeck et al., 1975.)

One possible configuration for a figure sensor is shown in Fig. 13. The probe beam from the laser is matched in diameter to the deformable mirror by a collimating telescope. The beam reflected from the deformable mirror back through the telescope is divided and imaged by optical elements on two detectors, one scanned and one a reference detector. It is combined with a uniphase reference beam which is derived from the same laser as the beam that probes the deformable mirror. The laser output is divided by a beam splitter into two paths, a probe beam and reference beam. Each beam is expanded and spatially filtered by separate but identical beam expanders. An optical delay line matches the reference and sample path lengths so that the laser can be used in a multimode configuration at higher power. The probe beam and reference beams are combined with a beam splitter and imaged onto the two detectors. Interference filters are used to remove background illumination.

B. Modal versus Zonal Correction

As discussed in Section I.B.1, the optical phase error can be expressed as

$$\phi_e(\mathbf{r}, t) = \sum_{n=1}^{\infty} a_n(t) f_n(\mathbf{r}) \tag{2}$$

An adaptive optical system attempts to generate a phase correction, ϕ_c, such that $\phi_c = -\phi_e$ so that the net optical phase error is zero. The system does this by adjusting the coefficients in an expansion similar to that in Eq. (2):

$$\phi_c(\mathbf{r}, t) = \sum_{n=1}^{N} b_n(t) f_n(\mathbf{r}) \tag{20}$$

The summation uses the same basic set of spatial functions as for the phase error, but the finite sum represents the number of servocontrol channels or number of degrees of freedom in the COAT system.

8. ADAPTIVE TECHNIQUES FOR WAVE-FRONT CORRECTION

One choice for the functions $f_n(\mathbf{r})$ is a division of the optical aperture into N independent (or nearly so) regions or "zones." Each servochannel then controls the phase of each individual zone. For obvious reasons, this type of system is known as a "zonal" system, and is the type of COAT system studied most thoroughly to date (Hardy et al., 1977; Cathey et al., 1970; Bridges et al., 1974; Pearson et al., 1976a; Pearson, 1976; Pearson and Hansen, 1977). A schematic representation of this type of system using a deformable mirror (Section III.C) is shown in Fig. 14a.

A second choice for the f_n is a set of functions, each of which is defined over the entire aperture. We call such functions "modes" and the resulting control system "modal." An implementation of a modal system is shown in Fig. 14b. Each control system now drives the amplitude of one mode, which is generated by an appropriate weighting of the drives to each actuator of a multiactuator active mirror. Figure 14c shows one particularly attractive choice of aperture functions, the Zernike polynomials, which can be used to represent the primary Seidel aberrations. Table IV gives the mathematical

TABLE IV

ZERNIKE POLYNOMIALS FOR MODAL COAT SYSTEMS[a]

n	f_n	Aberration name	β_n[b]
1[c]	X	Horizontal tilt	2
2[d]	Y	Vertical tilt	2
3[c]	$2R^2 - 1$	Refocus	1.5
4[c]	$X^2 - Y^2$	Astigmatism	3
5[d]	XY	Astigmatism	3
6[c]	$X(3R^2 - 2)$	Comma (Y symmetry)	4
7[d]	$Y(3R^2 - 2)$	Coma (X symmetry)	4
8	$X(X^2 - 3Y^2)$	120° (5th order, Y symmetry)	4
9[d]	$Y(Y^2 - 3X^2)$	120 degree (5th order, X symmetry	4
10[c]	$6R^4 - 6R^2 + 1$	Spherical aberration	2.5
11	$(X^2 - Y^2)(4R^2 - 3)$	Astigmatism (5th order)	5
12	$XY(4R^2 - 3)$	Astigmatism (5th order)	5
13[c]	$X(10R^4 - 12R^2 + 3)$	5th order coma (Y symmetry)	6
14[d]	$Y(10R^4 - 12R^2 + 3)$	5th order coma (X symmetry)	6
15[c]	$20R^6 - 30R^4 + 12R^2 - 1$	5th order spherical aberration	3.5

[a] Where $X = x/\sqrt{2} = R \cos \theta$, $Y = y/\sqrt{2} = R \sin \theta$, and $R = x^2 + y^2 \le 1$.

[b] Relative dither amplitude for constant effective modulation index in a multidither COAT system.

[c] Principal polynomials needed for blooming compensation.

[d] These five polynomials are not required for compensating only thermal blooming because of the symmetries in the blooming distortions; the transverse wind is in the X direction. See Bradley and Herrmann (1974). They are used for turbulence compensation.

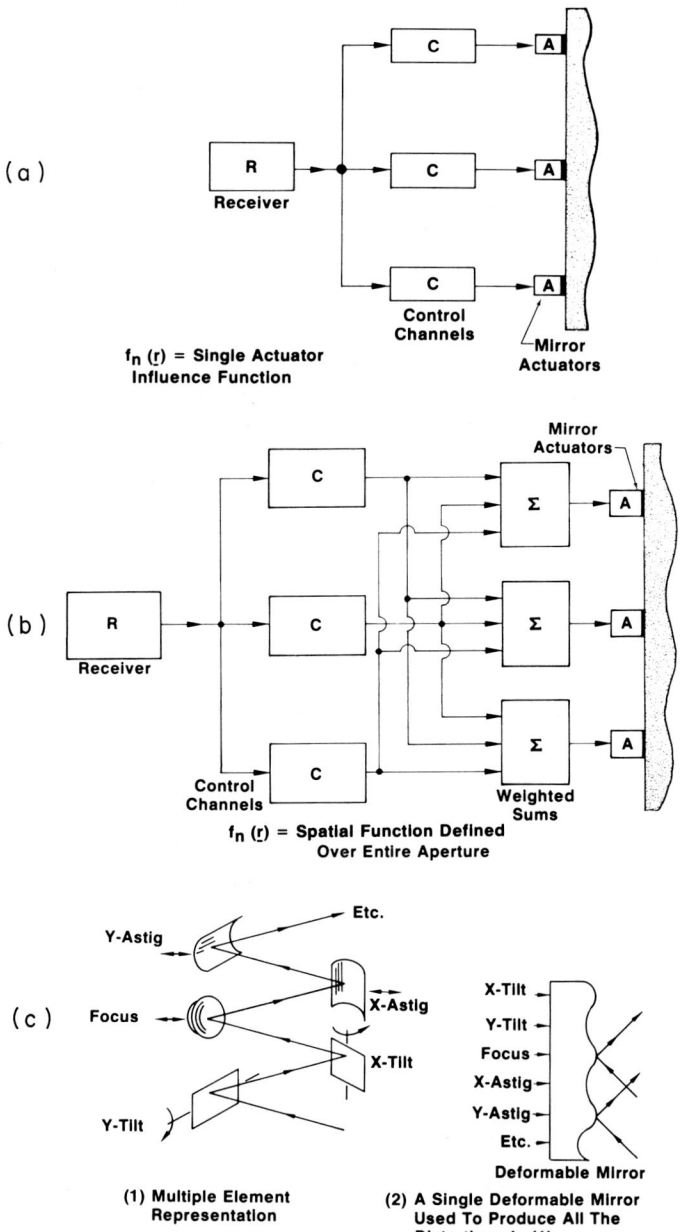

FIG. 14. Classification of adaptive optical systems by the spatial functions used to represent the correction phase: (a) zonal control where each control channel drives a single mirror actuator; (b) modal control where each control channel adjusts the amplitude of a spatial function defined over the entire aperture; (c) one choice for the spatial modes: the Zernike polynomials. Underscores denote vectors.

expression and the Seidel aberration name for the lowest-order Zernike polynomials. This type of system has been implemented in several forms with one (Erteza, 1976a), two (Erteza, 1976b), and five (Lind et al., 1977) modes.

The advantage of the zonal system is its ability to correct for arbitrary phase errors with relatively high spatial frequencies. In principle, the modal approach can also correct high spatial frequency errors, but in practice it is very difficult to construct a phase correction device that will accurately produce more than a few modes. If several corrector elements are used as illustrated in Fig. 14c, the system complexity rapidly approaches that of a zonal system.

The advantage of a modal system is its ability to produce a substantial improvement in beam quality with a minimally complex system. As an example, consider a five-channel modal system that acts on the seven lowest-order Zernike modes (ignoring the constant phase term) of tilt (2 modes) focus, astigmatism (2), and coma (2). Correction of these distortions will remove much of the common optical train errors, plus effect substantial correction for both atmospheric turbulence (Fried, 1965; Noll, 1976) and thermal blooming (Bradley and Herrmann, 1974). The correction for atmospheric turbulence is plotted in Fig. 15 using the results of Noll (1976). For comparison, the correction produced by 61-element zonal system is also plotted using the results of Brown (1975) [see Eq. (45) in Section IV.B]. As can be seen, correction of the first seven modes gives an improvement in

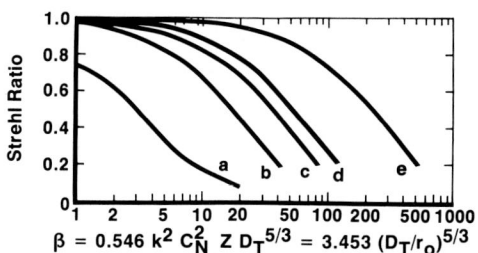

FIG. 15. Analytical comparison of turbulence compensation with modal, Zernike polynomial control, and zonal deformable mirror COAT control. Perfect phase correction is assumed for each indicated order (no servocontrol errors or deformable mirror mismatch errors). The transmitter diameter is D_T. The Δn parameter is defined by Noll (1976).

Curve	Error removed, order n, Δn	Aberrations removed
a		No correction
b	$\Delta 3$	Tilt
c	$\Delta 6$	Tilt, focus, astigmatism
d	$\Delta 8$	Tilt, focus, astigmatism, coma
e		61-Element zonal COAT

Strehl by a factor of (Noll, 1976)

$$\frac{S_{\text{COAT}}}{S_{\text{No COAT}}} = \frac{\exp[-0.0525(D/r_0)^{5/3}]}{\exp[-1.032(D/r_0)^{5/3}]} = \exp\left[0.98\left(\frac{D}{r_0}\right)^{5/3}\right] \quad (21)$$

It is also clear from Fig. 15 that increasing the number of degrees of freedom always gives improved turbulence compensation. Thus a 37-channel modal system and a 37-channel zonal system should have identical performance for turbulence compensation.

C. Error Sensing Algorithms

1. *Phase Conjugate*

The system commonly referred to as a "phase conjugate" COAT system is shown in its simplest form in Fig. 7a. This was the first type of COAT system to be demonstrated (Takken and Cordray, 1974) and has been extensively studied (Lavan *et al.*, 1976; Hayes *et al.*, 1977). Conceptually, this system is the easiest to understand. The concept is illustrated in Fig. 16. The initial distorted wave front from the transmitter array illuminates the target, which is assumed to contain at least one relatively bright "glint" point. If this glint point is small enough to be unresolved by the transmitting aperture, then the spherical wave emanating from the glint can be used as an effective phase reference for the optical system as long as the intervening atmospheric aberrations, optical system vibrations, etc., do not change significantly in the loop time (see Section VI.E). Loop time is defined as the optical transit

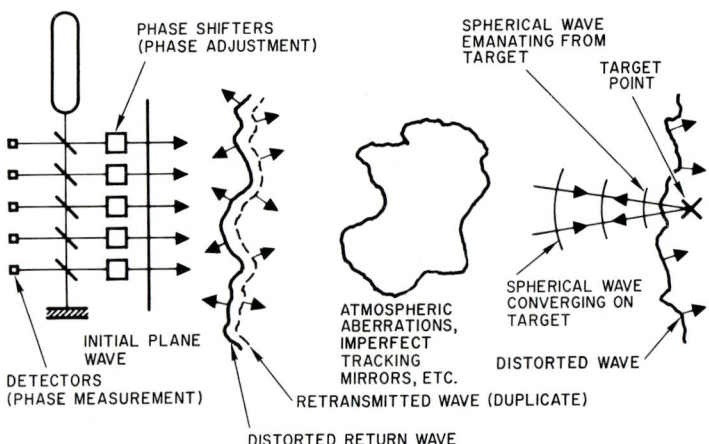

Fig. 16. Illustration of the operation of a phase conjugate adaptive array. (After Hayes *et al.*, 1977.)

time from the aberration source to the array and return, plus the time required for information processing.

The spatial phase distribution of the distorted return wave is measured by the detector array shown in Fig. 16 and a control loop is closed between each detector and its corresponding phase shifter. In this way, the retransmitted wave forms a good reproduction of the distorted return wave. Under ideal conditions (infinite signal-to-noise ratio, no turbulence, no noise or measurement error, etc.), rapid convergence should occur so long as the principle of reciprocity can be applied to the intervening medium. Fortunately, most optical media of interest are reciprocal. This includes phenomena such as weak or strong turbulence, vibrating optical systems, aberrations in optical surfaces, and weak thermal blooming.

The phase conjugate system, although conceptually straightforward, has several practical difficulties in actual implementation. Some of the potential problems are listed in Table III, such as backscatter interference, a complex receiver requiring N receiver elements for N array elements, and required doppler tracking in the receiver for moving targets. In addition, the receiver relies on the coherent interference of the return wave and a local oscillator beam so that operation with multiple-line laser sources (such as the HF–DF chemical laser) is very difficult. One final problem is the lack of phase shifters that can handle high optical powers so that the technique is limited to low-power applications or master-oscillator power-amplifier (MOPA) configurations. The system operation is by far the easiest to simulate on a computer and has formed the basis for most of the work done to date (Herrmann, 1977; Bradley and Herrmann, 1974; Brown, 1975). It can be analytically demonstrated that for linear distortions, *every* adaptive optical system will attempt to impress the phase conjugate of the distortion on the outgoing beam. Many COAT system features can thus be studied by analyzing the phase conjugate system.

2. *Interferometric*

Since adaptive optical systems operate on the phase of a beam, interferometric methods of measuring the error come naturally to mind. An interferometric method for figure sensing was discussed in Section II.A. A very important interferometric technique in adaptive optics uses a shearing interferometer (Wyant, 1973) or variations (Wyant, 1974, 1975) on this concept. In this concept, a lateral-shear ac interferometer is used to measure the local *slope* (derivative) of the wave front of interest. The operation of this device can be understood by reference to Fig. 17 (Fig. 17 and the following discussion are adapted from Hardy *et al.*, 1977). The input lens L_i forms a Fourier transform of the pupil containing the wave-front disturbance onto the grating G. The output lens L_o then forms the convolution of the pupil

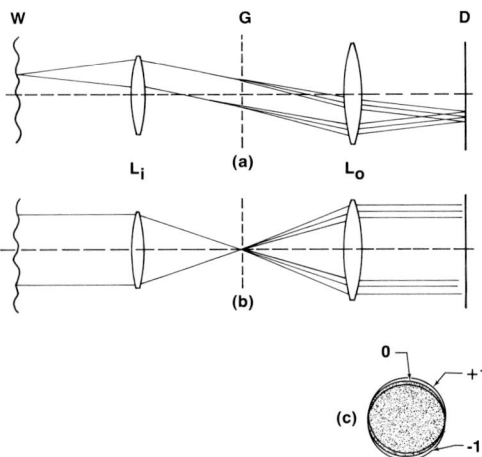

Fig. 17. Operation of a grating interferometer: (a) wave-front deformation W is imaged on to the detector array plane D so that each detector location corresponds to a specific zone of the wave front; (b) reference source focused on to the grating produced multiple images of the input aperture at the detector plane due to diffraction; (c) representation of lateral sheared images at the detector plane, showing zero order, +1 order, and −1 order sidebands which produce interference over the shaded area.

with the Fourier transform of the grating onto the detector plane D. The resulting overlapping pupil images interfere with each other at the detector plane, converting the wave-front phase variations into amplitude variations. If the grating is rotated (Hardy et al., 1977), each detector output is modulated, producing a signal whose fundamental component ω, is the number of grating cycles passing a given point per second, independent of the optical wavelength. The frequency ω corresponds to the interference between the −1 and 0 diffraction orders coherently added to the interference between the +1 and 0 orders. Proper selection of the grating transmittance functions insures that no other orders contribute to the modulation at ω, which is separated by electronic filtering.

The signal at any point on the detector plane at a Ronchi grating ac interferometer can be expressed as

$$I(x,t) \propto \tfrac{1}{2} + \gamma(2/\pi)\cos[\omega t + \phi(x)] \qquad (22)$$

where ω is the fundamental modulation frequency, γ is the modulation determined by the partial coherence of the reference source and the amount of path length distortion, and $\phi(x)$ is the phase angle of the ac signal. The phase angle may be expressed

$$\phi(x) = \alpha(x)S/\lambda \qquad (23)$$

where $\alpha(x)$ is the wave-front slope (dimensionless) at the corresponding point in the aperture. The shear distance S in a grating interferometer is proportional to wavelength, so that $\phi(x) \propto \alpha(x)$, i.e., the temporal phase angle of the signal at frequency ω is proportional to the slope of the wave front, independent of the wavelength of the light.

To reconstruct the wave front from the local slopes measured by the shearing interferometer, the optical path differences (OPD) are evaluated at an array of sampling points. The OPD at two points separated by distance L is then

$$\text{OPD} = \phi \lambda L / S \qquad (24)$$

To provide the maximum utilization of the light collected from the reference source, the entire area of each detector plane is divided into contiguous zones or subapertures. The phase of the ac signal from each detector then represents the average optical path length difference, in the direction of shear, over that subaperture. The subaperture dimension is dictated by the spatial frequency of the wave-front errors to be measured and by the required precision of measurement. Normally the spatial resolution of the wave-front measurement and correction (deformable mirror) systems will match each other.

The features of this type of shearing interferometer which are of particular importance in an adaptive optical system are as follows:

(a) It is a relatively simple device in which the number of optical components does not depend on the size of the aperture or on the spatial resolution of the wave-front measurement. These requirements determine the size of the detector arrays.

(b) Since the interferometer interferes the incident wave front with a displaced replica of itself, no local oscillator is required and it can operate with white-light sources, effectively measuring optical path length differences independent of wavelength.

(c) The optical efficiency is high; 45% of the incident light on the grating appears in the -1, 0, and $+1$ diffracted orders transmitted to the detector plane and is modulated at the fundamental frequency ω. By using a reflecting grating with an additional detector array, 90% of the incident light may be used for wave-front measurement.

(d) Variable shear allows maximization of contrast or modulation for various size sources and optimization of residual error in the corrected wave front.

(e) It is a common path interferometer and thus relatively insensitive to vibration and optical misalignment.

One disadvantage of the shearing interferometer is its requirement of a complete N-element detector array for each shear channel. Two shear channels are required (one for x and one for y shear), and the reflected energy (with a Ronchi grating, see Hardy et al., 1977) requires two more so that a total of four N-element arrays are necessary. Also, since the detector signals are measures of wave-front slope, a wave-front reconstruction algorithm must be performed before adaptive corrections to the optical path can be made. This reconstruction is really no more than a matrix inversion, however, which can be done in a fast, analog data processor (Hardy et al., 1977) or can be handled in a digital processor (Hudgin, 1977a). Fried (1977) and Hudgin (1977b) have discussed optimal techniques for performing this wave-front estimation.

A similar type of interferometric technique uses the interference between a pair of small circular apertures to measure the phase difference between two points on a wave front (Skolnick et al., 1974). The measurement technique is based on measuring the shift in location of the focal-plane intensity peak of the pair of circular apertures. An unambiguous measurement of the phase difference can be made if the phase difference does not exceed 2π. The entire wave front of interest is measured by scanning the sampling apertures over the wave front so that a phase-difference (phase-derivative) map is generated, as long as the spacing between the apertures is smaller than the highest spatial frequency in the phase distribution. In a sense, this technique is the serial-readout version of the parallel-processing shearing interferometer. Here only one detector is required, while $2N$ are required by the shearing system for $2N$ phase differences.

The accuracy of the phase-difference measurement is related to the contrast ratio of the diffraction pattern of the pair of circular apertures and the noise level. The contrast ratio is related to the intensity of the two adjacent circular areas that are being sampled in a given measurement. Thus the spatial uniformity of the laser intensity can have an important effect upon the accuracy of the wave-front measurement and must be included in the model. The measurement bandwidth of the sensor is limited by the finite dwell time of the scanning apertures; in general, system performance will be degraded by the finite framing time of the sensor if the phase of the laser is fluctuating rapidly. The technique finds its main application as a wave-front analyzer rather than as a wave-front error sensor for adaptive optics.

A model for the wave-front analyzer measurement process can be expressed in a few simple equations. For a given scan direction, say the X axis, the temporal variation of the detector output signal is given approximately by

$$I_D(X, Y, t) = I(X + \tfrac{1}{2}X_0, Y, t) + I(X - \tfrac{1}{2}X_0, Y, t) \\ + (2/\pi)I(X + \tfrac{1}{2}X_0, Y, t)I(X - \tfrac{1}{2}X_0, Y, t) \\ \cdot \cos[2\pi f_c t + \Phi_1(X + \tfrac{1}{2}X_0, Y, t) - \Phi_2(X - \tfrac{1}{2}X_0, Y, t)] \quad (25)$$

where $f_c = V_x/2X_0$ is the scan carrier frequency in the X-direction, V_x the velocity of the scanning slit, X_0 the separation of the sampling aperture centers, and Φ_i the average phase across the ith sampling aperture. The phase difference $\Delta\Phi$ is measured by using synchronous detection at the carrier frequency f_c:

$$\Delta\Phi_X = \Phi_1(X + \tfrac{1}{2}X_0, Y, t) - \Phi_2(X - \tfrac{1}{2}X_0, Y, t)$$
$$\Delta\Phi_Y = \Phi_1(X, Y + \tfrac{1}{2}Y_0, t) - \Phi_2(X, Y - \tfrac{1}{2}Y_0, t)$$
(26)

The phase is then found using a procedure similar to that discussed for the shearing interferometer. The complete model must also include the dependence of signal strength on aperture size so that tradeoffs between aperture size, aperture spacing, and phase accuracy can be determined.

Another class of interferometric sensors employs the principles of complex Fourier filtering. Included in this class are tests associated with the names of Foucault and Schlieren, but all such tests can be classed under the heading of Zernike Phase Contrast (ZPC). These tests all perform some operation on the wave front in the Fourier transform plane following a lens. The Zernike approach (Dicke, 1975a) places a phase-changing plate at the position of the zero order of the diffraction pattern. Thus, the constructive and destructive wave-front interferences formed in the reimaged pupil plane are modified, producing an intensity map of the phase differences in the pupil. The Foucault and Schlieren tests delete half of the diffraction pattern, thereby modifying the interferences in the reimaged pupil. The ZPC test is preferred because its results can be directly related to pupil coordinates whereas other embodiments are more difficult to analyze.

The advantage of such a system over other approaches is the smaller amount of data processing required to derive wave-front information from the intensity pattern. In the case of small phase errors, the formula

$$I(x', y') = |C|^2[a^2 + 2a\phi(x, y)] \tag{27}$$

applies, where C is intensity in the reimaged pupil, ϕ is the pupil phase function, and a is the transmission coefficient of the retarding plate. Thus, no reconstruction of phase is necessary, implying that more frequent sampling may be performed or that more prolonged integration time is permitted. The system also has the advantage of mechanical stability and simplicity once the phase plate has been installed, although boresighting with the system pointing axes is a later requirement.

The primary disadvantage is that the zero diffraction order is actually an image of the source. If the source is extended and resolved by the aperture, then the Zernike disk must be fabricated to match this image. Further, this measurement technique requires a limited spectral region to produce a useful range of optical path difference. Thus, unless multiple Zernike sensors are considered, each working in a restricted spectral band, the efficiency is low. Also, since the phase variation is being measured as an intensity variation, intensity variations across the wave front can introduce phase measurement errors.

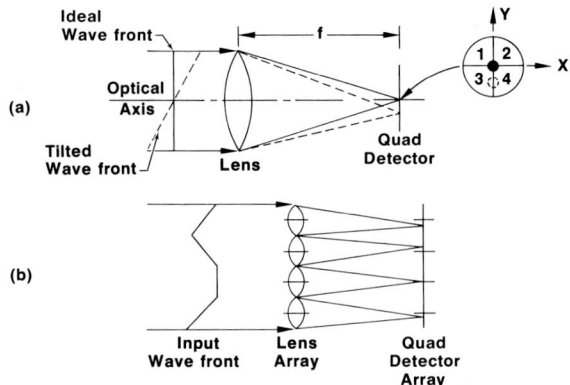

FIG. 18. "Hartmann" wave-front sensor: (a) single element illustrating how a lens and quadrant detector can be used to sense wave-front tilt; (b) multiple element version that senses localized tilts of a larger wave front.

3. *Hartmann Test*

The classic Hartmann test commonly used in evaluation of optical components can also be adapted to wave-front sensing for adaptive optics. Consider first the single lens with quadrant detector in its focal plane as shown in Fig. 18a. The lens and detector are aligned so that all four quadrants are equally illuminated when the incoming phase front is perpendicular to the optical axis. When a tilted wave front enters the lens, the focal-plane distribution shifts. The tilt angle of the wave front can be found from

$$\Delta\theta_X = [(V_2 + V_4) - (V_1 + V_3)]C, \qquad \Delta\theta_Y = [(V_1 + V_2) - (V_3 + V_4)]C \qquad (28)$$

where V_i is the voltage output signal of the ith quadrant and C is a constant that depends on the detector responsivity, the shape of the subaperture, the total optical power, and in some instances, on the shape of the target.

For adaptive optics applications, the aperture of the optical system is divided into many subapertures, the number being determined by the spatial frequency content of the wave front to be measured. A lens and detector in each subaperture is used to make a direct measurement of the wave-front tilt in each subaperture as indicated in Fig. 18b. Since the Hartmann system measures local phase slope, it must construct the actual phase front in much the same way as required by the shearing interferometer. Techniques for doing this are identical to those for the shearing system. The Hartmann system and the shearing interferometer system are basically image-compensation sensors since they can work with broad-band sources and extended targets. When used with a laser, they form the sensor for a "TRIM-COAT" system as discussed in Section I.C.

The Hartmann sensor has three advantages over the shearing interferometer: (1) it is a noninterferometric technique and thus has no 2π ambiquity problem; (2) it requires only one detector array to four for the shearing for the same optical efficiency; and (3) it is mechanically less complex, replacing optical and mechanical hardware with electronic processing. Its primary disadvantage is stringent alignment and manufacturing tolerances on the lens and quad-detector arrays. Under low signal conditions, as are often encountered in compensated imaging applications, it can be more sensitive to target shape than the shearing interferometer. A recent patent (Feinleib, 1979), however, discloses ways to overcome some of these problems.

4. Multidither

Dither systems have long been used for a variety of closed-loop servo-control applications. The basic error measurement scheme of a dither system involves making trial perturbations on the variable to be controlled and determining which perturbations (or which slope of the perturbations) improve the performance measure of the system. A variety of dither waveforms are possible, but here we consider only sinusoidal dithers.

To understand the operation of a multidither COAT system, consider the simplest possible system, a two-element phased array, as illustrated in Fig. 19. One element has a fixed phase, but the optical phase of the other one can be adjusted. At the target, the far-field diffraction pattern for a two-element array varies sinusoidally with distance along the direction along the array. A low-amplitude phase modulation, the "dither," is applied to the adjustable array element at frequency ω_d. This modulation causes the diffraction pattern to move back and forth. If the target contains a small reflecting

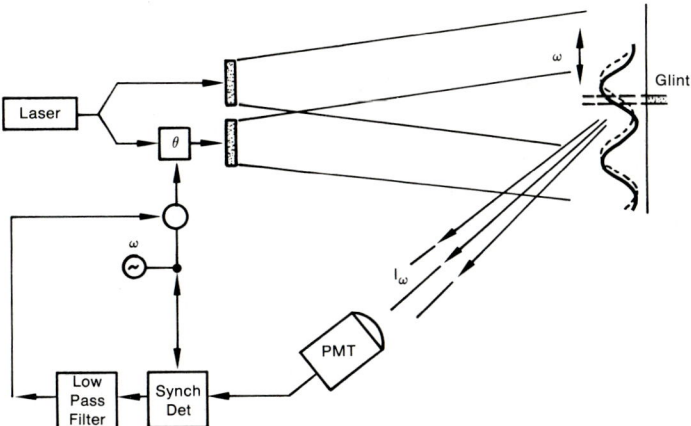

FIG. 19. Two-element dithered COAT system. The sinusoidal phase perturbation, applied to one element, scans the far-field diffraction pattern back and forth on the target glint. The reflected energy contains amplitude modulations that are sensed by the receiver.

region—a "glint"—that is unresolved by the transmitting aperture, the light reflected back toward the receiver is amplitude modulated at the dither frequency. This amplitude modulation is sensed by a single intensity detector and fed to a synchronous detector that is driven by the dither oscillator. A low-pass filter removes all but the slowly varying (much less than ω_d) signals. This "error signal" is then applied back to the array element phase shifter. The net result is a "hill-climbing" servo that adjusts the array element phase to position the diffraction pattern maximum on the target glint. This maximization will occur even if there are time-varying phase distortions between the transmitter and the target.

Although this type of simple explanation is intuitively appealing, the extension to many array elements is not obvious. It is not intuitively obvious that an N-channel COAT system, with N different dither frequencies ω_i feeding N different phase shifters, and N different synchronous detectors each extracting its appropriate signal from a common intensity detector, will work in the same fashion, but indeed, it does. Any one dithered element essentially forms a sinusoidal (spatial sense) interference pattern with the ensemble carrier of the remaining elements. This pattern scans back and forth at its characteristic frequency; and if it is not at a pattern maximum, it generates an amplitude modulation at this frequency. This operation has been proven both by experiment and by computer simulation (Bridges *et al.*, 1974; Pearson *et al.*, 1976a; Pearson, 1976; Pearson and Hansen, 1977) and can be demonstrated analytically (O'Meara, 1977a).

The analytical representation of the multidither system for "piston" elements (see Section II.B) as derived by O'Meara (1977a) is instructive. He derives the irradiance at a point (the glint point) as

$$I_p = I_0 \left| \sum_{n=1}^{N} A_n e^{i\Gamma_n} \right|^2 \tag{29}$$

where A_n is the relative field amplitude of the nth array element and Γ_n is the total optical phase (error, correction, dither) given by

$$\Gamma_n = \beta_n + \psi_0 \sin(W_n t) \tag{30}$$

The residual phase error is $\beta_n = \bar{\phi}_{en} - \phi_{cn}$, where $\bar{\phi}_{en}$ is the average phase across the nth element and ϕ_{cn} is the correction phase for that element. The dither amplitude is ψ_0, assumed equal for all elements. By rewriting Eq. (29) and keeping only terms through the second harmonic in ω_n, we find

$$\frac{I_{pm}}{I_0} = HJ_0^2 + \sum_{n=1}^{N} A_n^2 + 4J_0 J_1 B_m \sin(\beta_m - \beta_{mc}) \sin \omega_m t$$
$$+ 4J_0 J_2 B_m \cos(\beta_m - B_{mc}) \cos 2\omega_m t \tag{31}$$

where $J_i = J_i(\psi_0)$ is the ith integer Bessel function and

$$B_m = A_m \left[\sum_{\substack{l=1 \\ l \ne m}}^{N} (A_l^2 + H_m)^{1/2} \right], \qquad \beta_{mc} = \tan^{-1} \left(\sum_{\substack{l=1 \\ l \ne m}}^{N} A_l \sin \beta_l \Big/ \sum_{\substack{l=1 \\ l = m}}^{N} A_l \cos \beta_l \right) \tag{32}$$

$$H_m = \sum_{\substack{k=1 \\ k \ne l}}^{N} \sum_{\substack{l=1 \\ l \ne m}}^{N} A_k A_l \cos(\beta_k - \beta_l), \qquad H = \sum_{k=1}^{N} \sum_{l=1}^{N} A_k A_l \cos(\beta_k - \beta_l)$$

The parameter β_{mc} is the phase reference for the mth element, which approaches the weighted average of the remaining $m - 1$ elements as the system converges. Since the servosystem acts to drive the magnitude of the fundamental frequency (at ω_m) to zero, convergence occurs when all $(\beta_m - \beta_{mc})$ approach zero. The parameters H and H_m are convergence parameters that maximize at convergence, with H_m ranging from zero to $A_m^2 (N - 1)(N - 2)$.

Each dither frequency in a multidither COAT system can be considered as a subcarrier that carries the phase error information as amplitude modulation sidebands. Figure 20 illustrates this point. Clearly, the width of the disturbance frequency spectrum (or equivalently, the closed-loop control bandwidth used to correct it) will determine the minimum dither spacing that can be used. After detection at the receiver, the signal passes through a band-pass filter to eliminate signals outside the dither passband such as the dc component* (unmodulated return and background signal). This filter

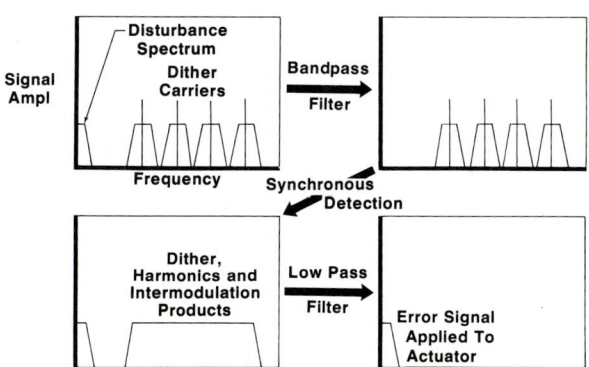

FIG. 20. Signal frequency domain representation of the signals in a multidither control loop.

---

* The near-dc signals can be used for an automatic gain control (AGC), a function that is essential for proper system operation with a variety of targets.

must have good low-frequency rejection without introducing phase shift at the lowest dither frequencies, and it must also pass the highest dither frequency without phase shift.

The dither signals are translated to dc (demodulated) in the synchronous detectors (analog multipliers in some systems) and then put through a low pass filter to obtain the correction signal. The low pass filter must perform two functions: (1) reduce feedthrough of residual signals at the dither frequencies and their intermodulation products below the level required for regenerative feedback (servoinstability); (2) keep the loop gain well below unity at any strong phase shifter resonance peaks. This must be accomplished while maintaining net gain and negligible phase shift below the servocontrol bandwidth frequency. Pearson et al. (1976a) discuss one of many possible designs for this type of servocontrol.

There are many factors that enter into the design of a multidither COAT system, some of which are listed in Table V. Some design rules of thumb are discussed in Sections III and IV. The most useful tool, however, for the design and evaluation of this type of nonlinear, multiple-loop servocontrol system is a detailed computer simulation. This type of simulation, shown schematically in Fig. 21, has been successfully used (Pearson et al., 1976a; Pearson, 1976; Kokorowski et al., 1977; Gurski et al., 1977; Radley et al., 1976) to design and study multidither control systems. A typical output from this type of simulation is shown in Fig. 22 along with an experimental comparison (Pearson et al., 1976a).

TABLE V
Required Parameters for Definition of a Multidither COAT System

Subsystem or device	Parameters
Servosystem	Number of channels Control bandwidth and error rejection (gain) Filters: number, break frequencies AGC concept Range compensation
Phase corrector mirror and dither mirror	Actuators: number, configuration, excursion Operating bandwidth minimum Acceptable stress, distortion, coupling Drive requirements
Receiver	Aperture configuration Type of detector (wavelength)
Options	Sample and hold Modulation index control Speckle noise reduction

8. ADAPTIVE TECHNIQUES FOR WAVE-FRONT CORRECTION

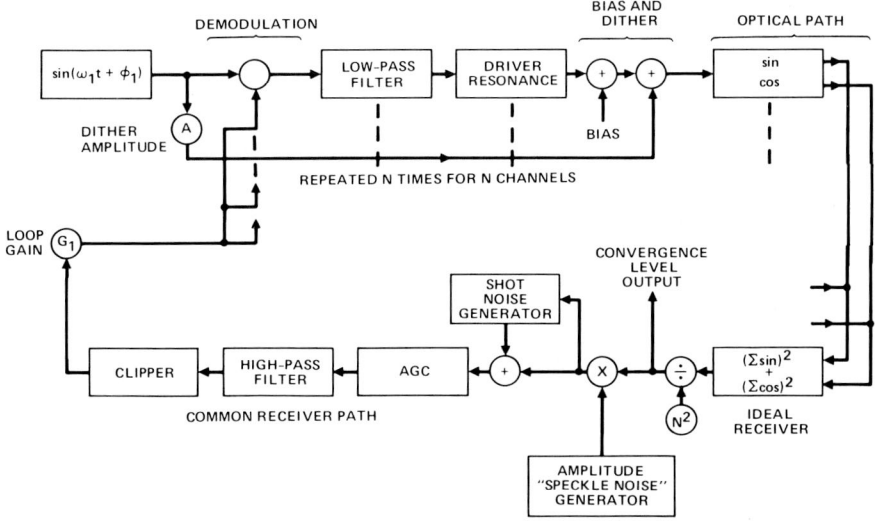

FIG. 21. Schematic diagram of the control loop for an N-channel multidither COAT system. There is a single common receiver path and N identical (except for the dither frequencies) control channels.

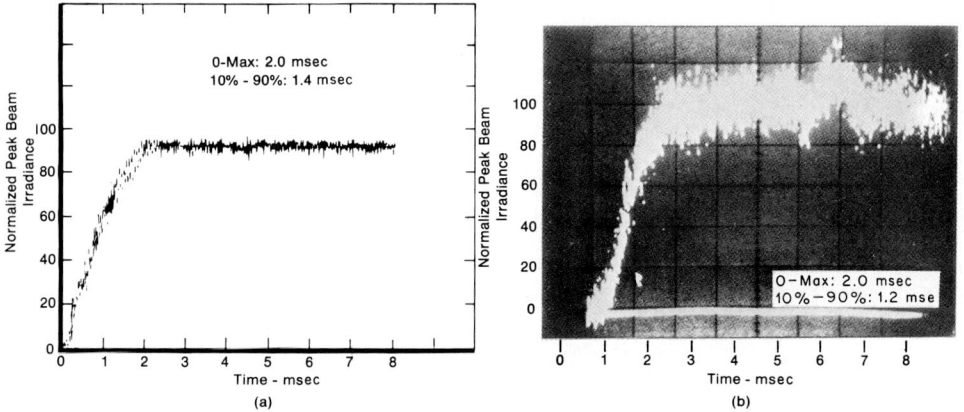

FIG. 22. Typical multidither COAT system convergence sequence for an 18-channel system with a 20 dither amplitude: (a) computer simulation; (b) experimental observation.

Multidither COAT systems can be used as outgoing-wave, return-wave, local-loop, or target-loop systems. Figure 23 shows these various implementations schematically. Note that in the outgoing-wave system, the output beam is dithered, which results in an unavoidable loss in Strehl ratio. This can also be seen in Fig. 22, is present in the expressions in Eqs. (31) and (32), and is further discussed in Section IV.D. The local-loop and return-wave systems, however, have the dither outside the outgoing laser path. The dither

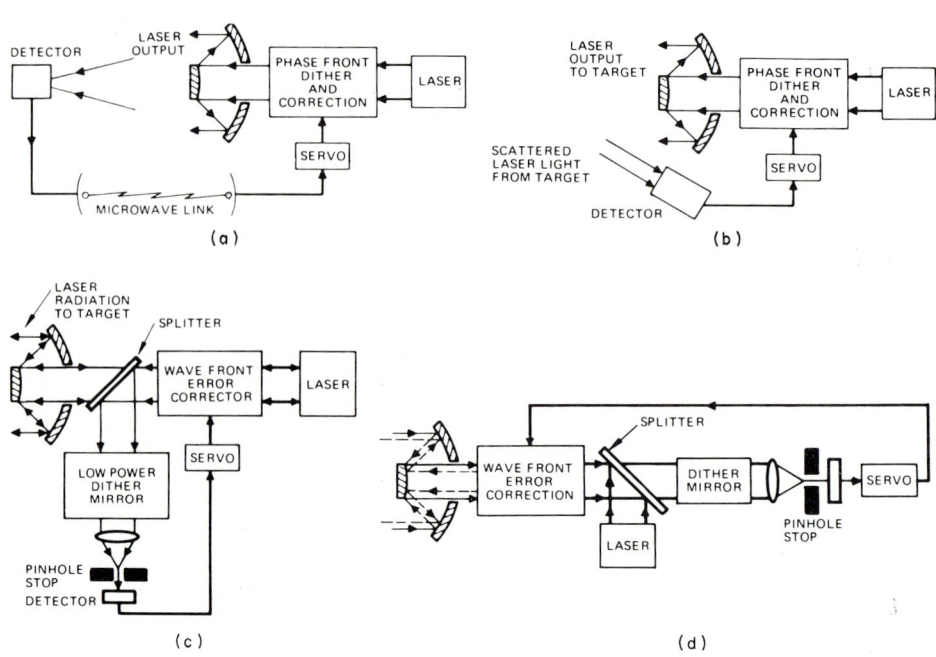

FIG. 23. Four types of adaptive optical systems that may be implemented with multidither servotechniques. Diffraction gratings would most commonly be substituted for the beam splitters at high power levels: (a) multidither outgoing-wave system transmitting optical power to a cooperative target, such as a satellite; (b) multidither outgoing-wave system operating as an optical radar; (c) local-loop multidither system that removes the wave-front (phase) distortions from an aberrated source; (d) multidither return-wave system.

amplitude in these cases can be increased to increase the signal-to-noise ratio without degrading the system performance.

One interesting feature of multidither COAT systems is their ability to operate with a variety of references. The usual mode of operation for an outgoing-wave system requires a localized target highlight—a glint. The system will form the beam on the brightest glint in the receiver field of view (Bridges *et al.*, 1974; Pearson *et al.*, 1976a; Pearson, 1976). By simply reversing the phase of the feedback signal, however, the beam will form on the *darkest* local target area. This mode of referencing is often called (Bridges *et al.*, 1974; Pearson *et al.*, 1976a) "black-hole tracking" and may be particularly useful when the incident laser creates absorbing regions on the target. The system can also use an edge, or a light–dark boundary, as a reference in some cases.

The return-wave system can also use glint tracking (pinhole stop as in Fig. 23d) or black hole by using a centrally obscured stop. The latter mode may have signal-to-noise advantages in some cases since shot-noise contribution is minimized by elimination of the strong dc signal. The "hot spot" generated on the target can also be used, just as it can be used by a compensated imaging system.

If one briefly compares dither systems to alternative COAT systems, one finds that dither systems have the advantages of simplicity, low cost, and versatility and disadvantages of low signal-to-noise (S/N) ratio and speed of response compared to phase conjugate systems or other interferometric techniques. It should be noted that while the simple dither systems are hopelessly noncompetitive in S/N for long-range imaging applications such as ground-based astronomical adaptive optics, the S/N is entirely adequate for a very large class of short-range applications. More complex dither systems are known wherein the speed of response and S/N ratios can be appreciably improved. However, one must abandon the simplicity of single-detector operation; these systems are beyond the scope of this discussion. A comparison of the various types of COAT systems based on the wave-front sensors discussed here is presented in Table VI.

III. PHASE SHIFTING ELEMENTS

A. Introduction

All adaptive optical systems contain elements for introducing the desired phase correction (phase shift) into the beam and, for multidither control systems, elements for phase modulating the beam to allow detection of the phase errors as discussed in Section II. These elements can be grouped into three general categories, (1) acousto-optical/electro-optical, (2) segmented mirrors, and (3) continuous surface mirrors as illustrated in Fig. 24. These elements can perform either or both the correction and modulation functions, depending on the design. When used for phase correction, the elements must produce phase changes corresponding to at least ± 2 wavelengths of path difference at low to moderate rates (10–1000 Hz), Dither modulation of the beam requires small amplitude ac motions of at least $+20°$ of phase shift at moderate to high frequencies (1–100 KHz). Since these two requirements place conflicting demands on the mechanical design, normal practice is to

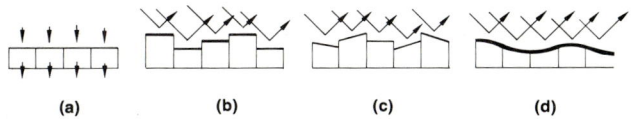

FIG. 24. Alternate configurations for the phase shifter element in an adaptive optical system. The continuous-surface deformable mirror has received the greatest attention to date: (a) electro-optical array; (b) segmented "piston" elements; (c) segmented "piston-plus-tilt" element; (d) continuous deformable mirror.

TABLE VI
COMPARISON OF FOUR BASIC CLASSES OF COAT SYSTEMS

System	Characteristics						Advantages	Disadvantages
	Type of optical receiver	Number of detectors	Sensitivity to backscatter	Operation with broad-band sources	Relative S/N for a given transmitter	Referencing options		
Multidither outgoing wave	Separate annular, or fully shared aperture	1 (Direct)	Very low	Yes	Lowest	Glint; edge; black-hole; extended diffuse surface with image plane aperture	1. Simple receiver 2. Multiple reference capability 3. Blooming compensation 4. Wavelength independent hardware 5. Compensates all distortions simultaneously	1. Potential problems with glint hopping and speckle effects 2. Most useful at ranges less than 10 km 3. Speed limited by dither frequencies
Multidither return wave	Fully shared aperture	1 (Direct) or heterodyne for sensitivity	High	Yes	Intermediate	Glint; receiver focal plane aperture	1. Dither mirror in low-power path 2. Offers higher speed for long range application 3. Simple receiver	1. Blooming compensation doubtful 2. Potential problems with glint hopping and speckle effects 3. Shared aperture receiver 4. Speed limited by dither frequencies

Phase conjugate	Fully shared aperture	N (Heterodyne)	Very high	Difficult	Highest	Glint only	1. Fast servo 2. Excellent sensitivity for long-range operation 3. Offset pointing easily accomplished	1. Complicated receiver; shared aperture 2. Potential problems with glint hopping and speckle effects 3. Blooming compensation doubtful
TRIM-COAT; shearing interferometer	Fully shared aperture	N (Direct)	None with incoherent source	Yes	Not comparable	Extended target referencing. May require target illumination or imaging or laser spot	1. No speckle or glint effects 2. Works with long ranges; fast servo 3. Compensated target image available 4. Modest stability requirements	1. Blooming compensation doubtful 2. Shared aperture receiver 3. Phase reconstruction required 4. Variable shear to eliminate 2π ambiguity
TRIM-COAT; Hartmann	Fully shared aperture	N (Direct)	None with incoherent source		Not comparable (higher than shearing)	Extended target referencing. May require target illumination or imaging of laser spot	1. No 2π ambiguity 2. Works with long ranges; fast servo 3. No target effects 4. Compensated image available	1. Blooming compensation doubtful 2. Shared aperture receiver 3. Phase reconstructed required 4. High stability requirements

provide separate dither and phase correction elements. For low bandwidth systems, the two functions can be combined in a single element.

B. Acousto-Optical/Electro-Optical Elements

In this approach the phase shift is produced by subdividing the beam into many individual beams, and passing each beam through a separate Bragg cell (Cathey *et al.*, 1970; SooHoo *et al.*, 1972; Mahajan, 1975) or electro-optical device (Kaminow and Turner, 1966). The Bragg cell introduces a phase shift by altering the frequency as a function of time by

$$\phi(t) = \phi_0 + 2\pi \int_0^t \Delta f(t)\, dt \qquad (33)$$

where Δf = instantaneous frequency change applied to the cell. Large phase shifts ($\approx 100\pi$ rad) can be produced in this manner and the technique avoids frequency response and dynamic range limitations of segmented and continuous surface mirrors. After passing through the cells, the beams must be recombined prior to transmission. Acousto-optical and electro-optical cells can also be used to modulate a single leg of an electronic interferometer for a figure sensor application (Section II.A) or to provide a local oscillator signal to perform optical heterodyne detection of target return signals in a phase conjugate COAT system. The major disadvantages of these elements are the need to subdivide and recombine the beam after introduction of the phase correction and the limitation to low power levels. Also, since the correction is generally of the form of a constant phase for all wavelengths, rather than a constant optical path, the correction is limited to a narrow spectral bandwidth.

C. Segmented Mirrors

Conceptually, a segmented mirror is by far the simplest device to introduce phase control and phase modulation into a laser beam. The active aperture is comprised of N individual mirrors, each attached to an actuator system.

The most basic configuration employs rectangular, square, or hexagonal segments with plano surfaces, each driven by a single actuator to produce a piston type motion of the segment normal to the mirror axis (Fig. 24b). Additional actuators can be added to provide two axis tilt and translation motion for each segment (Fig. 24c). Segmented dither mirrors normally employ piezoelectric actuators operating only in a piston mode. Phase correction segmented mirrors have been built that employ both piston and piston plus tilt motion and have utilized piezoelectric and hydraulic actuators.

Segmented mirrors have been utilized in two different arrangements for COAT and adaptive phase correction experiments. In one approach (Pearson et al., 1976a), the laser beam is split into several subbeams, each of which is directed onto a dither (tagger) segment and then onto an associated corrector segment. All beams are then recombined into the desired output aperture. In the second approach (Gebhardt et al., 1974), the segments are tightly packed to define a total active aperture which is completely filled with the expanded laser beam.

Design considerations for all mechanical phase shifters (including the continuous surface designs discussed in the next section), should have two major goals: as stiff an actuator/mirror combination as possible, and a light and stiff reflective mirror segment to reduce the accelerated mirror mass and "bounce" frequency of the mirror mass/actuator spring system. If the laser flux is sufficiently high, the segments will require either heat sinking, heat pipe cooling, or liquid cooling (which requires flexible inlet/outlet cooling connections) and a method of cooling to the mirror edges. The size and number of elements determine the ability to match a given phase profile as discussed in Section IV. Obviously, if a number of segments are being used to define a larger aperture rather than as phase shifters and taggers with a divided beam, the gap between segments should be minimized.

Segmented mirrors may be the only approach for producing large aperture (>10 m) telescopes. While small segmented mirrors are conveniently used, large arrays of segmented mirrors can have major design problems such as alignment of the individual segments in a large telescope to avoid translation differences between adjacent segments which are multiples of 2π, and maintenance of the alignment. Most likely, combinations of edge-sensing and full-aperture figure sensors will be required to maintain the figure. Moreover, simple plane surfaces will produce too large a phase mismatch to the required primary mirror contour, requiring the use of curved mirror segment surface contours.

D. Continuous Surface Mirrors

For a given number of actuators, piston-plus-tilt segmented mirror and continuous-surface deformable mirrors provide the best ability to spatially conform to wave-front error distributions. A comparison (Pearson and Hansen, 1977) of the different types is shown in Fig. 25. The data presented in the figure uses a one-dimensional computer simulation. For the two-dimensional case, the total number of mirror actuators is $(12)^2 = 144$ for

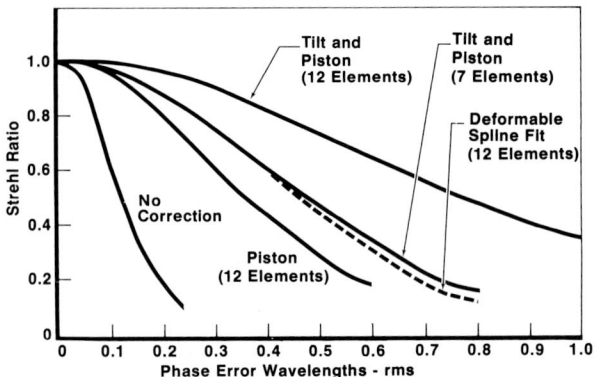

FIG. 25. Comparison of the phase-correction ability of three types of mirror systems. Piston-plus-tilt correction requires three actuators per mirror element whereas piston-only and deformable correction have one actuator per element. The curves are computed using a one-dimensional computer simulation. As a result, seven piston-plus-tilt correction elements correspond approximately to the same number of actuators (degrees of freedom) as 12 piston-only or 12 deformable-mirror elements.

the segmented and deformable mirror cases and $3 \times (7)^2 = 147$ for the piston-piston-tilt case. In general, two types of deformable mirrors have been developed to date: (1) monolithic piezoelectric mirrors (Feinleib *et al.*, 1974) (MPM) which consist of a monolithic block of piezoelectric (PZT) material with imbedded surface electrodes, to which is attached a thin reflecting surface; (2) thin-surface mirrors with discrete actuators that act against a massive substrate to deform the thin-front surface locally (Pearson

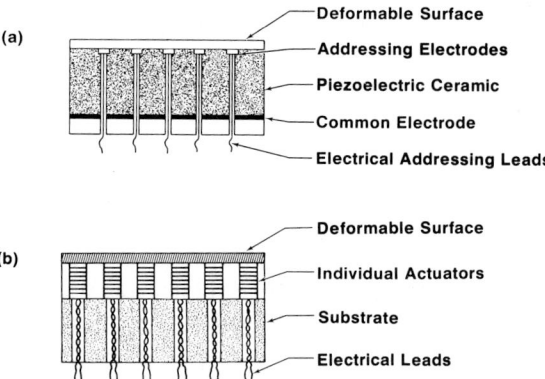

FIG. 26. Conceptualizations of continuous-surface deformable mirrors: (a) monolithic piezoelectric mirror (MPM); (b) discrete actuator mirror. The actuators can be piezoelectric, electromagnetic, hydraulic, or even driven by stepper motors.

and Hansen, 1977; Angelbeck *et al.*, 1975; Grosso and Vellin, 1977). These two types of mirrors are shown conceptually in Fig. 26.

In monolithic mirrors (Feinleib *et al.*, 1974), the surface of the mirror is deformed by applying bipolar voltages to the addressing electrodes with respect to the common back electrode. The resulting motion of the PZT is highly localized to the vicinity of the individual actuated electrode, with no significant interaction from electrode to electrode (Hudgin and Lipson, 1975). In essence, the electrode areas behave as linearly independent elements, with the displacement dependent only on the voltage applied to the electrode. The individual electrodes must be located at the front surface of the block. Reversal of the ground and common electrode results in significant attenuation of the motion at the active surface. Actuator packing density is limited only to the spacing needed to drill holes for the addressing electrode leads, with the practical spatial resolution being ≈ 1 mm. Mirrors have been built with PZT thickness ranging from 2.5 to 5 mm, and electrode spacings of 1 and $1\frac{1}{2}$ mm. For piezoelectric block thickness greater than the spacing between electrodes, all the deflection occurs at the mirror surface and effectively none at the ground (back) plane.

An important quantity for any deformable mirror is the actuator "influence function," $IN(r)$, defined as the local surface deformation produced when one actuator is energized and all others are restrained only by their internal spring forces. One interesting feature of the MPM is the ability to tailor the influence function simply by controlling the electrode geometry. This is possible with the MPM because the very thin faceplate has a negligible stiffness compared to the piezoelectric (PZT) substrate and thus will accurately reproduce the shape assumed by the substrate. Hudgin (1977c) has compared several different influence functions for relative ability to produce a desired surface figure. His results for atmospheric turbulence are shown in Fig. 27, where it can be seen that the rms fitting error has the form

$$E = \alpha(r_s/r_0)^{5/3} \tag{34}$$

where r_s is the interactuator center–center spacing, r_0 is the atmospheric coherence length defined in Eq. (12), and α depends on the influence function. As can be seen, any reasonably smooth function performs about equally well and, as previously observed, every deformable mirror function is substantially better than the piston (segmented) function. However, as Hudgin observes, a very broad influence function will require larger drive force (or power) to achieve the best fit since neighboring influence functions substantially overlap and interact with each other. In this sense, for a hardware design, the most localized influence function would appear to be optimal. There are other considerations, however, which are discussed below.

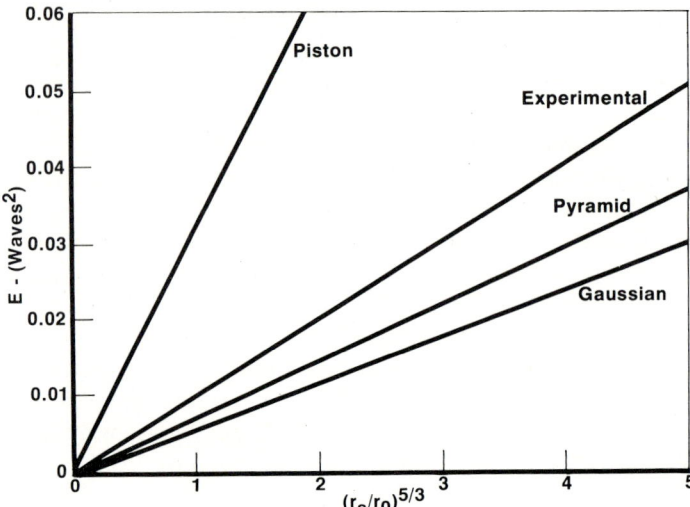

FIG. 27. The error variance of an active mirror correcting for atmospheric turbulence plotted as a function $(r_s/r_0)^{5/3}$, where r_s is the interactuator center-to-center spacing and r_0 is the coherence length of the turbulence.

The deflection (stroke) capability at an individual MPM actuator has been found to be higher than that predicted using the piezoelectric coefficient for the material (Radley et al., 1976). This is due to the fringing electric fields at each electrode increasing the lateral forces on the material, thus increasing longitudinal motion. Increases of approximately 1.4 in sensitivity (micrometers/kilovolt) have been observed. Nominal surface displacements of $+2$ waves visible have been obtained with $+1000$ V excitation. The absolute displacement possible is limited by the maximum field the piezolelectric ceramic material can tolerate before depolarization. For use as phase correctors, this limits these devices to visible and shorter wavelengths unless auxiliary amplification of the phase correction by means of multiple bounces (in a wedge, e.g.) can be employed. The excursion is sufficient, however, for dither mirrors in the infrared. A second limitation is the size restrictions (5–16 cm) imposed by obtaining thick isotropic monolithic discs with uniform properties.

Polishing of uncooled and cooled membranes bonded to the monolithic disk is somewhat easier than for thin membrane surfaces supported by individual actuators. However, contrary to intuitive feeling, analyses and experiments (Reynolds and Raymondo, 1977) show the resonant frequency characteristics for well-designed monolithic and discrete actuator cooled mirrors (i.e., low-amplitude motion) to be essentially the same. In both cases, the substrate structural responses dominate the resonance behavior for identical sized mirrors. Amplitude response for the discrete actuator mirror is stronger, however, unless oil damping between the faceplate and substrate is used.

In a discrete actuator mirror, expansion or contraction of the actuators relative to the stiff substrate produces the desired mirror surface motion. Actual mirrors can contain actuators sandwiched between the substrate and mirror as shown in the photograph (Pearson and Hansen, 1977) in Fig. 28 of a 10-cm diameter, 37-actuator mirror (mount deleted), or complex self-aligning, removable actuators such as those used in the 52-actuator mirror (Angelbeck *et al.*, 1975) of Fig. 29.

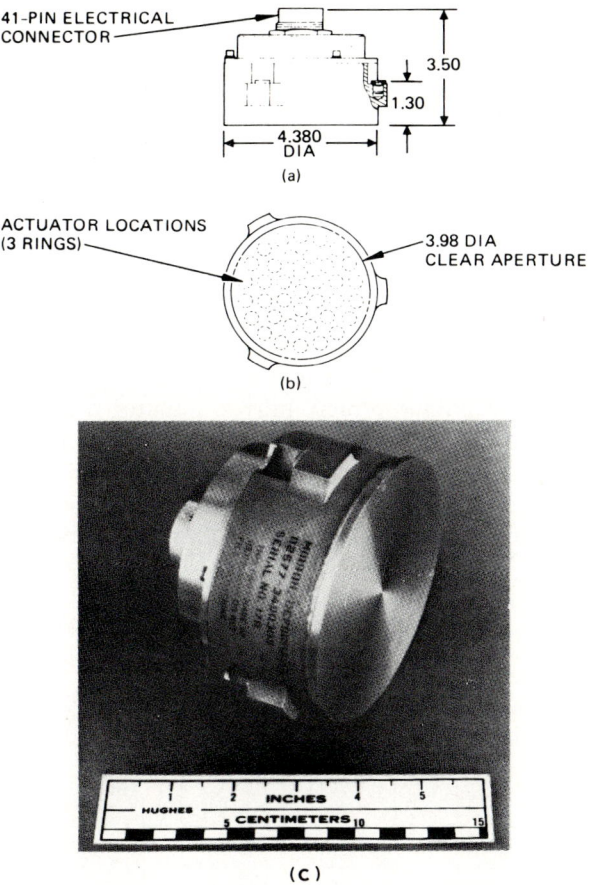

FIG. 28. All beryllium 37-actuator deformable mirror. Each preloaded-spring actuator cell has a **PZT** stack driver: (a) overall mirror schematic; (b) faceplate view showing three-ring, circularly symetric arrangement of actuators; (c) photograph of unpolished mirror.

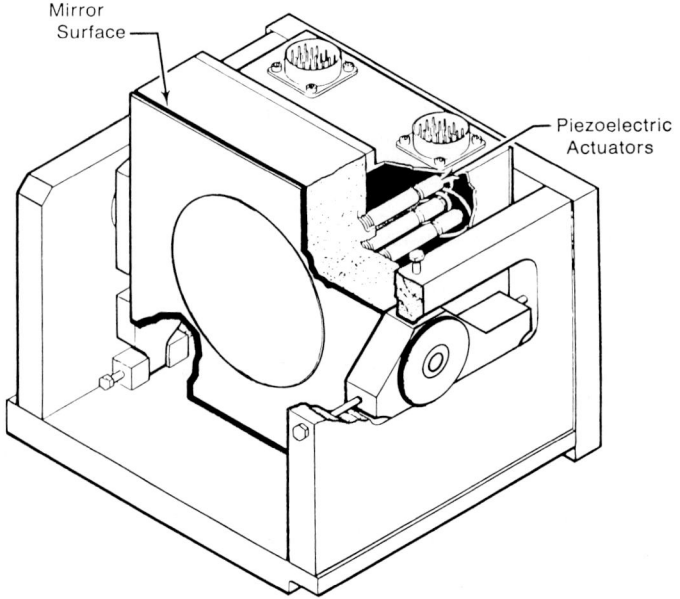

Fig. 29. Molybdenum, 21 cm active area, 52-actuator deformable mirror with fully removable actuators.

The important criteria in the design of individually actuated mirrors are the size, overall stroke, mirror structural resonance characteristics of the faceplate (stress and fatigue limits), mirror/actuator and substrate/mirror combinations, and the influence function for the number of actuators required to meet the correction goals. These parameters are governed in turn by the mirror faceplate thickness and material, interactuator spacing, actuator attachment geometry, and the stiffness of the mirror/actuator combination.

Actuation of the thin (on the order of 0.1 in.) mirror faceplate can be achieved with a variety of actuator types, with piezoelectric and hydraulic

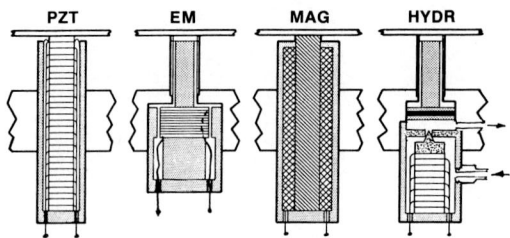

Fig. 30. Schematic cross section of typical discrete deformable mirror actuator designs: piezoelectric (PZT), electromagnetic (EM), magnetostrictive (MAG), and hydraulic (HYDR).

actuators being the most common. Figure 30 shows cross-sectional sketches of four actuator types. All are shown moving a section of a deformable mirror with respect to a thick reference backplate. The PZT actuator consists of a solid stack of ceramic disks of alternating polarity. An insulating jacket surrounds the stack and serves to transmit the force to the backplate.

The electromagnetic (EM) driver consists of a coil moving in a magnetic field and driving a piston against a spring. A cylindrical spring is shown here. A permanent magnet with pole pieces is used to transmit the force to the backplate.

The magnetostrictive (MAG) actuator consists of a solid cylinder of magnetostrictive ferrite surrounded by a wire-bound solenoid used to produce the driving magnetic field. The case surrounding the solenoid is used both as a pole piece for the magnetic return path and to transmit the force to the backplate.

The hydraulic (HYDR) actuator shown utilizes high pressure hydraulic fluid controlled by a PZT-driven valve (Hansen, 1975). The pressure is transmitted to the mirror surface by a piston working against a cylindrical spring. In the case of the hydraulic actuator, both hydraulic and electrical power are required.

Table VII compares key characteristics of these four actuator types in general terms. In many respects the PZT and magnetostrictive units are similar since they both involve dimensional change of the driver rod itself as a result of an externally applied field. The electromagnetic and hydraulic units are similar in that they use controlled force against an external spring.

The maximum extension that can be obtained in the PZT/MAG case is limited by saturation properties of the material itself under the applied field. For PZT-4 the maximum depoling field is 10^6 V/m, resulting in a strain of 3×10^{-4} m/m. For Vibrox-3 transducer ferrite, the maximum strain is 3×10^{-5} m/m at saturation. In addition to the difference in the numerical values of strain obtainable, note also that the electric field is applied to the PZT with little growth in the basic transverse dimension; only the connecting tabs and the surrounding insulator are needed. In the magnetostrictive case, the solenoidal winding and magnetic return path impose limitations in minimizing the transverse dimensions while maximizing the magnetic field. The lack of ability to pole ferrites also results in the requirement of an externally applied dc bias field if linear displacement is desired.

The maximum extension that can be obtained for the EM and hydraulic actuators is inversely proportional to the stiffness of the restraining spring. This spring, taken with the piston/mirror mass it controls, must have a resonant frequency substantially above the desired COAT control bandwidth ($\approx$ 1 kHz). Thus, the actuator must be capable of applying a large force to obtain the desired displacements. For the forces required to keep these

TABLE VII
Comparison of Actuator Types[a]

Actuator type	PZT	EM	MAG	HYDR
Maximum displacement limited by	$\Delta L/L$ limited by E_{max} (3×10^{-4})	ΔL limited by force against spring	$\Delta L/L$ limited by B_{max} (3×10^{-5})	ΔL limited by force against spring
Maximum pressure limited by	$\Delta L/L \cdot Y$ (2700 psi)	Limited by $B_{max} \cdot I_{max}$	$\Delta L/L \cdot Y$	Hydraulic technology 10,000 psi
Frequency response limited by	Mass plus bulk spring	Mass plus external spring	Mass plus bulk spring	Mass plus external spring, valve motion
Inherent stiffness	Y	External spring	Y	External spring
Power dissipation	Material loss tangent	Coil $I^2 R$	Coil $I^2 R$	Fluid turbulence; self-cooled
Drive requirements	Hi V, reactive load	Hi I, resistive load	Hi I, resistive load	Hi V, reactive load; plus hydraulic power
Transverse size constraints	Least	Large	Unknown	Somewhat larger than PZT

[a] Y is Young's modulus of appropriate material; ΔL the Linear displacement; and L the Actuator length.

resonances high enough, experience indicates that EM actuators would be much too large in transverse dimensions to attain the desired small inter-actuator spacing (2 cm or less). The hydraulic actuator, on the other hand, can achieve extremely high pressures with little growth in transverse extent,* since the fluid allows the force to be transmitted "around corners" so to speak. We may consider the hydraulic actuator as a *force* and *displacement amplifier* for the PZT valve control.

Other considerations such as power dissipation and drive impedance are also compared in Table VII. It is important to keep the amount of power

---

* This is only true as long as the piston displacements are small, as they are here; otherwise sufficient plumbing must be supplied to allow fluid flow.

8. ADAPTIVE TECHNIQUES FOR WAVE-FRONT CORRECTION

dissipated small in the vicinity of the back plate, since this is the stable surface to which the actuators are referenced. Excessive thermal distortion of the back plate would be undesirable. For the hydraulic actuator, circulation of hydraulic fluid effectively removes any heat generated by mechanical motion. PZT is an inherently low dissipation material. On the other hand, both magnetically actuated devices would have to have some cooling system applied to remove the I^2R losses.

These data lead to the conclusion that piezoelectric ceramics of the lead–zirconium–titanate (PZT) type are useful for the close actuator spacings and where a relatively small displacement (± 20 μm) is required against a relatively low force. For longer strokes against higher forces, the hydraulic actuator with a PZT controlled valve may be the preferred solution.

Achievement of large amplitude motions ($\gtrsim \pm 5$ μm) with all continuous surface deformable mirrors built to date has limited the mirror frequency response to less than 5 KHz due to the required length of the actuators and actuator/mirror stiffness. Significantly higher structural frequencies (40–80 KHz) can be achieved for nominal 20-cm diameter mirrors by using high stiffness to weight faceplate materials such as beryllium, and very stiff actuator/mirror/substrate combinations at the penalty of reduced stroke (± 0.5 μm) and limited duty cycles. The latter can be achieved using one or two thin piezoelectric crystals "sandwiched" between a beryllium mirror and a massive substrate. Consequently, current design approaches for multidither active systems utilize two mirrors, a large-amplitude corrector and a low-amplitude dither mirror. As mentioned earlier, for some applications with low bandwidths, it is possible to combine both functions on one mirror.

For a multidither COAT system, the dither mirror resonant frequencies should be kept above the maximum required dither frequency, if possible, to avoid servocontrol problems. As discussed in Section IV, the maximum dither frequency is given by

$$f_{\max} = [10 + C((N_a/2) - 1)]f_s \tag{35}$$

where N_a = number of actuators, f_s = the open-loop, unity gain servobandwidth, and C depends on the actuator-to-actuator mechanical coupling. The mirror resonance frequency increases with decreasing size; however, practical limitations exist in the minimum actuator centerline spacings due to the actuator size (a limit not imposed on monolithic mirrors). Current piezoelectric actuator designs for large stroke (± 40 μm) mirrors limit centerline-to-centerline actuator spacings to 1.5 cm. Accurate prediction of mirror structural response, especially for dither mirrors, can only be determined by structural analysis procedures using computer modeling. Typical responses for nominal 16–20 cm control and dither mirrors are shown in Fig. 31. Structural resonances limit the usable maximum frequency in current designs

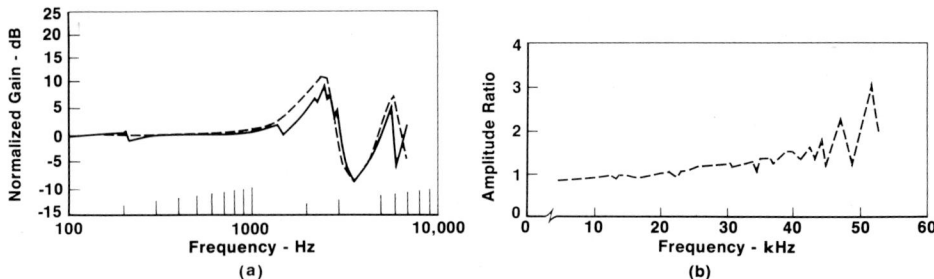

FIG. 31. Typical structural response characteristics of nominal 20-cm diameter deformable mirrors with drive applied to a central actuator: (a) molybdenum control mirror: (———) experimental frequency response, (b) beryllium (– – –) predicted frequency response; dither mirror.

to about 50 kHz. Pearson and Hansen (1977) have examples of the experimental phase and amplitude response observed in two other types of mirrors (one of which is shown in Fig. 28).

The choice of the mirror influence function can affect not only the ability to produce a desired phase front, but it can also affect two very important quantities: the servochannel cross coupling (when each mirror actuator is driven by a single servochannel) and the interactuator mirror surface "ripple" that occurs when several actuators are driven with near equal amplitudes (or when a linear tilt is desired, e.g.).

Following an analysis developed by O'Meara (1977b), the servocoupling C_s is defined as the ratio of the error signal in a given channel induced by the displacement of a single *neighboring* actuator associated with the given channel. For the special case where the influence function is of the form

$$IN(r) = \exp[-(r/r_e)^n] \tag{36}$$

where $r^2 = x^2 + y^2$, the servocoupling is (Pearson and Hansen, 1977)

$$C_{sn} = \frac{n 2^{[2/n-1]}}{\pi \Gamma(2/n)} I_n \tag{37}$$

where n is any number, $\Gamma(x)$ is the gamma function, and

$$I_n = \int_{\text{mirror}} \int \exp\{-[x^2 + y^2]^{n/2} - [(x-\beta)^2 + y^2]^{n/2}\} \, dx \, dy \tag{38}$$

The quantity β is related to the "mechanical" mirror actuator coupling C_M defined as

$$C_M = \exp[-(r_s/r_e)^n] \tag{39}$$

The value of β is found from

$$\beta = r_s/r_e = (-\ln C_M)^{1/n} \tag{40}$$

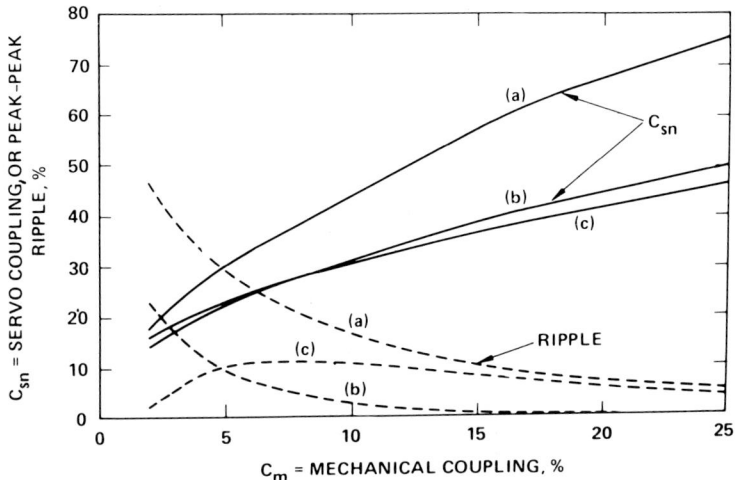

FIG. 32. Servocoupling C_{sn} [Eq. (37)] and peak–peak ripple as a function of deformable mirror mechanical coupling C_m [Eq. (39)]. Three influence function cases are shown: (a) "sub-Gaussian" ($n = 1.5$); (b) Gaussian ($n = 2.0$); (c) "super-Gaussian" ($n = 2.5$). (Further data are available in Pearson and Hansen, 1977.)

In order to estimate the mirror ripple effects, we make two simplifying approximations. First, we ignore the mirror edge effects and assume that all actuators have the same influence function. Second, we assume that the actual mirror surface is a linear superposition of the individual actuator influence functions for all possible drive amplitudes to the actuators. We then look at the surface produced when all the actuators in an infinite hexagonal array are given equal amplitude displacements. The results for the peak-to-peak ripple as a percentage of the maximum displacement are shown in Fig. 32 along with the servocoupling computed from Eqs. (37) and (39). Three cases are shown: $n = 1.5$ ("sub-Gaussian"), $n = 2.0$, and $n = 2.5$ ("super-Gaussian").

Based on these results for servocoupling and mirror ripple, we conclude that a nearly Gaussian influence function is a desirable characteristic for a deformable mirror and that a mechanical coupling coefficient of between 5 and 12% is a reasonable compromise between minimum ripple and servochannel cross coupling. Studies which are still in progress, however, indicate that for a particular value of C_M, there is an optimum value of n, n_{opt}, for minimum ripple. These studies also indicate that n_{opt} is larger than 2.0 for all values of $C_M < 20\%$. Work is currently in progress to further quantify these results and also to determine if there is an optimum value for n for minimum servocoupling C_{sn}.

Although we conclude that a Gaussian, or slightly super-Gaussian, influence function is desirable, we have not addressed the question of how to achieve this influence function. The influence function is controlled in discrete actuator mirrors primarily by three quantities: (1) the stiffness of the mirror faceplate, (2) the interactuator spacing, and (3) the geometry of the actuator attachment to the faceplate. These quantities also influence the maximum actuator motion as well as the surface ripple and frequency response of the mirror. The coupled relationships among all the various parameters means that the optimum mirror design is likely to vary with the intended application. Once the desired performance is specified, the various design tradeoff analyses can be performed.

Typical influence function profiles for the 37-actuator beryllium mirror shown in Fig. 28 and for the 52-actuator molybdenum mirror of Fig. 29 are shown in Figs. 33 and 34. The influence function for the beryllium mirror is approximately given by $\exp(-3.622 r^{1.5})$ and that of the molybdenum mirror is reasonably approximated by a Gaussian-type function in regions removed from the edge of the mirror.

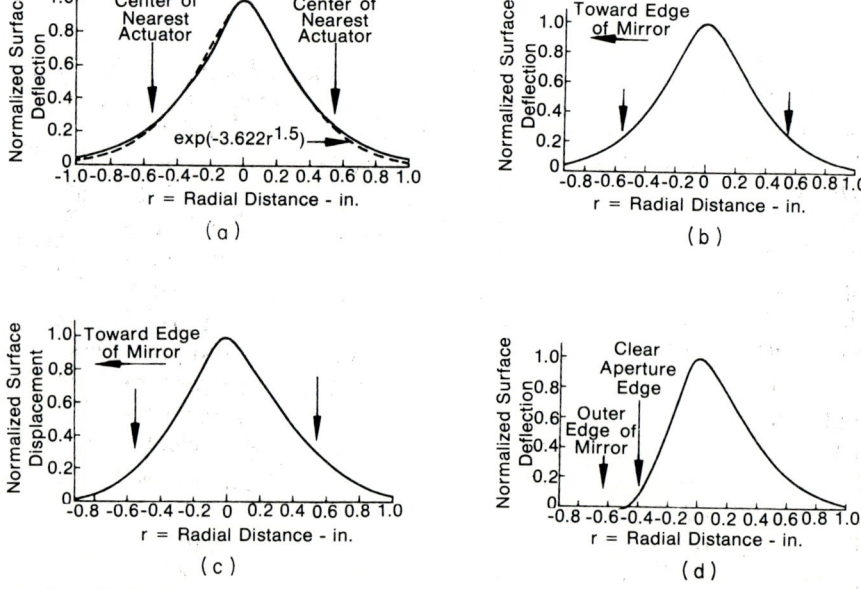

FIG. 33. Experimental influence function profiles of 37-actuator beryllium mirror: (a) center actuator; (b) first ring actuator; (c) second ring actuator; and (d) third (outmost) ring actuator. An empirical curve fit to the central actuator profile is shown as a dotted line in (a); the curve is is $\exp(-3.62 r^{1.5})$, where r is the radial distance from the profile peak (actuator center) in inches. The interactuator center–center spacing is $r_s = 0.55$ in.

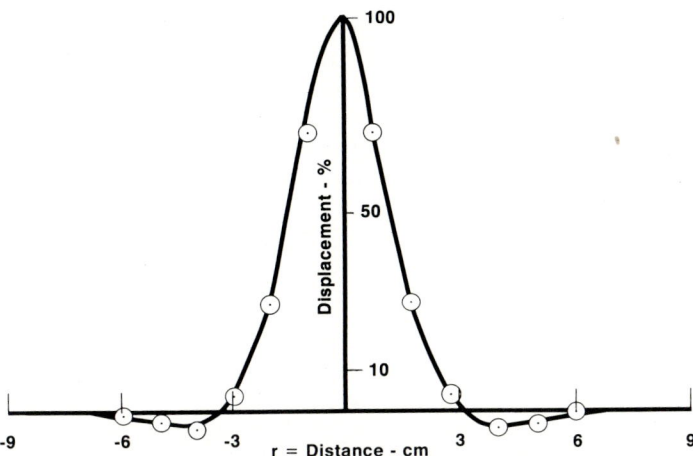

FIG. 34. Typical influence function for a central actuator in the 52-actuator molybdenum deformable mirror shown in Fig. 28. Near the peak the influence function is well represented by a Gaussian function. The interactuator spacing is $r_s = 3$ cm.

Multiactuated deformable mirrors can be used to implement "modal" correction using weighting matrices that produce the individual modes via coordinated motion of the individual actuators. However, this approach is more costly than using a mirror specifically designed to produce a given combination of "modes" and actuator nonlinearities, such as hysteresis and creep, require a local closed-loop control system to faithfully reproduce the specific mode shapes. Obviously, the design of a modal mirror is very dependent on the type of phase correction profiles required and the resulting mode shapes needed to match the correction profiles. This is one of the major weaknesses of a modal mirror: its capability of correcting only a given group of phase errors with reasonable accuracy. In addition, it is very difficult to design a mirror that can accurately reproduce a large number of terms of the series expansion, i.e., modes, that have been chosen to represent a phase correction. Consequently, a model mirror is useful only in situations where the general form of the required phase correction is well known, does not change character over the duration of operation, and can be represented by only a few dominant terms of the selected series expansion (modes).

The simplest form of a modal mirror is a tilt or steering mirror. The importance of this class of active mirrors for beam steering and autoalignment has resulted in this type of mirror being classified into its own category. Tilt mirrors are used either for autoalignment, fast pointing, or in combination with a deformable corrector mirror to relieve actuator dynamic range requirements. In the last application, the phase correction can be divided into the higher amplitude tilt error and the remaining low amplitude, higher spatial frequency terms. The key design features of a tilt mirror are a lightweight mirror moved by either two actuators about a central pivot point or two sets of actuators working in push–pull tandem. Both reaction designs, where actuator forces are predominantly absorbed and attenuated in the mount, and reactionless designs, where the actuator forces are dynamically balanced by reactive forces, have been utilized.

The next most common modal mirror is one that produces pure spherical or pure cylindrical correction via use of a single central actuator (spherical), or a limited number of edge actuators (cylindrical). Personnel at Lincoln Laboratory (Greenwood, 1976) have utilized a mirror which combines X and Y tilt plus a spherical correction. The spherical correction is produced by a single actuator that deflects a contoured mirror faceplate whose outer edges are attached to a solid frame. The frame in turn is tilted by tilt actuators, which are attached to the mirror mount. The design of such mirrors is a highly specialized field still under development and based on the use of structural theory to determine how a limited number of actuators strategically placed can produce a desired mode shape.

In summary, there are many interacting parameters that determine the performance of a deformable mirror. Tradeoffs among these parameters are necessary to arrive at a final design and detailed structural analyses are often required to establish the design before manufacture. Figure 35 shows some of the trends of various performance quantities as a function of actuator

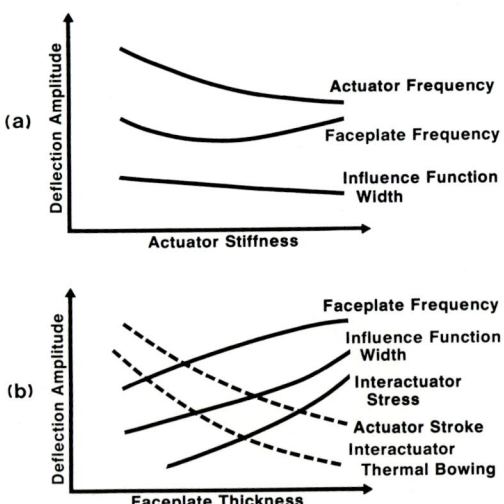

FIG. 35. Schematic representation of the trends in discrete actuator deformable mirror performance parameters as a function of surface deflection amplitude and (a) actuator stiffness and (b) faceplate thickness.

IV. SYSTEM DESIGN CONSIDERATIONS

A. Control Bandwidth Requirements

As discussed in Section I, the servocontrol bandwidth required in a COAT system depends on the temporal power spectrum of the various error sources. To set the bandwidth requirements, and the other requirements discussed in succeeding sections, we consider four error sources: (1) laser device; (2) optical train; (3) atmospheric turbulence; (4) atmospheric thermal blooming. Thermal blooming distortions are slowly varying since they are driven primarily by intensity variations, which are of small amplitude for most lasers of interest. The optical train produces low-frequency flux-induced errors and may have substantial fixed aberrations. Vibration-induced jitter of the components, however, may be substantial and can occur at rates up to 100–200 Hz.

The real bandwidth determining factor, however, is atmospheric turbulence. Several authors (Greenwood and Fried, 1976; Hogge and Butts, 1976a,b; Greenwood, 1977b,c) have considered the problem of bandwidth-limited adaptive compensation for turbulence. Following the results of Greenwood (1977c), we define the variance of the phase caused by the *uncorrected* portion of the power spectrum $P_t(f)$ as

$$\sigma_r^2 = \int_0^\infty |1 - H(f, f_c)|^2 P_t(f) \, df \tag{41}$$

where $P_t(f)$ is defined in Eq. (5), and $H(f, f_c)$ is the Fourier transform of the servoimpulse response function with f_c a characteristic cutoff frequency, such as a 3 dB point. For a simple RC filter,

$$H(f, f_c) = (1 + if/f_c)^{-1} \tag{42}$$

for which Greenwood calculates

$$\sigma_r^2 = 3.84 \times 10^{-2} \frac{k^2 C_N^2 V_T^{8/3}}{\omega f_c^{5/3}} \left[\left(1 + \frac{\omega R}{V_T}\right)^{8/3} - 1 \right] \tag{43}$$

The constant transverse wind speed is V_T, the beam slew rate is ω, the range to the target is R, and C_N^2 is assumed uniform along the path. The reduction in Strehl ratio due to a finite control bandwidth is then given by

$$S = \exp(-\sigma_r^2) \tag{44}$$

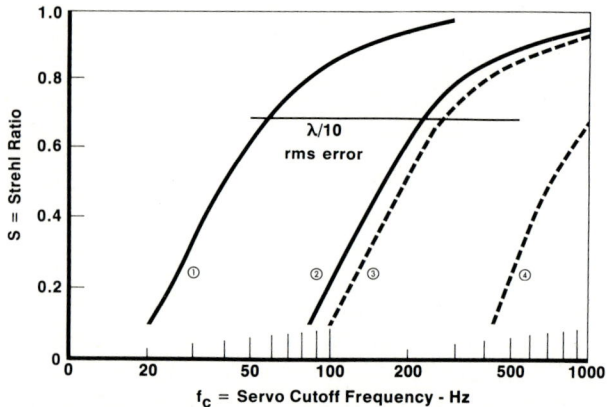

FIG. 36. Strehl ratio as a function of servo cutoff frequency [Eqs. (43) and (44)]. Two infrared laser wavelengths are used and two turbulence levels: moderate ($C_N^2 = 10^{-14}$ m$^{-2/3}$) and strong ($C_N^2 = 10^{-13}$ m$^{-2/3}$). Other values assumed are $R = 4$ km, $\omega = 0.01$ rad/sec, and $V_T = 4$ m/sec.

Case	$C_N^2 - 10^{-14}$ m$^{-2/3}$	λ, μm
1	1	10.6
2	10	10.6
3	1	3.8
4	10	3.8

Figure 36 plots S using Eqs. (43) and (44) as a function of f_c for two values of wavelength and turbulence level.

The results in Eqs. (43) and (44) make up a design rule of thumb which is subject to some approximations as discussed by Greenwood (1977c). As long as $\sigma_r \leq 0.5\pi$ ($\frac{1}{4}$ wave), however, the approximations are valid and the result is useful for all but perhaps the final stages of a servodesign.

The next consideration is what factors limit the servocontrol bandwidth in a COAT system. Ideally, it should be an electronic limitation, but in practice the limitation is mechanical: the frequency response of the deformable mirrors. For all nondither systems, this means a control bandwidth limit of 1–2 kHz, if control mirrors with frequency response such as the one shown in Fig. 31a are used. If smaller mirror surface motions can be used (e.g., visible-wavelength COAT systems), then the control bandwidths can be as high as 40–50 kHz (see Fig. 31b). At these large bandwidths, signal to noise may be a problem; Section IV.D discusses this consideration.

For a multidither COAT system, the choice of dither frequencies will also set the possible control system bandwidth, with higher bandwidths requiring higher dither frequencies. In a dither system, a conservative design rule of thumb requires the lowest dither frequency to be at least 10 times the control bandwidth,* f_s. By careful design, frequencies only 3 to 4 times f_s could be used. The dither frequency spacing must be at least $2f_s$ so that the signal spectra from adjacent channels do not overlap (see Fig. 20).

In practice, however, the required spacing depends on the channel–channel coupling, which can be produced, for example, by deformable mirror interactuator coupling [see Eq. (37)]. The spacing required also depends on whether one dither frequency is used for each channel or whether two channels are driven by the same frequency, one at $\cos(\omega_d t)$ and the other at $\sin(\omega_d t)$. Early work (Bridges *et al.*, 1973) in multidither COAT servodesign has demonstrated that such operation is possible and that it gives the same performance as one frequency per channel. It has the advantage that the total required dither bandwidth is reduced by $\sqrt{2}$ (instead of the intuitively expected factor of 2) from the one frequency per channel system. The same early work (Bridges *et al.*, 1973) has also shown that attempting to use one dither frequency for more than two channels is possible, but produces no further reduction in total dither bandwidth.

As stated earlier, computer simulations have shown that the total required dither bandwidth can be found from

$$f_{\max} = [10 + C(n-1)]f_s \tag{35'}$$

where n is the number of dither frequencies (one half the number of control channels in a "sine/cosine" system) and the conservative value of the minimum dither frequency has been used. The constant C depends on the servocoupling, which, if we assume all coupling occurs through the control and dither mirrors, depends on the influence functions of the mirrors. Table VIII gives values of C for several cases and Fig. 37 plots Eq. (35) for two values of f_s and C.

TABLE VIII

COEFFICIENT C IN EQ. (35) FOR MINIMUM DITHER FREQUENCY SPACING

Dither Mirror	Control Mirror	C
Segmented ($C_M = 0$)	Segmented	3.2
Segmented	Deformable, $C_M = 0.15$	4.2
Deformable, $C_M = 0.08$	Deformable, $C_M = 0.08$	4.2
Deformable, $C_M = 0.15$	Deformable, $C_M = 0.15$	5.3

* Defined as the open-loop unity-gain frequency.

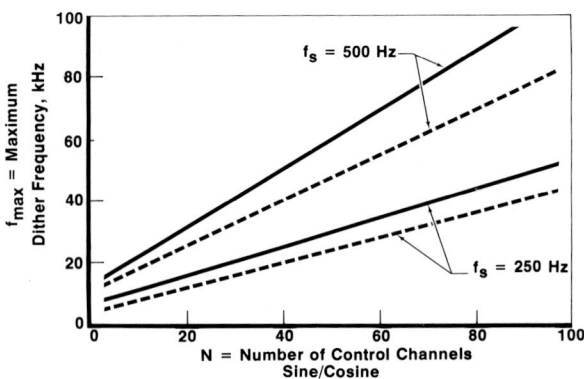

FIG. 37. Maximum dither frequency requirements as a function of the number of *control* channels for two cases: (1) (---), segmented dither and corrector mirrors, $C = 3.2$; (2) (——), both mirrors deformable with 8% mechanical coupling $C = 4.2$ ($C_m = 0.08$). Sine/cosine operation is assumed.

B. Phase Correction Requirements

1. Introduction

For a given aperture size, the number of actuators (or active zones) and actuator spacing (zone size) determine the ability to correct for a given phase error. In general, the required number of mirror actuators (active zones) is determined by the phase error spatial frequencies and the actuator stroke by one half the max-to-min phase amplitudes present in the distortion to be corrected. Actuator response requirements follow from the bandwidth considerations discussed in Section IV.A. Depending on the type of distortion to be corrected, the spatial frequency and amplitude present can vary significantly, with turbulence distortions requiring a large number of actuators and optical train component distortions requiring a minimal number. Several techniques exist for predicting a deformable mirror's ability to match a phase profile and for determining the required number of actuators. The most common ones, in order of increasing complexity are: (1) closed form analytical predictions, (2) statistical transfer function techniques, (3) influence function summation, and (4) spline fit approximation and finite element (NASTRAN) structural computer modeling. The last approach utilizes standard structural analysis to predict how a mirror surface moves for a defined actuator array, with each actuator at a defined position. The approach is costly, requires a new model each time the actuator pattern or spacing is changed, and normally is used only to evaluate a final design where the number of actuators have been selected by one of the other techniques. In all but the first technique, the results give the residual mismatch, either in terms of a phase distribution across the aperture or an rms phase distortion between the desired contour and the actual mirror. An accurate evaluation of the effectiveness of correction can only be obtained by computing the far-field pattern produced by the correction when propagated through the actual system. For the remainder of this discussion, we will consider zonal COAT systems and perfect wave-front sensing and control. Any residual phase errors then occur because of the finite number of deformable mirror actuators or because of their finite stroke. For a modal system, the number of modes required for turbulence compensation is discussed in Section II.B. Bradley and Herrmann (1974), Primmerman and Fouche (1976), Lind et al. (1977), and Pearson et al. (1976d) discuss the modes that contribute to thermal blooming.

2. *Actuator Number Requirements*

The actuator density requirement is determined by the spatial frequency spectrum $P_s(\kappa)$ of the phase error. Atmospheric turbulence and thermal blooming distortions dominate the requirement. For thermal blooming, the number of mirror elements is best determined by calculating the ideal phase correction surface (as was done in Bradley and Herrmann 1974) and then fitting this surface with a deformable mirror having a finite number of actuators using an expression such as that given in Eq. (20). The number of actuators is varied to bring the rms residual error below some value, usually $\lambda/10$.

The only accurate closed-form solution developed to date is for the correction of atmospheric turbulence, which when present usually dominates the requirement on number of actuators, even with other distortions present. This is particularly so at shorter wavelengths. Brown (1975a) has derived an expression for the Strehl ratio that can be achieved in the presence of turbulence when an ideal adaptive system is used with a deformable mirror having N actuators. The result is

$$N = \frac{(0.051 k^2 C_N^2 R D_T^{5/3})^{6/5}}{\ln(1/S)} \qquad (45)$$

where S is the Strehl ratio, $k = 2\pi/\lambda$, R is the propagation distance, D_T the aperture diameter, and C_N^2 the atmospheric structure constant. The result assumes $D_T/L_o = \frac{1}{2}$, where L_o is the turbulence outer scale. Equation (45) has been used to plot the Strehl ratio as a function of N in Fig. 38. As can be seen, 60–80 actuators will achieve a Strehl ratio of 0.6 or greater in all cases and increasing N further does not rapidly improve S.

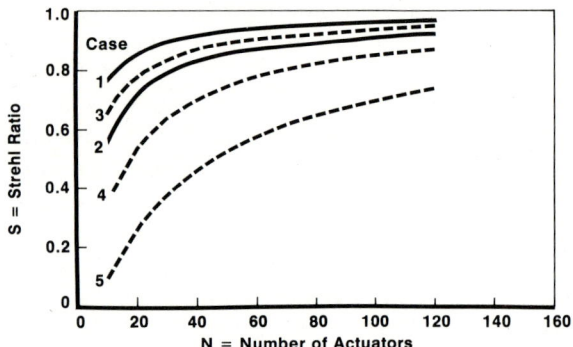

FIG. 38. Deformable mirror actuator number requirements imposed by atmospheric turbulence [Eq. (45)].

Case	D_T, m	λ, μm	$C_N^2 R$, 10^{-11} m$^{1/3}$
1	1	10.6	10
2	1	10.6	21.5
3	0.7	3.8	4
4	0.7	3.8	10
5	0.7	3.8	21.5

An alternate approach to determining N is to recognize the fact that a multiactuator deformable mirror acts as a high-pass spatial frequency filter which operates on the spatial frequency spectrum of the uncorrected wave-front error. Standard linear systems theory can then be used to characterize the deformable mirror performance (Harvey and Callahan, 1978). With this technique, the spatial frequency spectrum of the uncorrected wave-front error is given by the Fourier transform of the wave-front autocovariance function as shown in Fig. 39. The central ordinate theorem of Fourier transform theory requires that the volume contained under the wave-front error spectrum curve is equivalent to the variance of the wave-front error (Goodman, 1968). Multiplying this spatial frequency spectrum by the deformable mirror filter function results in a filtered spectrum whose volume is equivalent to the variance of the residual wave-front error after compensation. The inverse Fourier transform of this filtered spectrum yields the autocovariance function of the residual wave-front error. Frequencies with a spatial period less than twice the actuator spacing are effectively passed by this spatial frequency filter. For periodic errors, the deformable mirror will either pass or filter the wave-front error depending on the spatial frequency and the actuator spacing. For random wave-front errors (characterized by a continuum of spatial frequencies), a percentage of the error will remain after filtering.

8. ADAPTIVE TECHNIQUES FOR WAVE-FRONT CORRECTION

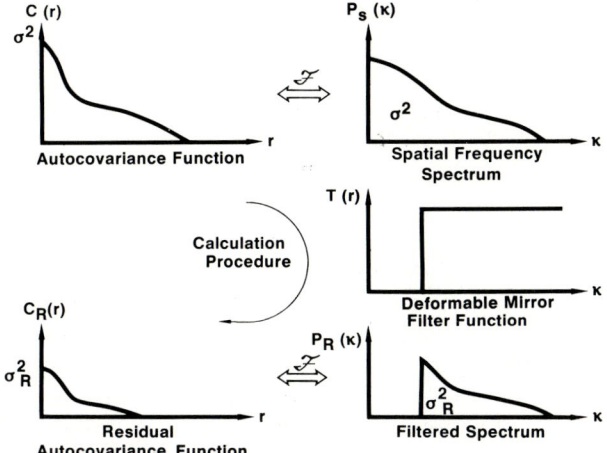

FIG. 39. Wave-front compensation by a deformable mirror viewed as a high-pass filtering operation.

The deformable mirror correction characteristics for random wave-front errors with Gaussian statistics is shown in Fig. 40 for several different actuator spacings as a function of the wave-front autocovariance length. These curves assume an idealized sharp cutoff characteristic as shown in Fig. 39. The larger the autocovariance width, the greater the percentage of errors represented by low spatial frequencies. The smaller the actuator spacing r_s, the more effective the deformable mirror is in eliminating wave-front errors. For

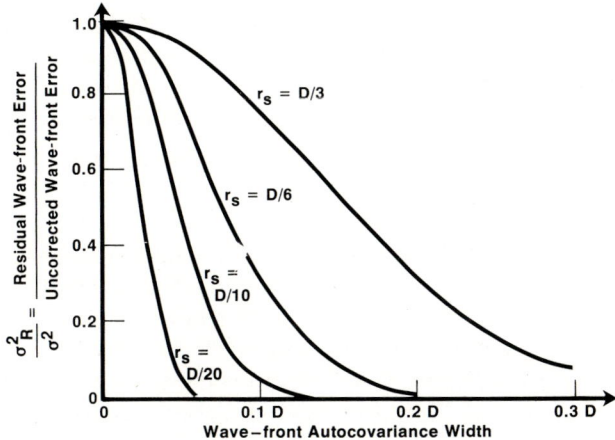

FIG. 40. Idealized (sharp cutoff) deformable mirror correction characteristics for random wave-front errors that have Gaussian statistics.

example, if the autocovariance length is equal to $D/6$ ($\approx 0.16D$) and the actuator spacing is equal to $D/6$, less than 10% of the uncorrelated wave-front error would be passed by the deformable mirrors. However, if the autocovariance length was $D/10$, the same deformable mirror would pass more than 30% of the uncorrected wave-front error.

A real deformable mirror with a finite number of actuators will not result in an idealized cutoff filter function as shown in Fig. 39 and assumed in Fig. 40. A realistic filter function can be constructed from measured actuator influence data for a typical deformable mirror. The deformable mirror can conform to any shape made up of a superposition of these influence functions located at the actuator positions and connected by a spline fit routine. Positioning the actuators to produce the best fit, in a least-square sense, to a sine wave of a given spatial frequency and calculating the rms error will characterize the deformable mirror performance in correcting wave-front errors of that spatial frequency. Repeating this operation for many spatial frequencies will result in the desired transfer function. This transfer function can be used to determine the residual wave-front error after compensation for any given uncorrected wave-front error incident upon the deformable mirror. For the special case of random wave-front errors with Gaussian statistics, a set of design curves similar to those illustrated in Fig. 40 provides a simple method of determining the actuator density and, as discussed below, stroke requirements (stroke $\approx \pm 2\sigma$) of a deformable mirror capable of achieving a specified degree of wave-front error compensation.

In addition to the ordinary fabrication errors, a deformable mirror will exhibit a periodic figure error associated with the actuator locations. This error results from the polishing process when the mirror faceplate is supported only at the actuator locations and is proportional to the required actuator stroke. This error has a period equal to the actuator spacing and hence cannot be corrected with the deformable mirror. Use of separate tilt and focus mirrors reduces the deformable mirror stroke requirement and hence this periodic fabrication error. This type of periodic error is not included in any of the discussions presented above.

3. *Actuator Excursion Requirements*

The total phase excursion required of a deformable mirror actuator depends on its location on the mirror, on whether or not the tilt error is removed by a separate mirror, and for turbulence correction, on the turbulence outer scale L_o. Unpublished results of Brown (1976a) indicate that $\pm 2\pi$ radians of phase shift should be sufficient for thermal blooming compensation. An interesting result for blooming compensation is that once the correct *shape* of the phase correction is found, then the *amplitude* of the correction can vary by as much as a factor of 2 with only a 10% change in corrected Strehl ratio (Primmerman and Fouche, 1976).

For turbulence compensation *without* a tilt mirror, it can be shown (Pearson and Hansen, 1977) that for an outermost actuator

$$(\Delta\phi)_{max} = \pm 0.566 k (C_N^2 R D_T^{5/3})^{1/2} \quad \text{rad} \qquad (46)$$

in order to insure that the required correction does not exceed the actuator excursion more than 0.3% of the time. For a center actuator, $(\Delta\phi)_{max}$ is reduced by a factor of 0.667 from that given by Eq. (46). Equation (46) also assumes near-normal incidence on the mirror and $D_T/L_o \approx 0.5$. If $D_T/L_o \approx 0$, $(\Delta\phi)_{max}$ is increased by a factor of 2.36 for an outermost actuator.

When an auxilliary tilt mirror is used to remove the tilt portion of the phase error, the excursion of a central actuator is unchanged, but that necessary for an outermost actuator is reduced (Brown, 1976b) by a factor of ≈ 0.67 for $D_T/L_o = 0.5$ and by a factor of ≈ 0.35 for $D_T/L_o \approx 0$ from the value given by Eq. (46).

The best estimate for the total required phase excursion when several distortions are present is given by a root-square summation (rss) of the individual requirements (Butts and Hogge, 1977). We thus have

$$(\Delta\phi)_{max, tot} = [(\Delta\phi)_{turb}^2 + (\Delta\phi)_{T.B.}^2 + (\Delta\phi)_{other}^2]^{1/2} \qquad (47)$$

Table IX summarizes the excursion and bandwidth requirements for two wavelength values and for selected parameter values. The excursions are given in units of microns of actual surface motion. In computing $(\Delta\phi)_{max, tot}$, we have assumed $(\Delta\phi)_{T.B.} = \pm \lambda$ and $(\Delta\phi)_{other} = \pm 0.3\lambda$, $D_T/L_o = 0.5$, $R = 4$ km, $\omega = 0.1$ rad/sec, and looked at an outermost actuator.

C. Signal-to-Noise Considerations

Given an ideal phase corrector element and sufficient control bandwidth for the phase errors being corrected, the final performance-determining criterion is control system signal to noise. Calculation of the signal to noise for a particular system can be quite involved and in general depends on the details of the system. Dyson (1975) has developed a generalized framework for the analysis of adaptive optical systems and other authors (Hudgin, 1977b; O'Meara, 1977a) have treated more specific systems.

McGlamery *et al.* (1975) have derived general results for a quadrant detector (Hartmann) sensor and for a shearing interferometer when the major noise source is photon noise. Visible-wavelength quantum-limited detection is thus assumed. They assume that the wave front across each subaperture in the sensor can be represented by tilt only and that the rms error in the tilt (slope) measurement σ_α is due to a limited number of photoelectrons available during a sampling period, which must be less than the atmospheric time constant. In all cases, σ_α is inversely proportional to $\sqrt{\bar{N}_T}$, where $\bar{N}_T$ is the average number of photoelectrons available to a single detector of the wave-front sensor. The brightness of the object, the optical transmittance of the

TABLE IX

MIRROR EXCURSION AND SERVOBANDWIDTH REQUIREMENTS FOR SELECTED PARAMETER VALUES

	Mirror excursion (μm), outer actuator						Servofrequency response (Hz) for Strehl ratio ≥ 0.67			
	$\lambda = 10.6$ μm, $D_T = 96$ cm		$\lambda = 3.8$ μm, $D_T = 60$ cm				$\lambda = 10.6$ μm		$\lambda = 3.8$ μm	
C_N^2, 10^{-14} m$^{-2/3}$	No tilt mirror	With tilt mirror	No tilt mirror	With tilt mirror			$V_T =$ 4 m/sec	$V_T =$ 10 m/sec	$V_T =$ 4 m/sec	$V_T =$ 10 m/sec
1 (Turb. only)	±1.7	±1.3	±1.3	±0.85			56	78	279	388
$(\Delta\phi)_{\text{tot, max}} =$	±11.2	±11.1	±4.7	±4.1						
10 (Turb. only)	±5.5	±4.2	±3.7	±2.8			224	311	1109	1541
$(\Delta\phi)_{\text{tot, max}} =$	±12.4	±11.8	±5.4	±4.8						

optical system, the quantum efficiency of the detector, and the integration time of the servocontrol all determine N_T.

Without derivation, we will state the results for the two types of detectors. For a Hartmann sensor,

$$(\sigma_\alpha)_{\text{Hartmann}} = \frac{1}{\lambda M_q \sqrt{\bar{N}_T}} \quad \text{(wavelengths/m)} \qquad (48)$$

where M_q is the slope of the image response function at the origin of the quadrant detector and $\bar{N}_T$ is the *total* number of photoelectrons from all 4 quadrants. For a point object and a rectangular subaperture of width W, $M_q = 1/2W$ (1/meters).

For a shearing interferometer,

$$(\sigma_\alpha)_{\text{shear}} = \frac{\sqrt{2}}{2\pi\lambda K[H(K,0)/H(0,0)]\sqrt{\bar{N}_T}} \quad \text{(wavelengths/m)} \qquad (49)$$

where $H(f_x, f_y)$ is the Fourier transform of the two-dimensional object irradiance map and the shear distance is $S = K\lambda$ (see Section II.C.2). In practice, the shear is varied so that $KH(K,0)$ is maximized, realizing that shear distances greater than the detector spacings should not be used and as a rule of thumb, the detector spacing is chosen about equal to the minimum value of r_0 [Eq. (12)] that will be encountered.

For many cases of interest, $(\sigma_\alpha)_{\text{shear}} \approx \sqrt{2}(\sigma_\alpha)_{\text{Hartmann}}$ unless four detector arrays are used in the shearing system, in which case the two systems give the same theoretical performance. The actual objects and ranges and receiver apertures must be investigated for each case of interest, however, to establish whether one system is ideally superior to the other. Since the shearing interferometer has an adjustable, object-dependent parameter, intuitively we would expect that it would give superior performance for some objects and ranges.

The signal to noise of a multidither COAT system has been discussed in some detail by O'Meara (1977a). We summarize here some of his principal results. In any multidither system, any residual phase error in a particular channel converts the dither modulation into an amplitude modulation. The servopower signal-to-noise (S/N) ratio depends on the square of the resulting amplitude modulation index M_e, which can be quite low in a multidither system. Specifically, it can be shown that for the mth channel,

$$M_{em} = [2J_1(\psi_0)/NJ_0(\psi_0)](\beta_m - \beta_{mc}) \quad \text{near convergence} \qquad (50)$$

and

$$M_{em} = [2J_1(\psi_0)/\sqrt{N} J_0(\psi_0)] \sin(\beta_m - \beta_{mc}) \quad \text{far from convergence} \qquad (51)$$

The various quantities in Eqs. (50) and (51) are defined in Section II.C.4, Eqs. (30)–(32). As an example, take $N = 60$, $\psi_0 = 20°$, and $(\beta_m - \beta_{mc}) = 0.2$ radians, giving $M_{em} = 1.2 \times 10^{-3}$.

The signal-to-noise ratio in a multidither system will depend on M_e, on the received optical power, and on the wavelength region of interest. For infrared operation, background and dark-current shot noise can be suppressed to negligible levels and detector electronic gains are usually near unity. The optical power at the detector required to achieve a given S/N is then

$$P_0 = \frac{2h\nu}{M_e^2 \eta}\left(B\frac{S}{N}\right) + \left[\left(\frac{2h\nu}{M_e^2 \eta}\right)^2 \left(B\frac{S}{N}\right)^2 + \frac{2kTh^2\nu^2}{M_e^2 \eta^2 e^2 R_L}\left(B\frac{S}{N}\right)\right]^{1/2} \quad (52)$$

where M_e is the amplitude modulation index, η the detector quantum efficiency, B the noise bandwidth (one phase correction channel), S/N the receiver output signal-to-noise ratio (power), T the noise temperature of the amplifier (including detector output circuit noise), R_L the detector load resistor, and h, k, and e are Planck's constant, Boltzmann's constant, and the electronic charge, respectively. Equation (52) divides naturally into three regions, as follows:

I. Thermal (amplifier) noise limited, in which

$$P_0 = \frac{h\nu}{M_e \eta e}\left[\frac{2kTB}{R}\frac{S}{N}\right]^{1/2} \quad (53)$$

II. A transition region, with about equal contributions from thermal noise and signal shot noise.

III. Signal shot noise limited, in which

$$P_0 = (4h\nu B/M_e^2 \eta)(S/N) \quad (54)$$

Figure 41 plots P_0 versus S/N for selected parameter values.

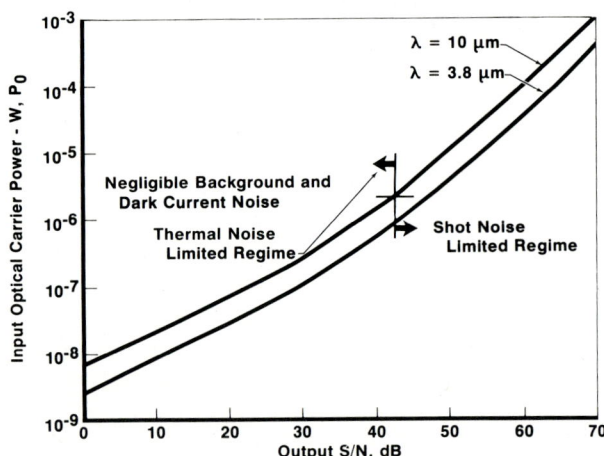

FIG. 41. Optical received power at the detector required to achieve a specified servopower signal-to-noise (S/N) ratio in a multidither COAT system. For direct detection receiver conditions: $\lambda = 3.8\,\mu\text{m}$; $\eta = 0.56$; $\lambda = 10\,\mu\text{m}$; $\eta = 0.4$; $B = 600\,\text{Hz}$; $M_e = 1.18 \times 10^{-3}$; $T_N = 300\,°\text{K}$; $R = 10^4\,\Omega$

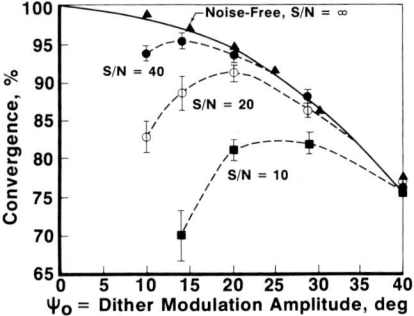

FIG. 42. Final maximum convergence values for an 18-channel multidither COAT system as a function of dither modulation amplitude and S/N ratio. The points are computer simulation values; the theoretical curve (———) is from Eq. (53) with $N = 18$.

One way to increase the S/N in a multidither COAT system is to increase the dither amplitude ψ_0. This is not without penalty, however, because the maximum converged level decreases as ψ_0 is increased according to

$$S = N^{-1}[1 + (N - 1)J_0^2(\psi_0)] \qquad (55)$$

This effect is plotted in Fig. 42 for $N = 18$. Note that $\psi_0 = 20°$ represents a good compromise between final converged level and adequate S/N.

There are two items which experience has proven to be essential for stable COAT control system operation with high energy lasers: (1) automatic gain control (AGC) and (2) compensation for laser power fluctuations, often called $P(t)$ compensation. An AGC network is used to stabilize the control loop gain in the presence of changing target returns and to provide automatic gain stabilization over a wide dynamic range of target return signals. Such a control can be implemented once in a common channel as shown in Fig. 21 or can be implemented for each control channel individually. The latter scheme or some combination of the two techniques, is most appropriate for Hartmann and shearing interferometer sensors.

A $P(t)$ compensation network is a type of AGC, but its function is to remove the deterministic laser power fluctuations that may interfere with proper control system operation. The usual implementation measures the laser fluctuations—$P(t)$—on a sample of the outgoing beam, delays the signal by the round-trip time to the target, and divides the received signal by the delayed, locally measured signal. This type of compensation is particularly important for multidither systems, where even small noise power within the dither band can adversely affect system performance. (See Section VI.B for further discussion on amplitude noise effects.)

The state of the art in COAT control system design has not progressed much beyond the fairly simple control systems described in this article and in the references. Modern estimation and control theory provides techniques for synthesizing systems that optimize some measure of performance such as minimum system response time subject to constraints on control magnitudes and rates or the maximum signal-to-noise ratio. Such techniques may have applicability to adaptive optical systems, but the benefits which might accrue from the use of optimal control synthesis methods and Kalman filter techniques are not immediately apparent. Asher and Ogrodnik (1977) discuss one formulation of the optimal control problem for N channels. They also apply their formulation to the problem of speckle noise cancellation in a one-channel dither control system (autofocus).

Optimal control synthesis methods rely on the knowledge of system states and control parameters and possibly state derivatives and the techniques are most readily applied for systems having a relatively few states and controls. Mirror-actuator dynamics, however, are characterized by multiple control points, with multiple resonances which give rise to many system states. In addition, optimal controllers are generally implemented on a digital computer (although it is possible to use analog realizations), especially if the system exhibits nonlinearities. Consequently, optimal controllers are not particularly compatible with large bandwidth, multiple-channel control systems required for many closed-loop adaptive optics applications.

Similar comments generally apply to modern estimation (generally Kalman filtering) techniques. However, there may be application for such methods in enhancing the return from target glint points or in overcoming spurious noise effects (Asher and Ogrodnik, 1977). Kalman filters include a model of the estimated dynamics of the plant whose performance is being measured. The filter predicts the system's expected behavior based on past information and also examines the latest sensed information which purports to describe its present situation. The two different sources of information are combined using weightings which are dependent on the uncertainty associated with each information source. Such a filtering scheme may be useful in the initial acquisition and tracking of a target glint point and in aiding a closed-loop A/O system to lock on a moving glint point in the presence of speckle effects and changing glint structure. A special issue (*Proc. IEEE*, 1976) contains several articles plus an extensive bibliography for those who wish to pursue further the use of advanced control techniques in adaptive optical systems.

V. COMPENSATION PERFORMANCE

A. Turbulence Compensation

Although the most commonly discussed application of a COAT system is atmospheric turbulence compensation, there are several other application areas. These include vernier pointing and tracking, removal of mechanical errors in an optical system such as distortions and mount jitter, and compensation for phase distortions occurring inside a laser device. The latter compensation and the removal of nonlinear propagation effects such as thermal blooming are discussed later in this section. This subsection deals with turbulence compensation.

The atmospheric correlation length r_0, defined in Eq. (12), can be interpreted physically as the distance perpendicular to the beam path over which index fluctuations are correlated. The reciprocal of r_0 is also a measure of the highest significant spatial frequency present in the index fluctuations. In order for a COAT system to correct adequately for the fluctuations, the turbulence must not have spatial frequencies that are comparable to the size of an element in the COAT transmitter. In other words, if the atmosphere is worse than about $\frac{1}{6}$ of a wave across an element, there will be some beam degradation that even a perfect COAT system cannot remove (except by reducing the element size). On the other hand, if the atmosphere is $\frac{1}{6}$ of a wave or better across the entire transmitter aperture, the COAT system has almost no atmospheric errors to correct for. These requirements can be stated mathematically as

$$D_T \geq r_0 \gtrsim D_e \tag{56}$$

where D_T is the transmitter diameter and D_e is the element diameter.

The first demonstration of closed-loop compensation with a multidither COAT system was reported by Bridges et al. (1974) using three-element and seven-element visible wavelength systems. The first demonstration of atmospheric turbulence compensation with this type of system was reported by Pearson (1976) using an 18-element piston-only type of system. Four criteria are commonly used to determine the degree of COAT compensation for atmospheric turbulence (Pearson, 1976): Strehl ratio, power on a glint with and without COAT compensation, power stability, and system convergence time. Table X presents typical observed values from Pearson (1976) of the time-averaged Strehl ratios as well as observations of the time-averaged power on a single glint P_g with and without COAT correction.

TABLE X
COAT TURBULENCE COMPENSATION PERFORMANCE

Experimental condition	S	$\bar{P}_g$ (Arbitrary units)	$(\bar{P}_g)_{\text{low turb}}/\bar{P}_g$
Low turbulence, COAT on	0.7	3.15	1.0
Low turbulence, no COAT correction	—	2.85	1.1
High turbulence, COAT on	0.8	1.34	2.3
High turbulence, no COAT correction	—	0.47	6.7

The last column in Table X compares the average glint power observed with COAT correction under low turbulence conditions to that observed for the other three cases. Without COAT correction, residual distortions at low turbulence levels reduce the average glint power by only a factor of 1.1, but with strong turbulence the power is reduced by a factor of 6.7. The COAT system, however, produces a factor of 2.9 correction. The COAT-corrected average intensity is still a factor of 2.3 below that observed in low turbulence, however, even though the apparent Strehl ratio is larger than in low turbulence. The lower corrected average intensity is caused by a combination of the effects of overall beam steering or wander, steering of the individual beam elements, and scintillation effects, all of which this system cannot remove. The major effect appears to be scintillation as further discussed below.

The stability of the target irradiance produced by the COAT system can vary greatly depending on the turbulence level. In a piston-correction system, beam wander introduced by the atmosphere will reduce the irradiance on a target glint in exactly the same way that moving glint will. The reduction occurs because the glint effectively moves off the phased array boresight axis. For the conditions in Pearson (1976), the array element diameter = $D_T/5$, so that the rms Strehl ratio reduction caused by steering is given approximately by

$$I_{\rm rms}/I_{DL} = \bar{R}_s[\sin^2(\alpha\pi/5)]/(\alpha\pi/5)^2 \tag{57}$$

The contribution of scintillation effects to the average Strehl ratio can be estimated by using Eqs. (10) and (11). For the experimental conditions of Pearson (1976), the scintillation contributed nearly twice as much as beam steering to reducing the average Strehl ratio in the experiment. The net expected Strehl ratio is $S = (0.65)(0.94) = 0.61$. The reciprocal of this value is 1.64, in fair agreement with the observed value of 2.3 given in line 3 of Table X. The beam steering contribution to the Strehl ratio reduction could be eliminated by employing a COAT-controlled tracking system, although scintillation effects will prove limiting even with a perfect tracking and phase control system (Bradley and Cheifetz, 1975). Since the amplitude fluctuations occur at about a 100-Hz rate, the tracking system would need to respond up to about 100 Hz. This conclusion is consistent with the results of Chase (1966), who considered the improvement for a heterodyne system when tracking is employed.

Two other important performance parameters for a COAT system are its ability to track a moving target and its ability to maintain the beam converged on the strongest glint present on a multiple glint target. Figure 43 shows the performance of an 18-element COAT system in removing turbulence for a fixed glint and in tracking a glint moving near a weaker one

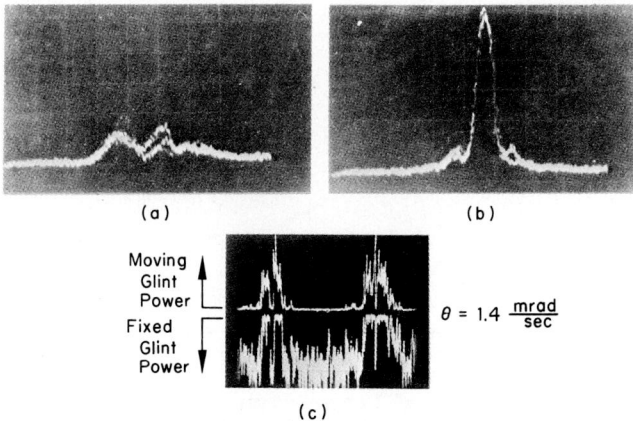

FIG. 43. COAT convergence in high turbulence ($C_N^2 = 6 \times 10^{-14}$ cm$^{-2/3}$). Each beam profile picture contains three traces, 30 msec apart: (a) 18-element array, COAT servoloop open; (b) COAT on (servoloop closed); and (c) power on each of two glints when one is stationary and one is moving; glint angular velocity is $\dot{\theta} = 1.4$ mrad/sec. The moving glint is 3 dB larger in reflectivity.

under heavy turbulence conditions. At its closest approach to the boresight axis, the moving glint has a 3 dB larger total return to the receiver.

Two things should be noted in Fig. 43c. First, the COAT system locks onto and follows the stronger glint as it moves across the transmitted beam. (The intensity envelope is just that of a single element of the transmitter array.) Similar behavior has also been observed with more than two glints. Second, although the strong turbulence occasionally causes the system to switch the beam from one glint to the other, the power is always on one glint or the other; the power is never shared between the two glints. The switching, in fact, is probably caused by the beam steering and scintillation discussed above. The measurements reported in Pearson (1976) indicate that one glint must be 2–3 dB larger than any other for the COAT system to form consistently the beam on only one glint without excessive switching from one glint to another nearly equal one. This conclusion on good glint discrimination remains true even if the receiver aperture has only one fifth of the diameter of the transmitter aperture.

Another cause for residual irradiance fluctuations occurs in deformable mirror multidither COAT systems: the tendency to lock up in nonoptimum $2N\pi$ states. O'Meara (1977b) has discussed this phenomenon in some detail. Briefly, the $2N\pi$ problem occurs because each channel in the deformable mirror dither system has multiple zero-error states of the servo that occur whenever the optical phase error in that channel is roughly a multiple of 2π as shown in Fig. 44. The maximum target irradiance occurs only for zero

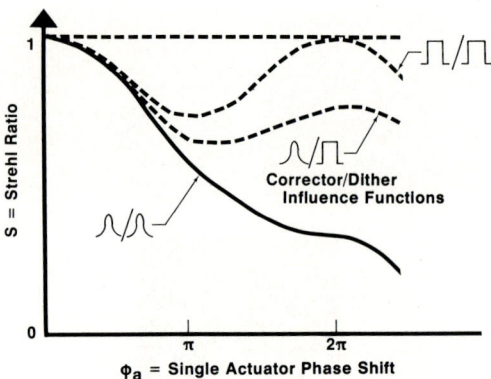

FIG. 44. Schematic illustration of $2N\pi$ ambiquities in multidither COAT systems. The influence functions of both the dither *and* corrector mirrors are important. For segmented mirrors, if the system locks up at $\phi_a = 2\pi$, the Strehl ratio is the same as at $\phi_a = 0$. This is not the case for deformable mirrors.

phase error, however, for deformable mirror systems. Depending on the state of the neighboring mirror actuators (O'Meara, 1977b) and on the mirror influence function, a $2N\pi$ condition in a given channel can be either an unconditionally stable state or a marginally unstable state where the servo has a finite error signal, but very low gain (gain is proportional to the slope of the curves in Fig. 44). In the second case, the servo will drive the channel phase error toward zero, but at a greatly reduced rate from the nominal small-error servoconvergence rate. In effect, the $2N\pi$ condition greatly reduces the control bandwidth in the affected channel. Rapidly varying errors such as turbulence would thus be poorly corrected and irradiance fluctuations could occur.

One type of evidence for $2N\pi$ states is shown in Fig. 45, which shows several time records of the error signal in a single servochannel. The channel observed drives a mirror actuator in the outer actuator ring and thus is more likely to encounter a 2π error. The fast transitions of over 2π in phase shift, indicated by arrows in Fig. 45, are attributed to the channel dropping out of a 2π lockup state. Since the transitions occur in roughly 4–10 msec, which is comparable to the servosystem convergence time, they are too rapid and too large in amplitude to be caused by the servo responding to normal atmospheric turbulence errors, which are normally slower at large amplitudes.

Clearly, optimum performance from a deformable mirror, multidither COAT system will be possible only when these detrimental effects can be removed or minimized. Numerous candidate techniques for alleviating the problem are currently under study. Any successful technique, however, must be able to distinguish between a $2N\pi$ servolockup error and a desired $2N\pi$ phase shift. Thus a simple limiting of actuator excursion to less than

8. ADAPTIVE TECHNIQUES FOR WAVE-FRONT CORRECTION 317

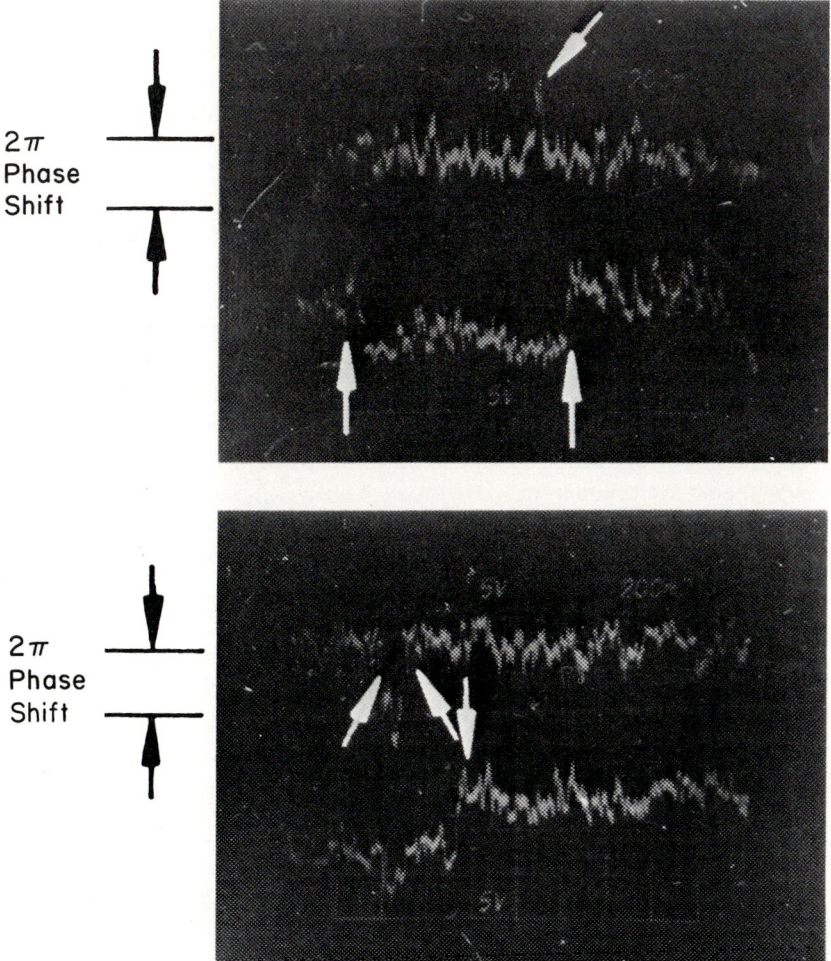

FIG. 45. Servoerror signals occuring in a single COAT control channel under conditions of high turbulence. The rapid transitions indicated by the arrows are attributed to the channel recovering from a $2N\pi$ lockup state.

$\pm 2\pi$ of phase shift will eliminate 2π lockups, but will also severely limit the error-correction dynamic range. Another simple technique is to limit the mirror actuator excursion between adjacent actuators to less than 2π phase shift. This technique limits the local wave-front error slope correction, which may not be an unduly severe restriction in many cases of interest. The technique has a further fatal difficulty, however, because a $2N\pi$ lockup occurs whenever the difference between the existing mirror state and the instantaneous wave-front error is $2N\pi$. In principle, a $2N\pi$ error condition could thus occur with a flat mirror. In addition, this technique will not rapidly relieve the "block" $2N\pi$ problem discussed by O'Meara (1977b). More sophisticated techniques that utilize dither second-harmonic information and automatic modulation index control are under study (O'Meara, 1977b).

In contrast to the multidither COAT system results, the performance of a phase conjugate adaptive array has been studied experimentally (Hayes et al., 1977) at $\lambda = 10.6$ μm over a relatively wide range of operating parameters. The experimental setup for this experiment is shown in Fig. 46a. The ability of the adaptive array to correct for turbulence-induced beam broadening was evaluated in a series of experiments at three different ranges. To demonstrate the correction capability for atmospheric effects, one test was conducted at 6 km range. Typical beam profile recordings in the adaptive and unadaptive mode are shown in Fig. 46b. Here the system has decreased the central lobe full width at half maximum from 15.2 to 12.5 cm. These measurements are taken in mild turbulence with $C_N^2 = 1.72 \times 10^{-15}$ m$^{-2/3}$. For these conditions $D_T/r_0 \approx 0.36$. The expected half-power spot size (Yura, 1973) for this turbulence level is $W_{1/2} = 15.9$ cm, in good agreement with the measured value of 15.2 cm, uncompensated.

Convergence was found to require approximately two loop times, following an initial hunting period, which increases with the number of elements and the complexity of the initial phase distribution. For high

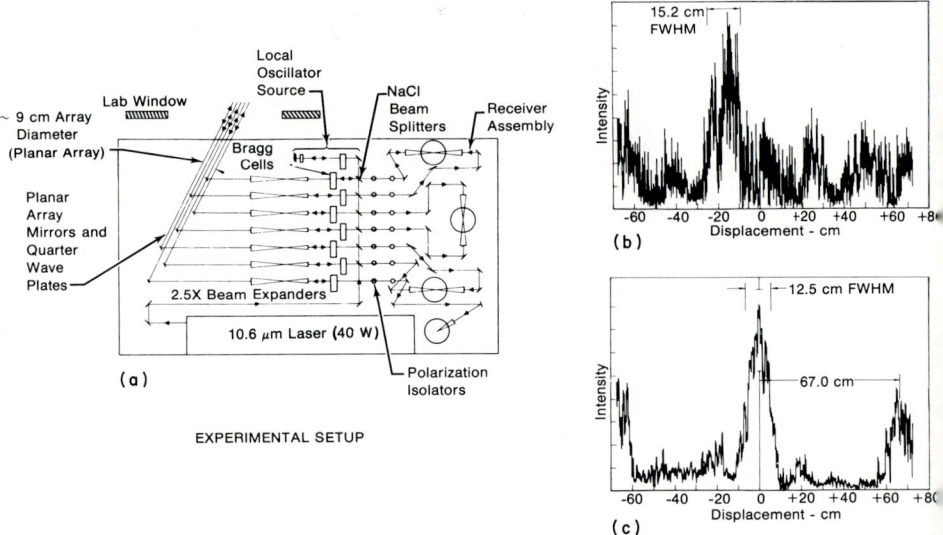

FIG. 46. Phase conjugate adaptive optical system tests: (a) optical configuration for the seven-element linear planar array; (b) focal-plane intensity distributions for the unadapted; and (c) adapted array cases for $R = 6$ km with a single glint target.

signal to noise, this initial hunting is completed in 20–100 μsec. However, for low voltage signal-to-noise ratios (≈ 2) at a 9.5 km range, the total convergence time lengthens to ≈ 2 sec. Essentially complete adaptive compensation for atmospheric turbulence and other fluctuation sources was obtained over a bandwidth of 2 kHz, with partial compensation out to 4 kHz.

For all situations tested, the phase conjugation array is found always to lock to the glint of highest reflectivity, as in the multidither experiments. In the phase conjugate tests, however, this occurs for reflectivity differences as small as 1.0 dB and for glint spacings either smaller or larger than the principal lobe null-to-null spacing.

A considerable potential improvement in performance for large ground-based astronomical telescopes may be achieved through the use of real-time atmospheric compensation (RTAC) for wave-front errors induced by atmospheric turbulence. The first reported use of an image compensation system was made by Hardy et al. (1974, 1977) using a shearing interferometer.

The basic concepts employed in the shearing RTAC are discussed in detail in Hardy et al. (1977). Other approaches to real-time compensation of atmospherically distorted images have been described by Muller and Buffington (1974; Buffington et al., 1977a,b) and Miller et al. (1974), using multidither image plane detection and by Dicke (1975b) and Mahajan (1976) using acousto-optical cells. A comparison of the efficiency of various approaches has been made by Dyson (1975).

The optical layout used in the RTAC system described in Hardy et al. (1977) is shown in Fig. 47. The first RTAC system with 21 deformable mirror correction zones was tested using an He–Ne laser as a reference source both in the laboratory using a simulated turbulent path and over an outdoor horizontal propagation range 300 m in length. Both static and dynamic test data were obtained.

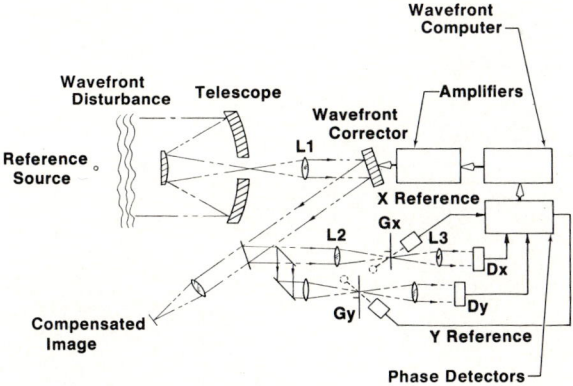

FIG. 47. Block diagram of a real-time atmospheric compensation system that uses a shearing interferometer to measure the wave-front error.

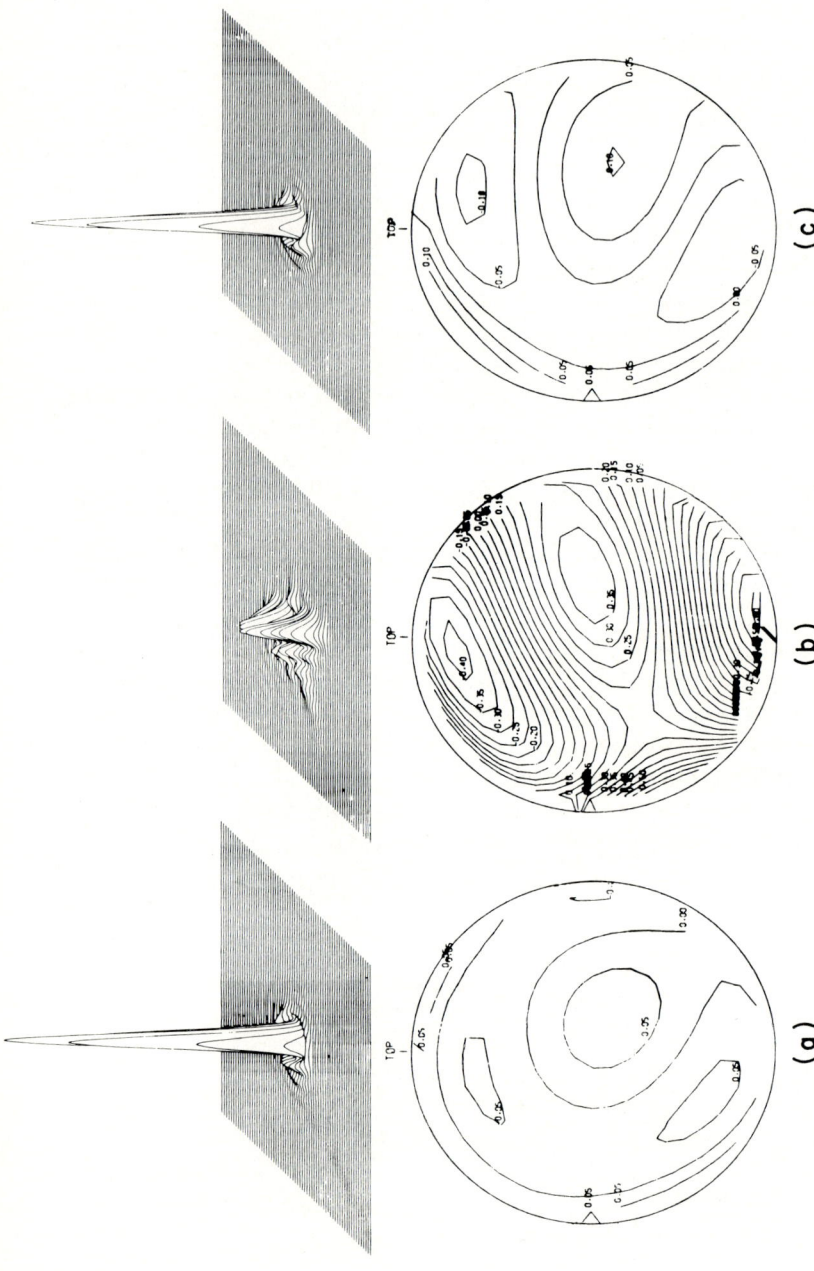

FIG. 48. Test data from experimental RTAC system (Hardy et al., 1977). The contour plots were made from interferograms of actual wave fronts at $\lambda = 0.633$ μm with contour spacings of 0.05λ. The corresponding point-spread functions were computed from these contour plots: (a) residual wave-front error of the RTAC with a plane-wave input, 0.04λ rms, 0.21λ peak to peak; (b) input wave-front distortion producing 0.27λ rms, 1.28λ peak to peak at system output with RTAC off; (c) same input as (b) with RTAC on. Residual wave-front distortion has been reduced to 0.06λ rms, 0.35λ peak to peak.

8. ADAPTIVE TECHNIQUES FOR WAVE-FRONT CORRECTION 321

Figure 48 shows the results of the static tests. The wave front at the output of the RTAC using a laser reference source with no added distortion is shown in Fig. 48a. The contours are spaced at 0.05 waves at 6328 Å, the residual error being 0.21 waves peak to peak and 0.04 waves rms. The corresponding pointspread function, computed from the measured wave front, is also shown.

A figured glass distortion plate introduced near the objective of the RTAC, with the correction loop nonoperating, produced the wave-front and point-spread function shown in Fig. 48b. The resulting wave-front error is 1.28 waves peak to peak, 0.27 waves rms.

When the RTAC was switched on, the wave front and PSF shown in Fig. 48c were obtained. The residual error has been reduced to 0.35 waves peak to peak, 0.06 waves rms, a reduction factor of about 4 times.

An example of the performance of this system with an extended image is shown in Figure 49. The bar-chart target, which was illuminated with white

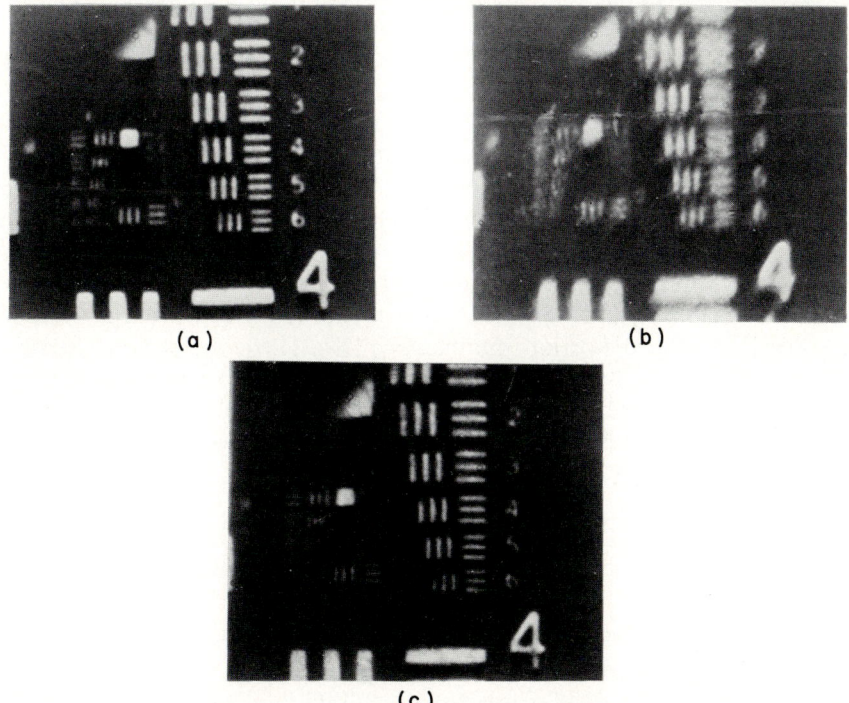

FIG. 49. RTAC performance with a three-bar resolution target and a He–Ne point reference. The resolution target was illuminated with white light and located in the same isoplanatic patch as the He–Ne laser reference source, which is masked out: (a) baseline RTAC resolution with no added wave-front distortion. The diffraction limit of the system is 130 cycles/mm; (b) distortion present, RTAC off; (c) resolution, with RTAC on. Target group 6–6 (114 cycles/mm) is just resolvable on the original negative.

light, was located within the same isoplanatic patch as a He–Ne laser. The laser provided a point reference source for the RTAC system control. Correction performance was to within about 80% of diffraction limit.

B. Thermal Blooming Compensation

The phenomenon of thermal blooming was described in Section I.B. Bradley and Herrmann (1974) were the first to demonstrate that one could compensate for thermal blooming by using an appropriate phase correction at the beam transmitter. More recently, an experiment was performed (Primmerman and Fouche, 1976) to apply the phase correction to a laser beam by means of a deformable mirror system. In the experiment, the deformable mirror phase profile was preprogrammed to match that calculated by Bradley/Herrmann for maximum correction of a truncated Gaussian beam undergoing forced convection-dominated thermal blooming. The profile included refocus and third-order spherical, coma, and astigmatism terms. The deformable mirror used in this experiment was a monolithic unit (Feinleib et al., 1974) similar to that described in Section IV.C. In this experiment the laser beam was slewed through the absorption cell by a variable speed slewing mirror. By detecting the light coming through the pinhole array located at the focal plane, both the intensity and the shape of the bloomed beam were measured.

As shown by Bradley and Herrmann (1974), the propagation of a slewed beam through an absorbing medium can be characterized by four dimensionless numbers: the distortion number N_D [Eq. (13)], the slewing number, N_ω [Eq. (14)], the absorption number, $N_A = \alpha R$, and the Fresnel number, $N_F = ka^2/R$. For laboratory experiments, an additional dimensionless number is required as a measure of the importance of conduction compared to forced convection:

$$\text{conduction number} = N_C = K/\sqrt{2}aV_0 \qquad (58)$$

where K is the thermal diffusivity. For convection-dominated heat transfer, $N_C \ll 1$.

In Fig. 50, the measured peak focal-plane intensity is plotted against input power for the uncorrected beam, the corrected beam, and the hypothetical situation of absorption with no blooming. Varying the power is equivalent to varying the distortion number, since $N_D \propto P$. The uncorrected curve was taken with the deformable mirror in the flat condition; the corrected curve was obtained by adjusting the amplitude of the mirror deformation to get the maximum possible intensity at each power. As expected,

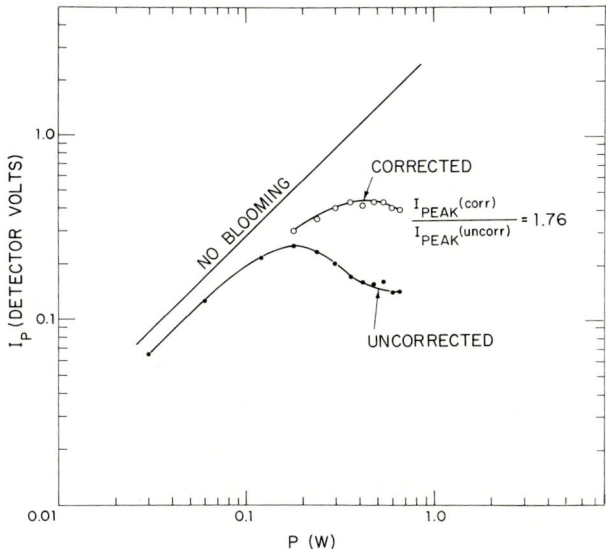

FIG. 50. Peak focal-plane intensity versus input power for corrected and uncorrected beams. The straight line would be the intensity if there were absorption but no blooming. $V_o = 1.65$ cm/sec; $N_c = 0.19$; $N_\omega = 10$. (After Primmerman and Fouche, 1976.)

the corrected curve shifts upward to higher intensities and outward to higher critical power. The maximum intensity increases 75% over the uncorrected case and that at certain powers there is a factor of 3 increase in intensity. This result is representative of consistently achieved improvements in maximum intensity of $\approx 70\%$ and in critical power of almost a factor of 2.5. A reasonable figure of merit in atmospheric propagation is the product of the critical power, P_c [Eq. (16)], and the peak intensity at the critical power: $I_p(P_c)P_c$. On the basis of this figure of merit, these experiments achieved a fourfold improvement.

The corrected curve in Fig. 50 was obtained by adjusting the mirror shape amplitude to give maximum intensity. But from a practical point of view, it is also important to know how sensitive this maximum is to changes in the deformation amplitude, with the correction *profile* fixed. In Fig. 51 the peak irradiance is plotted against peak-to-peak mirror deformation for a particular set of experimental conditions. The peak corrected intensity is a factor of 2.5 greater than uncorrected. But equally important, the correction curve is bell shaped with an extremely broad peak: the amplitude of the phase correction can vary $\pm 30\%$, while the irradiance decreases only 10%. This result is extremely encouraging, for it demonstrates that one does not have to apply phase corrections with great precision for them to be effective.

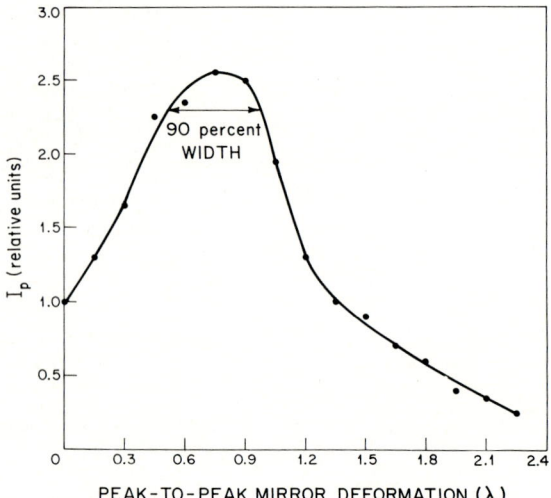

FIG. 51. Peak focal-plane intensity versus peak-to-peak mirror deformation amplitude. $V_o = 1.65$ cm/sec; $N_c = 0.19$; $N_\omega = 10$. (After Primmerman and Fouche, 1976.)

The previous experimental thermal blooming results were performed utilizing a predictive contour based on a theoretical phase profile. Herrmann (1977), however, has predicted that phase conjugate adaptive optical systems may be unstable if the thermal blooming becomes too strong. It has also been suggested (Pearson *et al.*, 1975) that all return-wave systems will suffer this type of instability. Finally, recent computer simulation studies of both multidither (outgoing-wave) and phase conjugate (return-wave) adaptive optical systems (Brown, 1975) have shown little, if any, improvement in peak focal-plane irradiance (Strehl ratio) for blooming-dominated propagation conditions.

Experimental measurements of adaptive phase compensation for thermal blooming have been made using an outgoing-wave 18-channel multidither COAT system. The adaptive system is the same one used in Pearson *et al.* (1976a; Pearson, 1976; Pearson and Hansen, 1977). For near-field thermal blooming, the data in Fig. 52 show that the optimum transmitter power (or "critical power") is increased by a factor of 2 to 3 and the peak focal-plane irradiance is increased by a factor of 4. The COAT system is this effective only because the phase distortion is so close to the transmitting aperture and thus it can be removed by a single-plane correction in the near field. In fact, a simple analysis (Bridges and Pearson, 1975) shows that the degree of correction in this case is limited only by the number of phase correction elements.

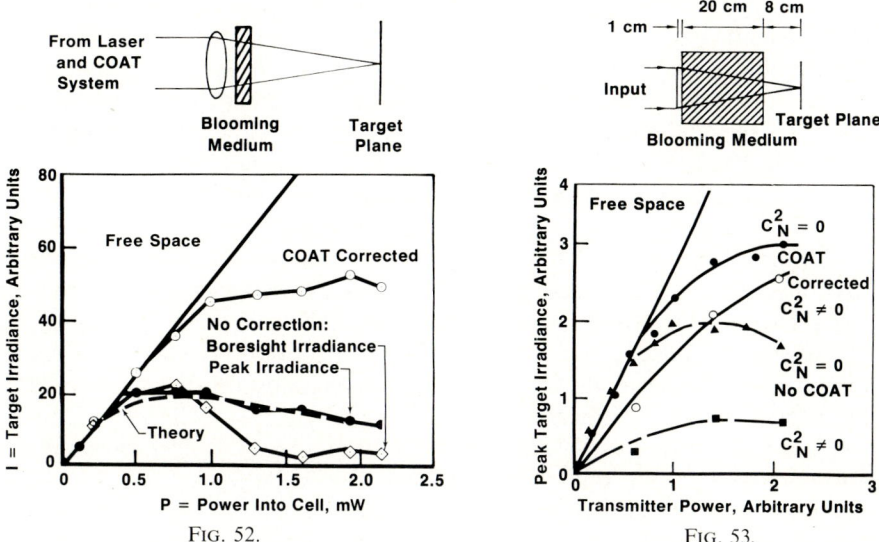

FIG. 52. Thermal blooming compensation with a thin, static-liquid blooming cell located in the near field of the transmitter: (---) the theoretical curve. (See Pearson et al., 1975.)

FIG. 53. Thermal blooming compensation with a flowing-liquid medium when the medium is in the first 72% of the path. Correction factor = 1.6. Correction for a single phase screen of artificial turbulence ($C_N^2 \neq 0$) is also shown. The lack of complete correction at low power for $C_N^2 \neq 0$ is a result of an insufficient number of COAT channels for the spatial frequencies in the turbulence screen used.

The more interesting scenario is one where the blooming occurs over the entire path with a focused beam. Such studies have been performed (Pearson, 1975, 1978) using the same 18-channel multidither COAT system. Representative data are shown in Fig. 53. The geometry of the flowing-liquid (I_2 dissolved in methanol) blooming cell allowed only 72% of the focused propagation path to occur within the cell.

The correction factor for thermal blooming alone is 1.58, as indicated by the curves marked $C_N^2 = 0$ in Fig. 53. The lower correction factors in these experiments are attributed to the absence of slewing, an arrangement which increases the effective distortions that cannot be removed by the COAT system, namely, those far from the transmitter. With slewing beams, the correction factor is expected to approach a factor of 3.

Another important observation is that no servo instability was observed as the blooming strength is increased for all the experimental conditions studied. In each experiment, the servobandwidth was optimized with no blooming or turbulence distortions present. The distortions were then added and the laser power increased (for blooming). In no instance did the COAT servo become unstable. The blooming compensation may not be substantial, but a stable correction state is always achieved and the COAT system never reduces the target irradiance. This type of stable behavior is not unexpected for intensity-maximization systems such as multidither outgoing-wave COAT, but as Herrmann (1977) has suggested, it may not be characteristic of return-wave systems, independent of whether laser return energy or target-image information is used to derive the phase errors.

Several potential solutions can be offered to the limited correction shown here and discussed in Section I.B. First, the phase errors can be moved closer to the transmitter by using slewed beams (Hayes, 1974) or by employing pulsed lasers (Wallace and Lilly, 1974) with repetition rates low enough to allow only a few pulses during a wind-driven medium transit time across the beam. Second, transmitter intensity tailoring or control (Wallace *et al.*, 1974; Yeh *et al.*, 1976; Pearson *et al.*, 1976b) can be employed. In fact, the simplest approach is to make the transmitter aperture as large as possible and to make the intensity as uniform as possible (Pearson *et al.*, 1976b). This procedure minimizes the blooming at a given transmitted power level. It also increases the effect of turbulence dramatically, but adaptive control is quite effective in removing most turbulence distortions if enough control elements are used (neglecting scintillation effects). A third possibility is the use of multiple correction planes by imaging various portions of the propagation path onto different phase correctors. This technique is limited to return-wave systems and has been suggested (Hardy *et al.*, 1977) for compensated imaging.

Aside from the limitation set by the geometry of the blooming problem, a second limitation on adaptive optical compensation has been suggested by Brown (1975b). This limitation is related to the adaptive control system bandwidth and takes two forms. First, if the control system is faster than the blooming medium response time, the instantaneous phase error will be corrected. This correction refocuses the beam, which produces a new (and stronger) phase error distribution, which leads to a new correction, and so on. This process may be unstable and not lead to a convergent solution as Herrmann has observed. Even if the process is stable, the correction algorithm will not properly sample the true slope of the target irradiance versus phase error curve. The system may thus converge to a nonoptimum stable state rather than the global optimum phase correction.

If instead of a fast response, the COAT system is slower than the medium, it would act like a "step, look, decide, step" servosystem. In this case, a change occurs and a decision is made whether the change increased or decreased the focal-plane irradiance. If the irradiance increases, the direction of the next change is the same; if not, the direction reverses. The amplitude of the test changes (the dither amplitude in a multidither system) will also be important in this type of system if local maxima are to be avoided.

C. Other Compensation Techniques

Although atmospheric phase aberration is the principal reason for the existence of adaptive optical systems, misalignment and phase disturbance within a laser resonator are other major effects which degrade the performance of any laser system. Mode discrimination as well as optical beam quality are often compromised as a result of temporally varying refractive index perturbations within the gain medium, optical cavity misalignment, mirror figure errors due to manufacturing tolerances, and thermally induced mirror distortions. These effects result in undesirable wave-front distortions which compromise the far-field irradiance.

The "adaptive resonator" concept consists of introducing a deformable mirror within the resonator and employing a closed-loop servosystem to control the surface figure of the mirror. By appropriately monitoring the phase and intensity distributions of the laser output, the intracavity deformable mirror can be configured in real time to compensate for undesirable perturbations within the resonator.

The feasibility of compensating for intracavity phase front distortions using a multidither zonal COAT system has been demonstrated (Freeman et al., 1977, 1978; Lind and Stevens, 1978) inside an unstable resonator cavity using a CO_2 electric discharge laser. A typical adaptive resonator experimental configuration is shown schematically in Fig. 54. The deformable mirror can be employed as one of the end mirrors of an unstable resonator or serve as an intracavity folding mirror.

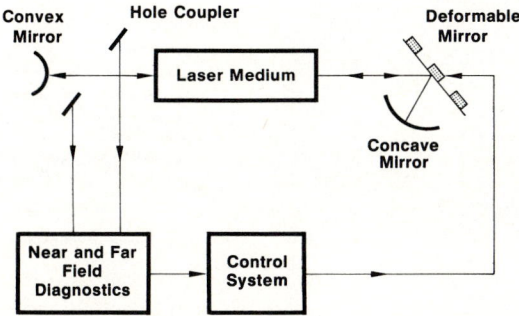

Fig. 54. Experimental configuration for an adaptive resonator. The deformable mirror is shown as a folding mirror inside the laser resonator.

In this experiment, a CW CO_2 electric discharge convection laser was used with a positive-branch confocal unstable resonator characterized by a geometric magnification of 1.35, a collimated beam diameter of 2.6 cm, and a cavity length of 2.84 m. The deformable mirror used was similar to the one shown in Fig. 29. Only the central five actuators were used to correct for undesirable phase perturbations originating within the gain medium or in the resonator optics.

Optimization of the far-field beam quality in real time was achieved by utilizing multidither techniques similar to those described earlier. The deformable mirror with its associated actuators in this experiment provided both phase dither and phase correction functions. The closed-loop bandwidth of the system was 65 Hz.

Static phase distortions (tilt and defocus) were introduced into the resonator in order to simulate typical phase perturbations encountered in

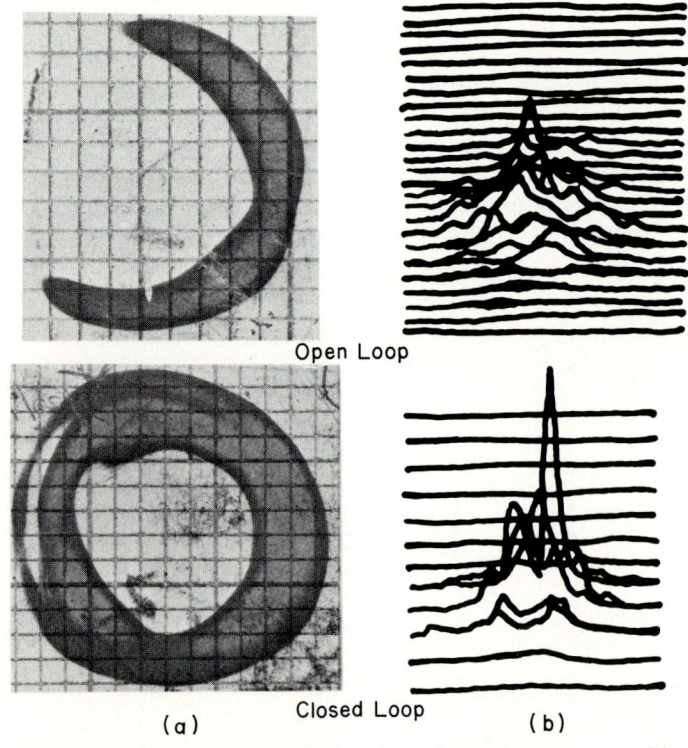

FIG. 55. Experimental performance of a five-channel adaptive resonator with multidither control: (a) near-field 10.6 μm patterns are recorded on burn paper; (b) far-field patterns are observed with IR scanning camera.

practical laser devices. These phase distortions were introduced in the vicinity of the convex mirror to allow the perturbation to propagate an appreciable distance within the resonator prior to encountering the deformable mirror. In all cases the intracavity induced degradations of the laser's performance were almost completely corrected by activating the closed-loop multielement COAT system. Figure 55 illustrates some typical results associated with a case in which a linear phase distortion was introduced into the resonator. The near-field data were obtained from near-field burn patterns and the far-field intensity scans were obtained using an AGA infrared scanning camera. The open-loop uncorrected laser performance and the closed-loop active resonator enhancement are depicted in the top and bottom rows, respectively. Although only five active actuator sites were utilized, the resonator performance improvement is substantial. The resulting near-field pattern closely approaches an ideal annulus. Moreover, the central lobe of the corrected far-field intensity distribution noticeably decreases in width and more than doubles its on-axis peak intensity with respect to the degraded far-field intensity distribution. In addition, the total power out of the laser was increased.

VI. TARGET CONSIDERATIONS

A. INTRODUCTION

There are four major target-related problems that can occur when an adaptive optical system attempts to focus a transmitted laser beam on a target. These four problems are listed in Table XI. The first three problems are particularly important for coherent-light multidither systems, although

TABLE XI
TARGET RELATED PROBLEM AREAS FOR ADAPTIVE OPTICAL SYSTEMS

Glint hopping
    Beam motion as COAT system attempts to select strongest glint on a dynamic, multiglint target

Speckle modulations
    Spurious signal modulation induced in receiver by moving speckle pattern returns from target

Target reflectivity changes
    Destruction of glint references by high intensity of incident beam

False aimpoint selection
    Aimpoint selected by system may not be desired point or does not fall on target

they are potential problems for any type of COAT system that uses coherent laser target return energy to obtain its phase error information.

Multidither systems that have been studied to date require a target highlight or bright feature (a localized region that has higher reflectivity than surrounding regions—a "glint") for proper operation. The system will always act to place the beam on the illuminated glint that has the strongest net return. (The net return is the glint directional reflectivity multiplied by the COAT array element pattern.) Since glint structures are known to evolve, replicate, shift, and/or disappear as the target changes aspect angle, the beam may not be stable on a dynamic target, and with featureless targets, standard outgoing-wave multidither systems cannot converge. In addition, if the laser beam directed toward the target succeeds in burning or in some way altering the target surface, even a stable glint reference may disappear. Offsetting the beam from a selected reference has been proposed as a solution to the glint destruction problem, but this technique is severely limited by the finite isoplanatic patch size as discussed in Section VI.C.

The final problem listed in Table XI arises mainly with systems that use noncoherent radiation either reflected from or originating on the target and attempt to focus the beam on the brightest point (image intensity centroid). One goal of an adaptive optical system design is to find a system that either does not experience these target-related problems or is minimally affected by them.

B. Potential Solutions to Target Problems

There are two possible approaches to eliminating the target-connected problems discussed above: avoid the use of coherent light returns or employ advanced techniques in the coherent light systems. One approach is to employ reflected sunlight or self-irradiance at infrared wavelengths as a return-wave technique for probing the propagating path and compensating the path errors. As discussed in Section II.D, these techniques are outgrowths of compensated (passive) imaging studies. Initial enthusiasm for this approach was quite high since this type of system offered attractive features. It appeared that appropriately chosen implementations would not require glints or other point references and could use the entire extended target as a reference. Furthermore, a variety of wave-front error sensors has been demonstrated or conceived that are capable of extracting the error information from extended-target returning wave fronts.

With further study, however, the initial enthusiasm dimmed somewhat. It was realized that this technique, like others, has its limitations. For example, it was realized that with cool targets operating at night, both self-irradiance and reflected light would sometimes be too weak to permit effective operation of the system. Furthermore, laser floodlight operation of the target cannot substitute for the lack of broadband emission, since self-interference of the return from spatially extended targets generates a spatial phase distribution that is intrinsically inseparable from the propagation path phase error distribution.

The major blow comes from another direction, however—the isoplanatic problem. For homogeneous paths, Fried (1971) has shown that when two source elements are separated by more than one half the atmospheric coherence length, they fall outside the isoplanatic patch; that is, they produce spatial phase distributions at the receiver aperture that are essentially uncorrelated at any one instant. A consequence of this fact is that if one attempts operation with information extracted from both points simultaneously, one obtains a composite phase measure that is incorrect for compensating the path to either of these points or to any other point for that matter. The isoplanatic problem is further discussed below. Given that TRIM-COAT systems may just trade one set of problems for another, an alternative approach to solving the target problems is to modify existing coherent-light systems so that they are relatively immune to the problems in Table XI.

C. Isoplanatism and Extended Target Referencing

Isoplanatic effects are expected to be important when the target size becomes comparable to the atmospheric coherence length. The quantitative relationship between the coherence length r_0 and turbulence strength is plotted in Fig. 4 for a 4 km range with wavelength ($k = 2\pi/\lambda$) as a parameter. For ground level (horizontal) operation, C_N^2 can easily exceed 10^{-14} m$^{-2/3}$, while for airborne operation it would be presumptuous to count on C_N^2 below 2×10^{-16}. The magnitude of the isoplanatic problem can be studied by looking at a technique called "offset pointing." Offset pointing is a technique where one point (glint) on a target is used as a reference, but the transmitted beam is deliberately pointed at some offset point. Offset pointing can be accomplished in practice either by time sharing (Pearson *et al.*, 1976a) the beam between two points using sample-and-hold circuitry or by adding a fixed offset to the transmitted beam (practical only with return-wave systems). The technique has been suggested as one solution to the problem of glint destruction with multidither COAT systems.

The isoplanatic problem will limit the distance that a beam can be offset from the reference point. Some results of computer simulation studies (Brown *et al.*, 1977) are shown in Fig. 56 to illustrate this point. The reference is at $d = 0$ and Fig. 56 plots the Strehl ratio when this fixed correction is used to point the beam through a stationary atmosphere a distance d from

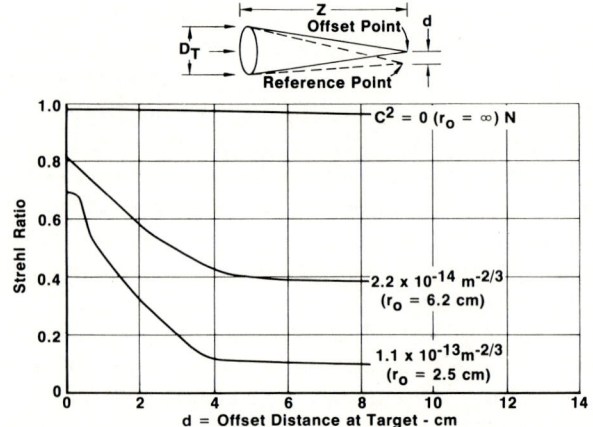

FIG. 56. Strehl ratio versus offset pointing distance d found by computer simulation (Brown et al., 1977). The r_o values are for plane-wave propagation and the following parameters were used: $\lambda = 3.8$ μm, $D_T = 70$ cm, $R = 4$ km.

the reference. As can be seen, the Strehl ratio drops rapidly with d for $C_N^2 \neq 0$, reaching about 50% of its value at $d = 0$ when $d \approx r_0/2$.

The overall conclusions which result from Figs. 4 and 56 are reasonably clear. First, at all wavelengths the isoplanatic patch diameter, which we define as $r_0/2$, will almost never be as large as 1 m (with 4 km paths). Second, infrared (IR) wavelengths are much better (by more than an order of magnitude) than the visible wavelengths in tolerating the isoplanatic problem. Third, referencing on truly extended (≈ 1 m) targets is not allowed by isoplanatic problems. Finally, analytical and computer simulation studies (Brown et al., 1977) have shown that the following general conclusions apply:

(a) Phase errors that are near the target do not result in erroneous measurements.

(b) Path effects that arise from phase errors near the receiver transmitter aperture present no problem.

(c) In general, the average path error estimators may be badly in error as a result of intermediate path phase errors.

(d) The only cases for which accurate correction occurs for the transmitter path are those wherein the reference is confined to a small source and the transmitter focuses back to this point.

D. Speckle Modulation

Any system that uses coherent light can encounter speckle effects. The effects of speckle in a COAT system are usually detrimental since speckle will obscure or confuse the information that the COAT sensor uses to perform its compensation function. Speckle will affect each of the systems discussed in Section I.C. and shown in Fig. 7 in different ways. The phase conjugate system in Fig. 7a spatially samples the return energy and uses heterodyne detection to determine the actual phase error by interfering a portion of the transmitted beam with the radiation returned from a target glint. Since the return radiation is coherent, it forms a speckle pattern across the receiver with both amplitude and phase variations that are unrelated to the phase distortions on the outgoing beam. If the speckle pattern variations are rapid (spatially) across the COAT receiver, the COAT system can get an erroneous result for the required correction to be applied to the transmitted beam, depending on the amplitude of the variations. If there are many target glints, a phase conjugate system will converge the energy on the strongest glint. This convergence is accomplished using phase fluctuations across the receiver (interference pattern) produced by reflections from the several glints. Speckle-induced variations will act to mask or confuse the glint-produced effects so that convergence may not occur on the desired glint point or may not occur at all.

The second system shown in Fig. 7b, uses an incoherent, image-compensation system to determine the optical path phase errors. This system uses broadband, noncoherent return radiation to derive the correction signals and thus is immune to speckle effects. From the standpoint of freedom from speckle, it is thus the most desirable system. It does have other problems, unfortunately, that may limit its usefulness as discussed previously.

The third system, multidither outgoing COAT is shown in Fig. 7c. The essential feature for our discussions here is that the phasing or error-correction information is contained as AM sidebands on the dither frequencies; the low-frequency or quasi-dc signals in the detector either have no effect on system operation or their effect can be eliminated by an automatic gain control (AGC). In effect, each dither frequency is a carrier for the error information in its associated servoloop. If there are other sources that can produce amplitude modulations in the receiver at or near the dither frequencies, these spurious modulations will compete with the impressed dither modulations for control of the servo, resulting in reduced performance (Kokorowski *et al.*, 1977; Pearson *et al.*, 1976c).

The existence of speckle as a source of such spurious modulations is illustrated (Pearson *et al.*, 1976c) in Fig. 57. The laser beam from a transmitter aperture of diameter D_T is focused on a target, which translates at velocity V_T and rotates at an angular rate Ω_T. If the target is anything other than a very small point specular, the reflected radiation forms a speckle pattern at the receiver. This speckle pattern contains a continuous spectrum of spatial frequencies, which for many target surfaces with a focused Gaussian beam is given by (Goldfisher, 1976)

$$S(f_x) \propto B \exp[-[(2\sigma/N)f_x]^2] \tag{59}$$

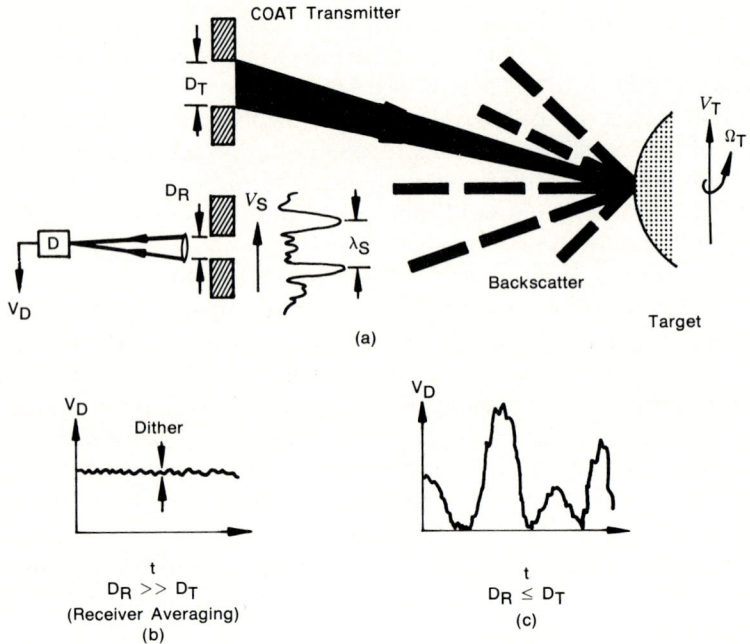

FIG. 57. Speckle effects in a multidither COAT system: (a) Definition of various geometrical quantities such as D_T and D_R, the transmitter and receiver aperture diameters; λ_s, the speckle coherence length; V_T and Ω_T, the target translational and rotational velocities; and v_s, the translation velocity of the speckle pattern in the receiver plane. (b) Receiver signal in the absence of speckle effects (e.g., when a very large receiver is employed); the only modulations are the dither control modulations. (c) Receiver signal when the receiver aperture is comparable to the transmitter. Large amplitude fluctuations can occur.

where $S(f_x)$ is the power spectral density at spatial frequency f_x, B is a constant, 2σ is the $1/e^2$ Gaussian beam radius at the transmitter, and N is the ratio of the $1/e^2$ beam diameter on the target to the diffraction-limited beam diameter. We have ignored a dc term (δ function) in the spectrum of Eq. (59) since it is not important to our conclusions here (it affects only the dc signal in the receiver and its amplitude is tracked out by an AGC).

The correlation length of the speckle is given by

$$\lambda_s = 2(2\sigma/N) \approx D_T/N \tag{60}$$

where we have chosen $D_T = 4\sigma$. If the target is stationary or if the receiver diameter is very much larger than the transmitter ($D_R \gg D_T$), the receiver signal consists only of small dither modulations on top of a dc level as indicated in Fig. 57b. The receiver averages over many speckle lobes and slow variations in the dc level are removed by an AGC network in the COAT servo system.

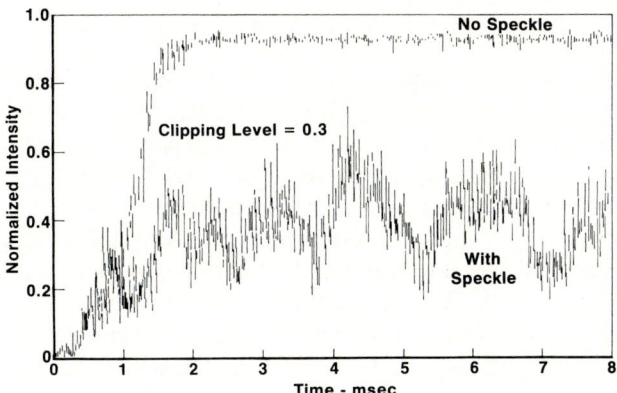

FIG. 58. Computer simulation results of speckle effects. With no speckle modulations, the 18-channel COAT system converges in about 2 msec. When strong speckle effects are present, the COAT system convergence is limited to an average level of about 40% of the "no speckle" maximum and the intensity fluctuates about this average level. Simulated Target: pyroceram sphere (10-cm diameter, 1-m radius) uniformly illuminated with 10.6 μm light at 2 km from transmitter/receiver. Target rotating at 2 rad/sec. Receiver: 1.2 × 1.0 m annulus.

The case most encountered in practice, however, has a receiver roughly equal in size to the transmitter. (Resolution, or maximum aperture dimension, is the most important parameter, not receiver area.) When the focused beam at the target has a diffraction-limited spot size ($N = 1$), the most probable speckle lobe size is equal to the transmitter diameter D_T. If $D_R \leq D_T$, large amplitude modulations can occur as the speckle pattern moves by the receiver. The motion of the speckle pattern is produced by translation or rotation of the target. When the effective speckle velocity at the receiver V_s divided by the speckle "size" falls in the dither band, the multidither COAT system can be affected. Thus potential problems occur when the $D_R \lesssim D_T$ and when

$$V_s/\lambda_s \approx (V_T + 2\Omega_T Z)/D_T \approx f_d \tag{61}$$

where f_d is within the dither frequency band of the system [see Eq. (35)]. An example of the effects of speckle in a multidither COAT system is shown in Fig. 58.

In summary, the performance of a COAT system in the presence of speckle is strongly dependent on COAT system parameters such as servo-correction bandwidth, number and value of dither frequencies, signal conditioning subsystems such as AGC, dither amplitude, and active noise suppression subsystems. Further results of studies on this phenomenon can be found in Kokorowski *et al.* (1977), Gurski *et al.* (1977), and Pearson *et al.* (1976c).

E. Range Considerations

Long-range adaptive optics applications introduce problems due to the optical transit time involved. For the two generic classes of systems, return wave and outgoing wave, the return-wave system can in general work against targets that are further away since only a one-way transit delay occurs. For outgoing multidither systems, the long path delays cause a problem in the synchronous detection of the modulation that provides the phase error information and in laser device power fluctuation correction. A fundamental range limitation exists for all adaptive optical systems; however, if the system is attempting to correct a time-varying disturbance located over part, or all of the propagation path, the "equivalent" delay time must be shorter than the rate of change of the disturbance or correction performance is reduced. For an outgoing system, this means, regardless of the location of the time varying disturbance, the round trip delay time τ must be shorter than the rate of change of the disturbance. For a return-wave system, the equivalent delay is only that portion of the total delay associated with the transmission path from the point of first contact with the time-varying disturbance to the receiver. It is this phenomena which allows astronomical adaptive "imaging" systems to successfully operate over very large distances while correcting for the atmospheric turbulence effects located in the earth's atmosphere.

We now consider further the multidither outgoing-wave system. With short ranges there is no time delay and therefore no phase error between the reference and return dither waveforms compared in the synchronous detection process. The effect of the round trip optical time delay τ is to modify the form of the sine $(\omega_m t)$ terms in Eq. (31) to the form $\sin \omega_m (t - \tau)$. Consequently, the result of synchronously detecting the expression in Eq. (31) with the time delay against a reference signal $\sin(\omega_m t)$ produces an error signal of the form,

$$S_{D_m} = A_m \sum_{j=1}^{N} A_j \psi_0 \sin(\beta_m - \beta_j) \cos(\omega_m \tau) \tag{62}$$

Several important points follow from this expression.

If $\omega_m \tau$ becomes as large as $\pi/2$, the detected component vanishes independently of the phasing errors. Worse, if $\omega_m \tau = \pi I$ (where I is an odd integer), the sign of the error signal reverses and the system will lock on minima rather than maxima of the composite beam. This identifies that such a system must either be restricted to ranges such that $\omega_m \tau$ is less than $\pi/4$ radians to avoid these effects, or compensation techniques must be implemented to correct the time delays. For an uncompensated system, the modulation frequency f_m is thus limited to

$$f_m = \omega/2\pi < c/8R \tag{63}$$

where c is the velocity of light and R is the range to the reflecting target.

High modulation frequencies offer advantages in minimizing noise interference problems and in accommodating large frequency errors. Moreover, turbulence correction requires bandwidths on the order of 100–300 Hz, which can be accomplished only with high dither frequencies. As an example, for a 300 Hz bandwidth system with 70 actuators operating in sine/cosine manner (corrector mirror and dither mirror with 15 and 0% coupling, respectively), the maximum dither frequency will be approximately 46 KHz. This equates to a maximum allowable range, without time delay compensation, of approximately $R = 1.6$ km. Obviously, range compensation is necessary if large ranges are to be considered.

Two possible classes of compensation techniques can be utilized (O'Meara, 1976). The first class corrects the time displacement between the reference and the return envelope by knowing or measuring the range (or radar pulse time τ) and thereby computing the associated phase shift (in the range $\pm \pi$ radians). This phase shift is then introduced as a phase delay correction in the reference path in advance of the synchronous detection operation.

The second basic class of system transmits the reference dither frequency, in the form of a modulation on some carrier, and detects the associated return signal, employing it as the reference in the synchronous detection operation. Since both the signal and the reference experience the same path delays, a purely synchronous detection operation thereby results independently of $\omega_m \tau$.

The propagation time-delay effects can also be addressed using modern estimation theory. The return signals can be recognized to contain errors and cross-channel coupling. This information can be incorporated into the design of Kalman or adaptive Kalman filters to yield the desired measurements.

REFERENCES

Angelbeck, A. W., Pietro, R. C., Hasselmark, E. D., Gagnon, E., Greiner, A. F., McClurg, W. C., Wisner, G. R., and Freeman, R. H. (1975). "Development of a Cooled Metal Mirror with Active Figure Control," Intern. UTRC Rep. (Dec.). United Technologies Corp., East Hartford, Connecticut.

Asher, R. B., and Ogrodnik, R. F. (197:). *J. Opt. Soc. Am.* **67**, 350.

Bradley, L. C., and Cheifetz, M. G. (1975). *J. Opt. Soc. Am.* **65**, 1212A.

Bradley, L. C., and Herrmann, J. (1974). *Appl. Opt.* **13**, 331.

Bridges, W. B., and Pearson, J. E. (1975). *Appl. Phys. Lett.* **26**, 539.

Bridges, W. B., Hansen, S., Horwitz, L. S., Kubo, R. M., Lazzara, S. P., O'Meara, T. R., Pearson, J. E., and Walsh, T. J. (1973). Coherent Optical Adaptive Techniques (COAT), RADC-TR-74-38 (Oct.). (Available from Natl. Tech. Inf. Serv. or Def. Docum. Center, Washington, D. C.).

Bridges, W. B., Brunner, P. T., Lazzara, S. P., Nussmeier, T. A., O'Meara, T. R., Sanquinet, J. A., and Brown, W. P., Jr. (1974). *Appl. Opt.* **13**, 291.

Brown, W. P., Jr. (1975a). "Computer Simulation of Adaptive Optical Systems," Final Rep., Contract No. N60921-74-C-0249 (Sept.). (Available from Natl. Tech. Inf. Serv. or Def. Docum. Center, Washington, D.C.).
Brown, W. P., Jr. (1975b). Private communication.
Brown, W. P., Jr. (1976a). Unpublished observations.
Brown, W. P., Jr. (1976b). Unpublished observations.
Brown, W. P., Jr., Yeh, C. W., Pearson, J. E., and O'Meara, T. R. (1977). *Conf. Laser Eng. Appl. (CLEA)* Pap. 17.4.
Buffington, A., Crawford, F. S., Muller, R. A., Schwemin, A. J., and Smits, R. G. (1977a). *J. Opt. Soc. Am.* **67**, 298.
Buffington, A., Crawford, F. S., Muller, R. A., and Orth, C. D. (1977b). *J. Opt. Soc. Am.* **67**, 304.
Bufton, J. L. (1973). *Appl. Opt.* **12**, 1785.
Butts, R. R., and Hogge, C. B. (1977). *J. Opt. Soc. Am.* **67**, 278.
Cathey, W. T., Hayes, C. L., and Davis, W. C. (1970). *Appl. Opt.* **9**, 701.
Chase, D. M. (1966). *J. Opt. Soc. Am.* **56**, 33.
Dicke, R. H. (1975a). *Astrophys. J.* **198**, 605.
Dicke, R. H. (1975b). *J. Opt. Soc. Am.* **65**, 1206.
Dowling, J. A., and Livingston, P. M. (1973). *J. Opt. Soc. Am.* **63**, 846.
Dyson, F. J. (1975). *J. Opt. Soc. Am.* **65**, 551.
Erteza, A. (1976a). *Appl. Opt.* **15**, 2095.
Erteza, A. (1976b). *Appl. Opt.* **15**, 656.
Feinleib, J. (1979). Sensor system for detecting wavefront distortion in a return beam of light, U.S. patent 4141652.
Feinleib, J., Lipson, S. G., and Cone, P. F. (1974). *Appl. Phys. Lett.* **25**, 311.
Freeman, R. H., Freiberg, R. J., and Garcia, H. R. (1977). *Conf. Laser Eng. Appl. (CLEA)* Pap. 17.3.
Freeman, R. H., Freiberg, R. J., and Garcia, H. R. (1978). *Opt. Lett.* **2**, 61.
Fried, D. L. (1965). *J. Opt. Soc. Am.* **55**, 1427.
Fried, D. L. (1966). *J. Opt. Soc. Am.* **56**, 1380.
Fried, D. L. (1971). *Appl. Opt.* **10**, 721.
Fried, D. L. (1977). *J. Opt. Soc. Am.* **67**, 370.
Gebhardt, F. G., and Smith, D. C. (1969). *Appl. Phys. Lett.* **14**, 52.
Gebhardt, F. G., and Smith, D. C. (1972). *Appl. Phys. Lett.* **20**, 129.
Gebhardt, F. G., Kendall, J. F., Jaminet, J. F., Hasselmark, E. D., and Bar-Tal, G. S. (1974). "Final Report on Multi-Segmented Mirrors." UTRC Rep. No. N911771-11, Contract No. N00014-72-C-0470 (Feb.). (Available from Natl. Tech. Inf. Serv. or Def. Docum. Center, Washington, D.C.)
Goldfisher, L. I. (1976). *J. Opt. Soc. Am.* **55**, 247.
Goodman, J. W. (1968). "Introduction to Fourier Optics." McGraw-Hill, New York.
Greenwood, D. P. (1976). Private communication.
Greenwood, D. P. (1977a). *J. Opt. Soc. Am.* **67**, 282.
Greenwood, D. P. (1977b). *J. Opt. Soc. Am.* **67**, 282.
Greenwood, D. P. (1977c). *J. Opt. Soc. Am.* **67**, 390.
Greenwood, D. P., and Fried, D. L. (1976). *J. Opt. Soc. Am.* **66**, 193.
Grosso, R. P., and Vellin, M. (1977). *J. Opt. Soc. Am.* **67**, 399.
Gurski, G. F., Nomiyama, N. T., Radley, R. J., and Wilson, J. (1977). *J. Opt. Soc. Am.* **67**, 345.
Hansen, S. (1975). "Hydraulic Actuators for Active Optical Systems," Final Rep., Contract No. N60921-75-C-0067 (June). (Available from Natl. Tech. Inf. Serv. or Def. Docum. Center, Washington, D.C.)
Hardy, J. W. (1978). *Proc. IEEE.* **66**, 651.

Hardy, J. W., Feinleib, J., and Wyant, J. C. (1974). *Dig. Tech. Pap.*, *Opt. Soc. Am. Top. Meet. Opt. Propag. Through Turbul.* Pap. ThB1.
Hardy, J. W., Lefebvre, J. E., and Kolipoulos, C. L. (1977). *J. Opt. Soc. Am.* **67**, 360.
Harvey, J. E., and Callahan, G. A. (1978). *J. Opt. Soc. Am.* **67**, 1367.
Hayes, C. L., Brandewie, R. A., Davis, W. C., and Mevers, G. E. (1977). *J. Opt. Soc. Am.* **67**, 269.
Hayes, J. N. (1974). *Appl. Opt.* **13**, 2072.
Hayes, J. N., and Ulrich, P. B. (1975). Private communication.
Herrmann, J. (1977). *J. Opt. Soc. Am.* **67**, 290.
Hogge, C. B. (1974). *In* "High Energy Lasers and Their Applications" (S. Jacobs, ed.), Chap. 4, pp. 230–234. Addison-Wesley, Reading, Massachusetts.
Hogge, C. B., and Butts, R. R. (1976a). *IEEE Trans. Antennas Propag.* **AP-24**, 144.
Hogge, C. B., and Butts, R. R. (1976b). "Analysis of Angle of Arrival Measurements at the Sandia Optical Range," pp. 8–17. AFWL-TR-75-311 (Feb.). (Available from Natl. Tech. Inf. Serv. or Def. Docum. Center, Washington, D.C.)
Hudgin, R. H. (1977a). *J. Opt. Soc. Am.* **67**, 375.
Hudgin, R. H. (1977b). *J. Opt. Soc. Am.* **67**, 378.
Hudgin, R. H. (1977c). *J. Opt. Soc. Am.* **67**, 393.
Hugdin, R., and Lipson, S. (1975). *J. Appl. Phys.* **46**, 510.
Kaminow, I. P., and Turner, E. H. (1966). *Proc. IEEE* **54**, 1374.
Knox, K. T. (1976). *J. Opt. Soc. Am.* **66**, 1236).
Kogelnik, H., and Pennington, K. S. (1968). *J. Opt. Soc. Am.* **58**, 273.
Kokorowski, S. A., Pedinoff, M. E., and Pearson, J. E. (1977). *J. Opt. Soc. Am.* **67**, 333.
Lavan, M. J., Cadwallender, W. K., and DeYoung, T. F. (1976). *Opt. Eng.* **15**, 56.
Lind, R. C., and Stevens, R. R. (1978). *Opt. Lett.* **3**, 79.
Lind, R. C., Minden, M. L., Hansen, S., and Nussmeier, T. A. (1977). "Multidither Adaptive Algorithms," Final Rep., Contract No. F30602-76-C-0022 (Mar.). (Available from Natl. Tech. Inf. Serv. or Def. Docum. Center, Washington, D.C.)
McGlamery, B. L., Silva, D. E., and Harris, J. L., Sr. (1975). "Computer Studies of Compensated Imaging Systems" (Sept.). Scripps Inst.f Oceanography, Univ. of California, San Diego, California.
Mahajan, V. N. (1975). *J. Opt. Soc. Am.* **65**, 271.
Mahajan, V. N. (1976). *Proc. Soc. Photo-Opt. Instrum. Eng.* **75**, 109.
Miller, L., Brown, W. P., Jr., Jenney, J. A., and O'Meara, T. R. (1974). *Dig. Tech. Pap.*, *Opt. Soc. Am. Top. Meet. Opt. Propag. Through Turbul.* Pap. ThB2.
Muller, R. A., and Buffington, A. (1974). *J. Opt. Soc. Am.* **64**, 1200.
Noll, R. J. (1976). *J. Opt. Soc. Am.* **66**, 207.
O'Meara, T. R. (1976). Private communication.
O'Meara, T. R. (1977a). *J. Opt. Soc. Am.* **67**, 306.
O'Meara, T. R. (1977b). *J. Opt. Soc. Am.* **67**, 318.
Pearson, J. E. (1975). "Coat Measurements and Analysis," RADC-TR-75-47 (Feb.). (Available from Natl. Tech. Inf. Serv. or Def. Docum. Center, Washington, D.C.)
Pearson, J. E. (1976). *Appl. Opt.* **15**, 622.
Pearson, J. E. (1978). *Opt. Lett.* **2**, 7.
Pearson, J. E., and Hansen, S. (1977). *J. Opt. Soc. Am.* **67**, 325.
Pearson, J. E., Brown, W. P., Jr., Kokorowski, S. A., O'Meara, T. R., and Pedinoff, M. E. (1975). *J. Opt. Soc. Am.* **65**, 1212A.
Pearson, J. E., Bridges, W. B., Hansen, S., Nussmeier, T. A., and Pedinoff, M. E. (1976a). *Appl. Opt.* **15**, 611.
Pearson, J. E., Yeh, C., and Brown, W. P., Jr. (1976b). *J. Opt. Soc. Am.* **66**, 1384.
Pearson, J. E., Kokorowski, S. A., and Pedinoff, M. E. (1976c). *J. Opt. Soc. Am.* **66**, 1261.

Pearson, J. E., Brown, K. M., Minden, M. L., Price, K. D., and Yeh, C. (1976d). "Multidither Adaptive Algorithms." RADC-TR-76-364 (Nov.). (Available from Natl. Tech. Inf. Serv. or Def. Docum. Center, Washington, D.C.)
Primmerman, C. A., and Fouche, D. G. (1976). *Appl. Opt.* **15**, 990.
Proc. IEEE (1976). Adaptive Systems. **64**, Spec. Issue (August).
Radley, J., Nomiyama, N., Wilson, J., and Gurski, G. (1976). "COAT; Modal–Zonal Comparison," Final Rep., Contract No. N60921-76-C-0122 (Aug.). (Available from Natl. Tech. Inf. Serv. or Def. Docum. Center, Washington, D.C.)
Reynolds, H. C., Jr. and Raymondo, P. J. (1977). Unpublished observations.
Skolnick, M. L., Gowrinathan, S., Harris, J. S., Medico, L. J., Roberts, F. E., Robertson, H. J., and Rosenzweig, D. N. (1974). "Laser Wavefront Analyzer Using Sliding Reference Interferometry," AFWL-TR-74-73 (July). (Available from Natl. Tech. Inf. Serv. or Def. Docum. Center, Washington, D.C.)
Smith, D. C. (1969). *IEEE J. Quantum Electron.* **QE-5**, 600.
Smith, D. C., (1977). *Proc. IEEE* **65**, 1679.
Soo Hoo, J., Hayes, C. L., and Brandewie, R. A. (1972). *Proc. Tech. Program, Electro-Opt. Syst. Des. Conf., Chicago, Ill.* p. 164.
Takken, E. H., and Cordray, D. M. (1974). *Appl. Opt.* **13**, 2753.
Tatarski, V. I. (1961). "Wave Propagation in a Turbulent Medium." McGraw-Hill, New York.
Wallace, J., and Camac, M. (1970). *J. Opt. Soc. Am.* **60**, 1587.
Wallace, J., and Lilly, J. Q. (1974). *J. Opt. Soc. Am.* **64**, 1651.
Wallace, J., Itzkam, I., and Camm, J. (1974). *J. Opt. Soc. Am.* **64**, 1123.
Whinnery, J. R., Miller, D. T., and Dabby, F. (1967). *IEEE J. Quantum Electron.* **QE-3**, 382.
Wood, A. D., Camac, M., and Gerry, E. T. (1971). *Appl. Opt.* **10**, 1877.
Wyant, J. C. (1973). *Appl. Opt.* **12**, 2057.
Wyant, J. C. (1974). *Appl. Opt.* **13**, 200.
Wyant, J. C. (1975). *Appl. Opt.* **14**, 2622.
Yeh, C., Pearson, J. E., and Brown, W. P., Jr. (1976). *Appl. Opt.* **15**, 2913.
Yura, H. T. (1973). *J. Opt. Soc. Am.* **63**, 567.

Index

A

Abbe number, 48
Absorption number, 322
Acrylic, 81
Actuator, 284
   discrete, 286, 289, 296
   electromagnetic (EM), 290–292
   hydraulic (HYDR), 290–292
   influence function, 287
   magnetostrictive (MAG), 290–292
   piezoelectric (PZT), 290–292
Adaptive resonator, 327
Aluminum, 103
Angular scattering, 216
   BRDF, *see* Bidirectional reflectance distribution function
Annealing, 56
Autocorrelation, 216
Autocovariance
   functions, 223, 226–228, 230–235, 239
   Gaussian, 195, 218, 223
   length, 217, 305

B

Beam propagation, 155
Beam waist, 183
Beryllium, 102
   lightweight mirror, 113
Bidirectional reflectance distribution function (BRDF), 223, 224
Bragg cell, 284
Brewster window, 164

C

CER-VIT, 102
   lightweight mirrors, 113
Coatings for plastics, 84
Coherent optical adaptive techniques (COAT), 247, 250, 251, 259, 276
   imaging, 256
   multidither, 256, 275, 280, 293, 309, 327
   phase conjugate, 256, 268, 284
   S/N ratio, 311
   target return, 255
   turbulence compensation, 312
   zonal, 266, 327
Confocal parameter, 181
Copper, 103
Correlation length, 252, 313, 334
Crown glass, 48
Crystals, optical, 68

D

Density, image, 126
Detective quantum efficiency (DQE), 128
   for electrons, 148
Diamond turning, 220
Dielectric multilayers, 236
Diffraction
   Fresnel, 249
   Helmholtz–Kirchoff, 194, 195, 199
   Kirchoff, 201, 224
   scalar, 193, 194
   Stratton–Chu–Silver (SCS), 196, 199
   vector, 193, 196
Diodes, light emitting, 34
Dispersion equation, 50
Distortion number, 322
DQE, *see* Detective quantum efficiency
Duran, 50, 102

E

Electroless nickel, 103
Electron exposure, photographic, 148
Expansion coefficient, 106

F

Figure sensor, 264
Films
  evaporated, 220
  sputtered, 220
Flint glass, 48
Fluorescent lamps, 11
Foucault, 273
Fourier filtering, 273
Fourier series, 225
Fourier transform, 223, 225, 227
Fraunhofer diffraction, 187
Fresnel diffraction, 185
Fresnel number, 250
Fused quartz, 65
Fused silica, 65, 102
  lightweight mirror, 110

G

Gaseous discharge lamps, 10
Gaussian laser beam, 176, 333
Germicidal lamps, 18
Glass, optical, 48
Glass-ceramics, 101
Glassy carbon, 103
Glint, 276
Graded refractive index, 174
Grain, photographic, 121
Granularity, Selwyn, 127
Graphite–epoxy, 103
Gratings
  Ronchi, 270, 271
  scattering, 218
  sinusoidal, 218

H

Hartmann test, 274, 308
Heterodyne detection, 255, 258
High pressure lamps, 30

I

Incandescent lamps, 2
  operation, 4
  polar distribution, 2
  tungsten–halogen, 3
Information content, 138

Information storage, 141
Information storage capacity, 145
Infrared materials, 71
Infrared optical materials, 71
Interferometer
  grating, 270
  shearing, 269, 271, 272, 308, 319
Interferometry
  differential interference contrast, 201
  Fizeau, 230
  fringes of equal chromatic order (FECO), 230, 232
  multiple beam, 230
  Nomarski, 201, 208, 221
  speckle, 262
Internal transmittance, 58
Isoplanatic patch, 331, 332
Isoplanatism, 331

K

Kalman filters, 312
Kinetic cooling, 249

L

Lambertian, 216
Lambert's law, 216
Laser
  damage, 211
  HF–DF, 269
Laser beams, 155
Latent image, 121
Light emitting diodes (LED), 34
Light sources, 1
  gaseous discharge, 10
  incandescent, 2
Lightweight mirrors, 110
  reflection under load, 114
  ribbed, 114
  triangular and hexagonal cell, 116
Low-expansion materials, 101

M

Matrix, ray transfer, 157
Measurements of scattering, 228
Medium pressure lamps, 18
Mercury lamp, 18
Metal additive lamp, 30

Metal additive (halide) lamps, 23
Methyl methacrylate, 81
Microirregularity, 200
Microscope, scanning electron, 229
Microscopy
  electron, 229
  stereo electron, 228
Mirrors
  continuous surface, 285
  mechanical loads, 98
  modal, 297
  monolithic, 287
  monolithic piezoelectric (MPM), 287
  piezoelectric, 286
  segmented, 284, 285
  thermal behavior, 118
  thermal effects, 99
Modal correction, 264
Mode matching, 179
Modulation transfer function (MTF), 132
Molybdenum, 103
MTF, *see* Modulation transfer function
Multidither, 275, 277
Multilayers, scattering, 236

N

Natural radiation, 38
Noise, image, 126

O

Optical crystals, 68
Optical glass, 48
  bubbles, 57
  chemical properties, 59
  optical homogeneity, 56
  physical properties, 62
  staining, 60
  strength, 64
  striae, 56
  transmittance, 58
  weathering, 60
Optical materials, reflective, 97
Optical properties of glass, 49

P

Paraxial rays, 156
Particulate scattering, 214

Phase conjugate, 255, 324
Photoflash lamps, 33
Photographic detectors, 121
  characteristic curve, 124
  detective quantum efficiency, 128
  information content, 138, 141
  latent image, 121
  quantum efficiency, 122
  spatial frequency analysis, 132
Photographic grains, 121
Plastics, optical, 70, 79
  manufacturing methods, 92
  mechanical design, 89
  mold design, 90
  optical design, 85
Polarization, 193, 237, 238
  grating, 237
  scattering, 237
Polystyrene, 81
Principal planes, 172
Profilometer, 232

Q

Quantum efficiency, 122
Quartz, fused, 65

R

r_0, 313
Radiation effects on materials, 109
Random roughness scattering, 239
Ray matrix, 157
Real time atmospheric compensation (RTAC), 319
Reflective materials, 97
  absorption and scattering, 100
  geometric stability, 98
  property comparisons, 101
  radiation effects, 109
  stability, 107
  thermal behavior, 118
Refractive index
  optical glass, 49
  optical plastics, 81
  stress effects, 53
  temperature effects, 52
  vitreous silica, 67
Refractive indices, 217
Responsive quantum efficiency, 123
rms roughness, 202

S

Sapphire, 69
Scattering
  angular, 216, 221, 222, 228
  bidirectional reflectance distribution function, 223
  diamond-turned surfaces, 233
  dielectric multilayers, 236
  diffuse, 223, 224
  dipole, 236
  infrared, 207
  microirregularity, 200
  Mie, 192, 193, 197, 214, 216, 217, 242
  numerical methods, 198
  particulate, 197, 214
  Rayleigh, 192, 197
  resonance, 197
  scaler theory, 194
  total integrated (TIS), 193, 199, 204, 213–216, 236, 241
  vacuum ultraviolet, 229
  vector theory, 196
  x-ray, 229
Schlieren, 273
Scratch/dig, 208
  specification, 208
Secondary spectrum, 52
Signal-to-noise ratio, 138
Silica, fused, 65
Silicon carbide, 103
Sodium lamp, 18, 27
Solar radiation, 38
Speckle, 332
Spectral distribution of lamps
  gaseous discharge, 13, 19
  incandescent, 7
Spectrum lines, 51
Stability of materials, 107
Strehl ratio, 247, 250, 252, 254, 299, 303
Stress–optical coefficient, 54
Superpolish, 202

Surface plasmon, 206
  excitation, 206
Synchronous detection, 272, 278, 336

T

Telescope matrix, 171
Thermal blooming, 249, 253, 269, 299, 303, 322
Thermal expansion, 62
  coefficient, 106
Thick lens, 167
Thin lens, 165
Total integrated scatter (TIS), 193, 199, 204, 213–216, 236, 241
Tungsten filament lamps, 2
Turbulence, 252
Turbulence compensation, 312

U

ULE, 102
  lightweight mirrors, 113

V

Vitreous silica glass, 65

W

Wiener spectrum, 136

X

Xenon lamp, 31
x-rays, detective quantum efficiency (DQE), 150

Z

Zernike phase contrast, 273
Zernike polynomials, 265–267
Zonal correction, 264